The App
Introducing the Philosoj
Empirical I

MW00881020

by Russell Hasan

Table of Contents

Chapter One: Introduction

Aristotle once said "all men by nature desire to know." This book is for readers who desire knowledge, specifically that understanding of reality which comes only from philosophy. This book presents two new ideas that have never previously been presented in the history of philosophy. These two new ideas are, first, the philosophical scientific method, and second, pure empirical essential reasoning. This Introduction will offer a summary of the problems that this book seeks to solve, while the rest of the book contains depth and detail in evolving the ideas into a comprehensive theory that solves these problems.

Before I define these questions, I want to briefly touch upon the areas of philosophy that they involve. There are three: first, epistemology, which is the study of knowledge and its methods and limits, second, the philosophy of science, which seeks to explain the justification for science and the limits of science as knowledge, and third, the philosophy of mind, which seeks to discover what is the mind and how the mind relates to reality.

These three areas are quite broad. But if you read this book in its entirety, you will find that I provide a brand new, complete, systematic theory which explains how knowledge works and what are its limits, how science works and why it is knowledge, and what is the mind and its relation to reality. If I am right, this is a really important book, on a par with Kant or Plato or Aristotle or Ayn Rand in the scope and magnitude of what it seeks to achieve within the realm of philosophy. If I am wrong, my philosophy is, at least, an interesting point of view to debate, one which should be added to the chorus of viewpoints in the history of philosophy. Whether I am right or wrong is for you to decide, hopefully after you have read my work and thought about it.

So let me get right to my summary of what sorts of problems this book seeks to solve. Does the external world exist? Does physical matter exist outside of our minds? Does reality exist objectively? How can we know the answers to those questions or prove that our answers are true? What is truth? What is knowledge? What is the scientific method, and can it be used to answer these questions? What is the being of things, and what is the essence of a

1

thing? Can we make a valid inference from the being of one thing to the beings of all things, in a way that is necessary and universal, for example, to infer "All X are Y" from one specific X being Y? Is there a principle that justifies induction? How does reasoning work, and is it based on perception? What does it mean to be objective or subjective? What is sensory experience, and what is consciousness? What is the mind? Does the mind see the external world, or does it merely interact with intuitions and subjective experiences?

This book will explore experimental philosophy, physicalism, mind-brain identity theory, science vs. atheism, empiricism, essentialism, and the scientific method, as well as the scope and limits of knowledge, the definition of truth, and the meaning of objectivity, in order to answer these questions. My analysis is always in the context of epistemology, the philosophy of science, and the philosophy of mind. My methodology is, ultimately, simple, in the sense that my arguments generally follow this pattern: I will ask you to observe something in your own experience, I will then argue that this observation leads to a certain conclusion as a matter of reason and rational deduction, and then, once you admit that you have seen these experiences too, and that my arguments make sense, you must then at that point accept my conclusion as rational, since it was based upon your own experiences, which you observed with your own mind, and upon inferences which make sense in the context of those experiences. The experiences I will draw upon are as simple as you observing yourself choosing what to eat from a menu of food in order to study free will, slicing open an apple and looking at its red color and tasting it to observe how perception works, or tossing a pebble into the air to study the details of ontological being by observing what it is about the rock that makes it be a thing that is flying through the air. The results, however, might be nothing less than a complete understanding of how the human mind is situated in reality and how reason and perception can achieve knowledge. Of course, to go from an apple to universal truth is no simple task, so you must forgive me for the long size of this book. I think you will enjoy reading it, as every page contains interesting thoughts.

The structure of my book is as follows. In Part One, I present a collection of essays which propound the theory of the philosophical scientific method. In Part Two-A, I provide a short essay which summarizes pure empirical essential reasoning, in the context of contrasting my philosophy with Objectivist epistemology. In Part Two-B I publish a longer book which offers the complete line of reasoning justifying pure empirical essential reasoning as a means of knowledge. At the end of this book I give some speculative ideas which are worth considering, and then in the conclusion I explain the meaning of the title of this book, and the metaphor of the apple of knowledge.

Won't you come and join me on this great intellectual journey of philosophy?

Part One: The Philosophical Scientific Method

Chapter Two: The Philosophical Scientific Method

As a threshold matter, we must define what we mean by "the scientific method." The scientific method, as a method for using the analysis of empirical data in order to confirm or refute a hypothesis, asks one question: "does the world actually look precisely the way that we would expect if our theory was correct?" It the answer is "yes," then we can continue to develop our theory. But if the answer is "no," then we must concede that our theory was incorrect and has been refuted by reality, and we must abandon our theory and move on to a new hypothesis. The scientific method as used by research scientists is based around controlled experiments to collect data which their theory can conform to or be refuted by, and their theories are typically based on physics, chemistry or biology. But the theory of the philosophical scientific method asserts that the scientific method can be applied to philosophy, and can be used with any theory or idea, in the realm of philosophy or in any other field, such as economics or history.

Any theory or idea can be supported or contradicted by an observation or experience of physical reality. In order to perform a philosophical scientific experiment, only three things are necessary: (1) the theory or idea to test, (2) the set of expectations that we would expect to see if our theory was true, based upon our thinking about expectations that takes place prior to our experience, and (3) the experience of reality or the observations of physical reality and objects in the world. Once we have these three elements, we can evaluate whether (3) our experience matched (2) our expectations. If they did, then (1) the theory is supported. If they did not, the theory is disproved. It is often said in university philosophy classes that the difference between science and philosophy is that scientific theories can be experimentally tested and philosophical theories cannot. I argue that this idea is untrue, and that the scientific method is not limited to science. The scientific method, reduced to the simple template of checking ideas by experimental verification against observable reality as seen by empirical perception, can also be used as a template for everyday thinking, and it can be applied to all philosophical ideas, or to any other type of idea.

By applying the scientific method to philosophical analysis, philosophers can discard and reject theories which do not conform to the results of their observations and experiences of the physical world, which would be a vast improvement over the philosophical method which currently prevails in most university philosophy departments, where academic professors get lost in a dream world of abstract theory that never gets verified against experience. By taking a philosophical approach to understanding the scientific method, science can move beyond its role as merely the producer of technological gadgets, and science can re-conceptualize itself as a system for achieving certain, absolute, provable knowledge of objective physical reality, capable of answering deep questions about the human experience, and as a body of ideas that can solve most of the problems in our quest to understand the world in which we live.

Note that in order for the scientific method to be accurate, we cannot make post-hoc rationalizations, which can also be called after-the-fact justifications, i.e. rationalizations that take place after the experiment has been conducted, to try to make our experience conform to what we wanted to believe. We must not make tortured justifications to squeeze our experience into what we expected to see. The scientific method, as applied to philosophy (and to natural science also) does not work in the absence of intellectual honesty in evaluating whether or not reality conforms to our beliefs, regardless of our emotions and feelings or political biases regarding what we desire to be proven true or false. Ideally, the set of expectations is decided prior to the experiment taking place, because if we form our expectations after the observations of data have already been made then we tend to shape our expectations to achieve the confirmation or refutation that we desire, i.e. this leaves us more vulnerable to confirmation bias. If the experience must be analyzed after the observation has already taken place, as will often happen due to practical necessity, then we must seek to reason what our expectations would have been "before the experiment," in order to do our best to avoid "after the experiment" post hoc rationalization.

Sometimes, different interpretations of the same data are plausible. When this happens, you must look to reality and use new experiments to determine which interpretation is true. This can be done by an experiment which will distinguish between the two interpretations on the basis of the difference between them and what difference it makes whether one or the other is true. Or one can look at reality more closely and in greater detail, with new data, to see which theory is true. One theory will be confirmed by the empirical data and all others will be refuted, if scientific experiments are run properly and thoroughly, because it is something existing in external reality which causes experiments to confirm or deny an idea, because the data and observations have as their source something in reality, and this thing in objective reality which the experiment explores will cause the results of the experiment to show whatever it is that it shows. So the scientific method, if applied correctly, will result in one's theory matching what exists in reality, which is how I define what it means for a theory to be "true."

I am not the first philosopher to argue for a philosophical scientific method. One thinks of, for example, Karl Popper's falsification theory, or A. J. Ayer's verification theory, as attempts to introduce an empirical scientific method into philosophy. My theory of the philosophical scientific method is new and unique, for several reasons, two of which I mention here.

First, in addition to asserting that a scientific hypothesis can be disproved and falsified by the scientific method, if the observations contradict the hypothesis, I argue that a scientific (or philosophical) theory can be proven true by the scientific method, if the experiment yields observations which are precisely what one would expect if the theory was correct. Thus, experience and experimental observation are capable of proving a hypothesis false, and they are equally capable of proving a theory to be true, and thereby establishing a belief that can properly be called knowledge according to the scientific method. Other philosophers of science, such as Popper, have stated that the difference between a scientific theory and a non-scientific idea is that a scientific theory can always be refuted by new evidence or data in the future, whereas a non-scientific idea cannot be disproved. If Popper were right, then science could never achieve knowledge, because every scientific belief would always be in danger of being refuted in the future. In contrast to Popper's falsification theory, the philosophical scientific method can prove that a scientific theory is true and correct, in which case it will never be refuted by new data in the future, by means of conducting an experiment where experience matches precisely what we would expect if the hypothesis were true. Science can confirm a hypothesis as well as refute a hypothesis.

For example, suppose that I am at home, and I hear a noise at the front door of my house. This noise makes me guess that a visitor is at the front door. I now have a hypothesis in mind, namely, that someone is at the door. I will then go to my front window and look out. If I do not see anyone, then my hypothesis was falsified, disproven, and refuted. But if I see that someone is there, then my theory was proven and confirmed. Whether the hypothesis was proved or disproved, it was tested in the same way, via empirical experience and observation. If both proof and refutation are based on experience, then proof that a theory is true should be understood to constitute knowledge, as known by the scientific method, to precisely the same extent that disproof and refutation can constitute knowledge. To say that science is based on guesses which can always be refuted in the future by new data does not reflect what actually happens in scientific analysis. When I see the person at my front door I know that someone is there, and the fact that a visitor is at the door was so clearly confirmed by visual observation, i.e. I am looking at the person there, that it could not reasonably be refuted by a method that claims to rely on experience for knowledge. The scientific method achieves knowledge to the same extent whether the hypothesis is confirmed or disproved, because either the positive proof or the negative refutation, a belief that either the thing is or is not, is based on whether you see a person at the front door.

Secondly, prior philosophies of science distinguished between ideas that admit of scientific experimentation and those which do not. Such theories claimed that religion, and, perhaps, philosophy, literature, etc., do not admit of experimental testing, whereas hard sciences, such as chemistry and physics or, perhaps, social sciences like economics and history, can articulate theories which can be proved or disproved by experience and data. Some pro-science philosophers have said that only the falsifiable, disprovable areas of thought can be called knowledge, and the areas where theories cannot be tested and disproved are unimportant or outside the scope of science and reason. Some people think that in math and science there are "right" and "wrong" answers demonstrable by proof and deduction, whereas in philosophy, the arts, and the humanities (and to some extent, the social sciences) no demonstration of right and wrong can exist, and every inference is open to debate and discussion and interpretation. This might be argued as a justification for the scientific method being inapplicable to philosophy and religion. If this were conceded, then it would also follow that religion can protect itself from the philosophical scientific method by claiming that religious ideas are not the sort of ideas that can be proved or disproved by observation and experience.

My argument is new and different, because I argue that in literally every area of human thought, including philosophy, literature, or religion, an intelligent and enterprising person can design experiments which will articulate a theory and then collect observations of experience which would either conform to or contradict the expectations entailed in the theory. I argue that right and wrong answers can be proven true or false, in every area of human thought, including philosophy and the arts and humanities. It follows from my claim that religious ideas, as well as philosophical ideas, can all be put to the test of experience and experimentation. In the upcoming sections of this book, I will show what it would look like to design an experiment to test the hypothesis that God exists, to test the theory that humans have free will, to test the idea that the mind and the brain are identical, and other philosophical scientific experiments.

Some thinkers believe that some issues are abstract theory with no basis in practical reality, such that they cannot be proved or disproved by means of experiments and empirical observation. For example, the theory that seventeen angels can dance on the head of a pin, or the theory that the Kantian distinction between the transcendental vs. the transcendent vs. the immanent is accurate, are thought to be untestable, because nothing in empirical observation will show its truth one way or the other. Philosophers of science like Popper and Ayer, as well as Pragmatists like William James, might have thought such theories to be meaningless, and neither true nor false. I argue that these ideas do make a difference, and something is at stake in them, such that experience can confirm or refute them.

Generally, what is at stake in the idea that seventeen angels can dance on the head of a pin, is that religion is true and God exists. If you can prove on the basis of scientific reasoning that God does not exist (and I will show later how this is possible) then you prove that zero angels can dance on the head of a pin. What is at stake in the

9

idea of the Kantian distinction is the truth or falsehood of the entire Kantian philosophy. Later I will show what is at stake in Kantianism, and particularly in the section on Objectivism vs. Subjectivism I will show that the idea that the world revolves around the mind can be tested by an experiment to establish whether perception and subjective belief can constitute reality on the basis of whether the mind can change or control our experiences of reality. If empirical experience refutes Kant, then the philosophical scientific method proves that the Kantian distinction of transcendent vs. transcendental is false.

If someone proposes a theory which is mere abstract theory detached from any basis in practical reality, then what is at stake in its truth or falsity is whether human thinking should focus on and believe in impractical useless ideas, or whether reason is a practical tool for engaging and interacting with reality. The difference it makes is how we as humans should go about thinking about reality. If reason and thought are practical, and this is proven by experience and observation, then each and every abstract impractical imaginary daydream that could be asserted is proved to be false. I argue that every theory which can be articulated, and every idea that has ever been debated, can be put to the test of experience and experiment, if someone actually wants to invest the thinking necessary to design an accurate experiment. Something will always be at stake, although some deep thought may be necessary to identify what is at stake and what difference it makes.

Chapter Three: Experimental Philosophy and External Contradictions

In this section I would like to introduce two related concepts: (1) experimental philosophy, and (2) external contradictions.

Many philosophers claim to want to introduce the "scientific method" into philosophy. Hume said this. So did Kant. But philosophers have been notoriously lazy in terms of actually using science in philosophy. The scientific method begins with a belief, theory, assumption, premise, hypothesis or postulate, and then looks at experience, usually in the form of empirical data from tests and experiments, and science asks: "was our experience exactly what we would have expected it to be if our theory was correct?" If not, throw out the theory and try a new one. If the evidence supports the hypothesis, then it is confirmed.

What would experimental philosophy look like? Let me offer two short oversimplified examples, both of which will be explored in detail later, in order to give you an introductory taste of my theory. First, if Kantian subjectivism is true and the mind creates the experience of space and time, then you could jump out a window and your mind could alter your experience of reality so that your experience of space would look like you were flying. As you fall out the window, your mind would constitute your experience of reality by imposing categories onto your experience, therefore your mind could create the experience of flying. If the world revolves around the mind, then we would expect the mind to have the power to control the world. If our observations contradict this, then Kant is disproven. Second, if Hume's skepticism were true, we could not know that the Sun will rise tomorrow morning. If you possess knowledge that the Sun will rise tomorrow, and this knowledge is confirmed, then it can't have been true that you did not know anything. But Hume believes that we could not know for certain that the Sun will rise tomorrow.

A related concept involving the use of empirical data in philosophy is external contradiction. Every great philosopher claims to have a "coherent" philosophy, that is, one which is internally consistent and has no internal contradictions. The only problem with these claims is that it is incredibly easy to think up a theory which is internally consistent. Hume's philosophy is. So is Kant's. So is Ayn Rand's, and so are the theories of most other great thinkers. Perhaps the only example I can think of a famous theory with an internal contradiction is Christianity, namely that God is omnipotent and loving but evil exists and God permits evil to exist, or that God is the only power in the universe but the Devil is also powerful, or that Christ is both human and divine.

Instead of seeking internal coherence, the far more difficult, but more important, task of a theory is to not have external contradictions, in other words, not to conflict with what our experiences teach us. For an example let me use Rand. As one Objectivist wrote in a reply to an article of mine, the idea that "emotions are robots programmed by a person" is internally consistent with Rand's philosophy. Indeed, Rand seems to have believed that her followers Nathaniel Branden and Barbara Branden could program their emotions like robots, to make Nathaniel Branden love Rand and to make Barbara love Nathaniel, because Rand (as she asserted) was Nathaniel Branden's "highest value" and therefore he could program his emotions of romantic passion to conform to the value judgments of his logic and abstract reasoning. Yes, this seems crazy, but no idea in Rand's philosophy actually contradicts this position. Therefore the Randian theory of love lacks any internal contradiction, so it is coherent.

But, even lacking internal contradiction, Rand's theory of love had an external contradiction. Experience proved that emotions are not robots and that love comes from preference and personality and "chemistry" and falling in love as well as from logical evaluations of compatibility and sharing abstract ethical values. As described in Barbara Branden's book "The Passion of Ayn Rand," the application of Rand's theory of love led to sexual dysfunction and irrationality, climaxing in the Branden-Rand sex affair explosion which destroyed Rand's movement in the 1960's. That series of events disproved Rand's "emotions are robots programmed by logic" hypothesis. If Rand had been paying attention and had applied the philosophical scientific method to her philosophizing, then she would have noticed that experience refuted her theory of romantic love. History indicates that Rand clung to her theory even after the Branden fiasco, instead of rejecting it when she faced the external contradiction. That was not a scientific attitude for her to have. If being scientific is a desirable trait, then we must challenge, and be ready to discard, any belief which is exposed as having an external contradiction.

Chapter Four: Life Experiments

The scientific method applied to human existence can take the form of "life experiments," in which you use the scientific method to test basic ideas, like whether honesty is good or bad, or whether God exists, or which career you should pursue. In a life experiment, you choose a theory to test, which is the hypothesis, you then define what you would expect to see if the theory was true, and you then live your life as if it was true, and apply and use it in your life, for a period of time. During this time you carefully observe and record your observations, and at the conclusion of the period of time, you evaluate your notes and determine whether the theory is confirmed or disproved by your observations. I recommend one month as the ideal time period for a life experiment, although as little as one day might yield useful observations.

History has examples of famous life experiments, such as in Benjamin Franklin's "Autobiography," where Franklin tested his ethical beliefs, and in Henry David Thoreau's "Walden," where Thoreau tested the naturalist approach to human existence. I myself can mention two life experiments that I have run. Both of these examples may seem eccentric or extremist, but this is only because most people are not used to the intellectual honesty of testing an abstract idea and taking seriously the task of seeing what it would look like if applied in practical reality. First, I tried to test whether prayers and wishes and casting magic spells can have a beneficial impact of helping you to get what you want. I found that they cannot. Second, I experimented with whether it is possible for the human brain to make all decisions consciously and for the conscious mind to manage the subconscious mind in every detail. I learned that it is not possible, because the human brain was designed for the subconscious mind to handle most mundane matters automatically, while the consciousness thinks about the big picture of the person's life and pays attention to the most important behaviors.

For life experiments, it is necessary for me to explain why if a hypothesis is true then it will be useful and beneficial and improve your life, such that if a theory causes suffering and agony then this indicates that it is false, and if applying a theory to your life makes you happy and helps your life then this shows that it is true. My theory disputes the Pragmatist idea that truth and usefulness are identical, or that an idea is true because it is useful. In contrast, I argue that an idea will tend to be useful because it is true. The explanation for why this happens comes from what I call the ethical theory of "pursuance." For thousands of years, philosophy and ethics have struggled to figure out how to go from "is" to "ought," in other words, how to translate our knowledge of what reality is and what things exist, into a set of beliefs about what we should do and what is right and wrong. Pursuance solves this problem, and also shows why a belief in the truth will almost always tend to improve a person's life. Pursuance means that the objectively right thing to do is to treat things in reality as what they are, and not to treat them as something different from what they really are. In other words, ethical behavior is behavior that is pursuant to objective reality.

For example, say that you have a hammer on your left and a pillow on your right. Say also that you need to hammer nails into wood to build a table, and after that you plan on taking a nap. You could hammer the nails with the hammer or the pillow, and you could sleep with your head resting on the hammer or on the pillow. The objectively right behavior is the one that is pursuant to objective reality. Thus, the objectively right thing to do, under an objective ethics, is to hammer the nails using the hammer, because the hammer is made of metal and is hard and is in a good shape for your hand to grip and swing to use for hammering nails. And the right thing to do is to sleep with your head resting on the pillow, because the pillow is full of feathers and soft and a good shape to go under your head. In contrast, it would be wrong under objective ethics to sleep on a hammer or hit nails with a pillow, because of what things are in objective reality.

You should hammer the nails into the table using the hammer, and you should sleep with your head on the pillow. These are both objective facts, but they are also "should" statements of ethics. Right and wrong can be translated into good and evil. If you are a carpenter, and you need to build tables in order to survive, and you need a good night's sleep for your body to stay healthy, then pursuance as an ethics can distinguish between right and wrong as good vs. evil because they will make you either happy or miserable. The good and right choice will treat reality as it is, which enables

15

you to build tables using the hammer. By being a successful carpenter you can make money, support your wife and children, and be happy. The evil and wrong choice would result in failing to build tables and sleeping with your head on a hard uncomfortable hammer, which will make you suffer in misery and die.

Note that the qualities of the hammer and the pillow which make the difference are absolute, objective, physical characteristics, as known by perception and reason observing the physical world, which I consider to be one of the hallmarks of science. Thus, pursuance is a scientific ethics, which derives right and wrong from what does or doesn't exist. Philosophy professors have argued that the fact that a hammer is useful for hammering nails is a subjective quality that comes from human intent, not from objective reality. They assert that the ability to hammer nails exists for humans, not in the hammer in itself. This is incorrect, since the qualities and properties that make the hammer useful for hammering nails, e.g. the shape of the handle and the hardness of the metal of the hammer head, are all objective physical properties as known by scientific empirical observation.

This example of the hammer and pillow captures the general principle that a person's life is almost always benefited by applying the truth, at least if we assume that a person's happiness comes from survival and success in achieving one's tasks and goals. In general it is good if a person treats reality as what it is, and not as what it is not. The few conceivable examples, like a criminal aiming at gun at our head and threatening to shoot us unless we loudly declare that hammers are softer than pillows, are uncommon enough to be worth ignoring. If we accept the ethics of pursuance as the basis of objective right and wrong, then we can extend the pursuance argument to the scientific method. If you test an idea in your behavior during your life experiment and it works, then it is probably true, and if it causes pain and harm then it is probably false. What works vs. what doesn't work won't tell you why the theory is true or false, but reasoning based on your observations, as combined with research and reading about the subject matter or discussing it with other people, should give you enough clues for you to figure out what it was about the hypothesis that was true or false.

Every opportunity to learn from experience can be structured as a life experiment. You identify the idea to test, define your expectations, and then experience reality and collect observations. Your theory will be either confirmed or contradicted. For example, you could test drive a Toyota and a Ford and see which one conforms to your expectation that a good car will corner well and have good traction. You could test the idea that the right amount of time to boil pasta is ten minutes by trying it and then tasting the pasta. You could experiment with the idea that your talents suit you for a career as an accountant by interning at a tax accountant's office and then evaluating feedback from your employer as to whether you did a good job. Or you could read the first thirty pages of "The Hobbit" in order to test out the theory that your reading all of Tolkien's epics would be a fun and enjoyable way to spend time. In all of these examples, the practical application of the philosophical scientific method means that you believe what is confirmed and you abandon those beliefs that were refuted. If you had thought that you would really love the Toyota and hate the Ford, but you do the two test drives and you actually like the Ford and find the Toyota annoying, then the philosophical scientific method would tell you to buy the Ford.

Unfortunately, sometimes the window of opportunity to verify an idea will be narrow and difficult, because simulations will lack the realism of a situation when it really happens. For example, the only good way to confirm that your friend is a true friend will be to have a crisis and see whether your friend is there for you or not, and a true crisis cannot really be modeled in simulations before the fact, although perhaps a previous minor problem or test drill will approximate it enough to model what would happen in a crisis. However, from the experience of a crisis, you will be able to verify from empirical experience which of your friends are good friends and which are not. Similarly, dating a person for three years might be sufficient to evaluate whether you are in love, but is it perhaps true that only the real experience of being married for several decades could confirm or refute the belief that your partner is the right person to spend the rest of your life with. In a sense, the rising rate of divorce since 1950 may be a sign of culture catching up to empirical philosophy.

It is useful to draw a distinction between "a posteriori" knowledge and "a priori" knowledge. This distinction was made famous by Kant but has been used by many philosophers throughout the history of philosophy. "A posteriori" knowledge is knowledge that comes from experience. Literally it is a Latin phrase meaning that it "comes after" experience. "A priori" knowledge comes "from before" experience, meaning that it is knowledge from intuition that the mind imposes onto the world. As each of these examples indicates, true knowledge can come only from experience a posteriori. Before-the-fact a priori guesses and intuitions have no basis for being trustworthy as knowledge of reality. Contrary to the beliefs of mainstream philosophy, which hold that a posteriori knowledge is impossible, the philosophical scientific method as deployed in life experiments is a means of achieving a posteriori knowledge.

Chapter Five: Objectivism vs. Subjectivism in Epistemology

First I will define what I mean by objectivism and subjectivism, and then I will explain how to run a philosophical scientific experiment in order to discover which is true. Let me begin by noting that my use of the term "Objectivism" for epistemology has a completely different meaning from the term as used in political philosophy to denote the libertarian capitalism of Ayn Rand. By objectivism I mean the belief that reality exists objectively, i.e. that a physical external world exists, and also that reality consists of objects which possess identity and can be studied by reason. By subjectivism I mean the belief that reality is subjective, which means that the human mind creates and controls reality, that mind exists but objective physical material does not, reality consists of subjects not objects, and the world is constituted by beliefs, feelings, and wishful thinking. Subjectivism in its most extreme forms becomes solipsism, panpsychism, or social construct theory, and I group these theories together as all being based on the claim that reality is either totally or mostly subjective.

To test a hypothesis, as a first step we must identify what we would expect if the theory were true, and then as a second step we must journey into the realm of experience and see if what we observe either conforms to or contradicts our expectations. If objectivism were true then we would expect to interact with reality by going out into the objective external world and physically engaging the objects that we find there. If subjectivism were true then we would expect to interact with reality by altering the subject, not the objects, i.e. by changing how we perceive something or what we believe about something in order to control it. Wishes, faith, willpower, and perhaps also magic spells, would be the tools to succeed.

Having defined the expectations, we can now set the parameters for an experiment. For example, one could test the ideas by placing a mug of coffee on a kitchen counter, and then trying to move the coffee from the counter to the kitchen table. If you must do this by physically moving the object, then this proves that objectivism is true. If you can move the mug to the table subjectively by believing that it is on the kitchen table or by your mind causing it to move or by changing your perception so that you see it as being on the table, then your mind controlled or constituted your experience, and subjectivism would be supported. Run the experiment and see for yourself what you observe.

It is worth noting that some things exist, for example a phobia or fear, where the problem would be solved by altering your beliefs, feelings, or mind. This could be interpreted as evidence in support of subjectivism. However, this example is not inconsistent with objectivism, to the extent that you must force your feelings to conform to and obey objective reality. For example, if you are afraid of squirrels then you must make your subjective feeling match the objective fact that squirrels are not dangerous. If subjectivism were true then we would expect the belief that squirrels are scary to cause squirrels to become monsters, because only the inside of the mind would exist and no external reality would exist outside the mind of the person with the phobia, so the fear would be real and true, and no outside reality would exist which could contradict or disprove the ideas within our minds.

Subjectivism asserts that our experience of reality is subjective. Objectivism says that the reality that we see is objective, and no difference exists between the external world and the world as we experience it. For example, suppose that I hold an apple in my hand. To say that the thing in my hand is my mind's representation of the apple adds nothing useful or beneficial to the analysis, because a representation or idea, concept, or perception in my mind is merely my subjective experience, which cannot be studied by science. In contrast, to say that this thing is the apple itself in external reality adds a lot to the analysis, because we can see that the apple is a fruit with seeds that can be planted or eaten, etc. If biology and botany study apples, and not a world of mind or spirit, then in order for science to be possible we must say that this apple is a thing in itself. Also, if the apple is in my mind then I can't be right or wrong about it in a way that I can demonstrate to other people. But if this apple exists in objective reality then statements about the apple can be true or false, and I can demonstrate the truth to other people by pointing to the apple and showing it to them and pointing out what it is about the apple that makes the statement true or false. For example, I can prove to you that this is a seedless apple by slicing it open and showing it to you. The apple would be the same for everyone and you and I could talk about the same apple. And empirical experience would provide knowledge of the apple. (This issue will be revisited in depth in Part Two, in the section on Adler and the Objectivity Proviso.)

I will conclude by explaining that objectivism is an epistemology which justifies science while subjectivism is an epistemology that justifies religion. Objectivism is based on the external world of objective reality, while subjectivism, in its various manifestations, is based on subjective perception, interpretation, belief, thought, faith, or language, or the structure of the mind as it experiences reality and pure reason, or relativism and inter-subjectivity. If existence exists objectively then science is our tool of controlling and coping with reality, because science is the method which looks at the external world and finds ways to control objects derived from an analysis of objects and experience. If existence is subjective then religion would be the proper tool for controlling reality and getting what we want, via our belief and faith and wishful thinking or our minds imposing our desires onto reality. If subjectivism were true then we could make problems go away by redefining our words so that we could no longer talk about the problem, because our language as a tool of thought would constitute reality, or we could make our minds impose our desires onto our experience in the act of constituting reality, or we could cause the existence of God by having faith in Him and believing in Him. The objectivism vs. subjectivism debate is not mere abstract theory and theoretical philosophy. It has a clear and obvious practical impact on how we live our lives as human beings and how we seek to interact with reality.

Chapter Six: The Mind as Brain vs. Soul Experiment

Is the mind a brain or a soul? The philosophical scientific method proves that the mind is the brain, because our experience is precisely what we would expect if the mind was the brain. How do we know that the mind is the brain? Let me point out the six main areas of our experiences which prove that the mind is the brain:

1. Brain damage. Science has learned, from over 100 years of experience, that brain damage to a specific section of the brain corresponds to a failure in a specific function of the mind. The following data is taken from Wikipedia. Damage to the hippocampus, which is a part of the brain, can cause memory loss, which in its mild form becomes Alzheimer's disease and in its most extreme form becomes anterograde amnesia, which is a total inability of the mind to form new memories. In the brain disease called prosopagnosia, brain damage to the fusiform gyrus can cause the mind to cease to recognize people's faces, so a person can see the face of a loved one and not know who they are, and be unable to remember faces. In left hemianopsia, damage to the right occipital lobe causes the person to be unable to see things in the left side of their field of vision while still seeing objects to their right. In the condition known as topographical disorientation, lesions to the posterior hippocampal gyrus and anterior lingual gyrus cause the person to be unable to mentally map their environment or to navigate locations by means of landmarks. In the developmental disorder known as Autism, a delay in the physical development of the brain of a child causes the afflicted teen or adult to have difficulty speaking and evaluating social situations, and in its most extreme form the person may be totally unable to speak. Vision, memory, speaking, and thinking are generally thought of as aspects of the mind, so the evidence shows that parts of the brain correspond to parts of the mind. This strongly suggests that the mind is the human brain.

Going slightly beyond the findings of research on brain damage, science has also studied brain activity while a person performs certain behaviors and learned that activity in specific regions of the brain corresponds to certain cognitive actions typically regarded as parts of the mind. For example, vision is done by the occipital lobe. Memory is done by the hippocampus. Making decisions is done by the amygdala. Processing language and choosing words for one's speech is done by Broca's area and Wernicke's area. Emotional reactions like fear and anger are done by the limbic system. And inhibiting emotions and imposing restraint and reason onto impulses is done by the prefrontal cortex. In each of these cases the alignment between function and brain region is not perfect, and many different parts of the brain all contribute to the end result. But the data is precisely what we would expect to see if the mind is the brain.

2. Drug effects and psychiatric medications. Drugs such as alcohol and marijuana have a specific effect upon the mind. Thinking and reasoning, which some philosophers see as the highest aspects of the spiritual being of the soul, are affected by such drugs. These drugs are physical substances which alter the brain, with alcohol acting as a downer and marijuana affecting neurotransmitters like dopamine. The drug called ecstasy, which alters mood and makes people happy, works by increasing serotonin in the brain. Joy and other emotions are generally regarded as existing in the mind, but the physical substances of drugs can alter them, which suggests that emotions have a physical existence in the brain. And, aside from recreational drugs, there are medical drugs such as antipsychotics used to treat schizophrenia. Such psychiatric medications are drugs which can make the difference between sanity and insanity for a mental health patient. But thinking sane or insane thoughts seems to be an attribute of the mind. This is what we would expect if the mind is the brain. If the mind were a soul then we would not expect drugs to alter or control it.

3. Sleep and Hunger. If the mind was a disembodied soul then we would expect it to remain awake while the brain sleeps. We would expect the mind not to be controlled by bodily urges such as hunger, etc. People can't think clearly while sleepy. People get irritable while hungry. And they make decisions which should be governed by reason and logic, such as what food to buy, differently based on whether their body is hungry.

4. Space and Time. In rejection of the Cartesian notion of dualism, that the mind is a res cogitans which does not exist in space and time, in contrast to the res extensa which exist physically in space and time, i.e. that mind and body are two different substances,

I will note that thinking a thought takes time, e.g. it might take you five minutes to think up your list of food to buy for a grocery list, and thinking often exists in a way that is spatially tied to the mind being in a specific place, e.g. your thinking about which can of soup to buy situates your mind physically in the location of the soup aisle at the grocery store. If thought were based on a disembodied nonphysical soul then we would expect thought to look like speculation regarding how many angels can dance on the head of a pin, whereas thought, as we experience it in our own lives, looks like practical thinking to engage physical reality in space and time, like thinking about how to save money on food when your brain and body are physically located in a grocery store.

5. Sensations. A person's primary means of interacting with reality is through sensory perception, such as sight, hearing, smell, etc., and if the seat of the mind was a "soul" then we would expect our minds to possess supernatural means of knowledge that would give us mystical insight into reality. No such thing is evident in our typical understanding of human behavior or human thinking. If the soul theory was true and the brain theory was not, then the physical senses would be an illusion or distraction, and we would not expect that we must rely on them to live. For example, we would not need our eyes and the vision parts of the brain to drive a car, and we could instead navigate roads while driving a car by mystical insight. Nor would we require the tongue's taste and the nose's smell and the brain's regions processing taste and smell to eat a sandwich and know whether the food we eat is good or bad. Our experiences as human beings clearly contradict the soul theory expectations, i.e. what we would honestly expect to see if the mind were a soul and not a brain.

6. Memory, Learning, and Childhood. If the mind were the soul then we would expect to have clear memories of our past lives or of where we were prior to being born into a human body. I have no such memories, and you must ask yourself whether you do. Also, if mind is soul then we would not expect to need to teach things to children, because knowledge would come from the soul. If the mind is the soul, and the soul attaches to the body at conception, then a baby would have a mind like an adult. If language and math are known by the soul, and they do not have their origin in physical reality, then we would not expect to need to teach them to children.

On the other hand, if the mind is the brain, then we would expect the development of the brain to parallel increases in knowledge and the stages of education as the body grows from a young child to a teen, which is what we can see happening in real children. Plato attempted to counter this observation by arguing that when we learn something we are really remembering what our soul already knew. But, if the soul has this knowledge of what is taught already in its memory, then the honest expectations that we would expect to see in observable reality would be that we would have some sort of access to this memory, so we could know things without learning them, and we would expect children to already possess a foundational knowledge of math, language, science, etc., before it is taught to them. Yet what really happens is that math in its entirety must be learned, from addition and subtraction through multiplication tables all the way up to calculus

One counterargument against me is people's anecdotes of out of body experiences and spiritual experiences. But this sort of anecdote is adequately explained by hallucinations, wishful thinking and self-delusion, combined with the brain's strong social desire in seeking acceptance from the socially dominant religious groups in which many people want to participate.

Chapter Seven: The Free Will Experiment

As another example of showing that the philosophical scientific method can achieve scientific answers for philosophical questions, I will here explain how to design an experiment that tests whether free will or determinism is correct, as shown by empirical experience. The first step is to define the expectations that we would expect if one or the other were true, and the second step is to look at experience and observe reality to see which theory is confirmed and which is refuted. If humans did not have free will then we would not expect to have to make choices, or you would expect your choices and decisions to be illusions which have no effect. For this experiment, go to a restaurant and look at the menu and then experience what happens between the time when you look at the menu and the time when you begin eating your food. What you will observe is that you had to make a choice about what food to order. You looked at the options in the menu, thought about it, made a decision, and then placed the order which you had chosen. This choice was meaningful and had an effect. For example, if you choose to order a salad then you will eat a salad, and if you don't choose to order a salad then you will not get a salad. Because it had a real, visible result, it was not an illusion.

So we have shown by means of the philosophical scientific method that people make choices. The determinist reply to this argument for free will might be that one's choices are controlled by the factors that one considers in making the choice. For example, one's craving for a salad forced one to choose to eat a salad. But experience refutes this, in a number of ways. First, behavior does not happen automatically, and choices are made as a result of thinking and choosing, as we can see via introspective observation if we examine our mind while we are making a choice, e.g. while we are choosing what to eat at a restaurant. The set of items, prices, and tastes on the menu, combined with our appetite and cravings, does not define one or the other decision. For example, if the chicken sandwich is cheap but the garden salad tastes good but you have recently had a craving for a baked potato, then this set of factors does not produce one determined result automatically. Instead we must make the choice between the sandwich or the salad or the potato, and, although we consider factors, we are capable of choosing one or the other, so the factors do not control or determine the choice and the behavior is not automatic.

Generally, if pleasure and pain controlled human decisions then we would not expect people to enlist in the army to fight for ideals, nor would we expect anyone to ever choose difficult or unpleasant behaviors. If the brain were controlled by DNA and genetic impulses then we would expect the genetic drive to procreate to render monogamy and marriage impossible and to force everyone to behave with sexual promiscuity. And if the mind did not play any role in making choices then we would expect to react to stimuli immediately, by instinct, like an animal does, whereas what we observe is that humans think before making a choice.

The argument from ethics also proves that humans have free will. If humans did not have free will then we would not need ethics and values to guide our choices, so the fact that ethics is useful and necessary proves that humans have free will. My argument here is not the Pragmatist argument that a belief is true because it is useful. Rather, my argument is that a belief will tend to be useful because it is true, so the usefulness of an idea strongly suggests that the idea is true, although utility and truth are not identical. On this basis, the human need for ethics is strong evidence that we possess free will.

The philosophical scientific method shows that humans have free will, but we still need to explain why we have free will, and how free will works. The philosopher Laplace offered an argument in support of determinism, that because only bodies in motion exist, if you know all the physical properties of every object in reality then you can predict what the future will look like with perfect accuracy, hence no room for free will exists in the Universe. Here I will offer a new argument, the "omniscient computer model" argument in favor of free will, which is the equal but opposite of the Laplace argument.

This argument is a thought experiment: imagine that humans build a computer that knows everything, including the existence of every subatomic particle and every vector of motion in the universe. If this omniscience computer were to model the universe and predict the future, say 10 years in the future, then, if everything is predetermined, and the model knows everything, then the model would be perfectly accurate. But, to extend the experiment further, if

27

the humans who run the computer look at the model and see what will happen in 10 years and they don't like what the future will be, for example if the model shows that a viral plague will break out in 10 years and kill everyone, then they could act to change the future, by analyzing the virus and creating a vaccine, and then the future would be different and the model would be no longer correct. The computer model could not have included what the humans would do after seeing the model's predictions, since that would then lead to a second model which would include the first model and the human response to the first model, and the human computer operators could respond to the second model, such as by seeing that their vaccine failed and developing a second vaccine, and then a third computer model would follow, and a fourth, and so on, and a model accounting for the human ability to choose futures from among alternatives known in the present reduces to infinity and absurdity.

Thus, if humans have the ability to predict the future, then they can choose among different alternative futures, and then total physical determinism reduces to a contradiction, which proves that humans have free will. Laplace said that if the future can be predicted then humans lack free will, but I say that if humans can predict the future then we can choose among competing alternative futures. Note that this is a philosophical argument, which could be expanded and improved by a scientific argument once science unlocks the puzzle of how the brain works and how awareness of the future and the brain's choices between alternatives exists in human brains.

Let me conclude by noting that my theory of free will extends only to the claim that a human being can make a choice which is meaningful and non-illusory and which has an effect in practical reality. I do not believe in the radical free will of Sartre and Szasz, which claims that a person is totally responsible for every action that they do even if something biological in the brain, such as mental illness, drugs, or some other controlling factor, interfered with their ability to make a choice. In a different section I offered proof that the mind is identical with the brain. In a properly functioning brain, i.e. in normal situations, ethical responsibility attaches to having made a choice, such that we deserve reward or punishment for good or bad choices. But if a biological problem in the brain prevents a freely made choice then no ethical responsibility can attach because no choice was actually made. Something similar would be true if external factors interfered with the freedom to choose, e.g. if a criminal points a gun at you and forces you to do something then you would not be morally responsible, although you might be partly to blame depending on the details of the situation.

Much has been made in philosophy recently of the theory that free will and determinism are compatible. They are not compatible in the sense that the claim that "humans have free will" clearly contradicts the claim that "human behavior is determined," and a contradiction is not coherent or plausible. But they are compatible in the sense that a complete understanding of human choice understands that the mind's choices can influence the brain, and also the brain's biology can influence choices. This could be called top-down causation vs. bottom-up causation. Compatibility extends to the observation that DNA and the brain's neurons define the range of one's natural talents and abilities, but decisions and choices then actualize one's potential, such as by working hard and being disciplined vs. being lazy and self-indulgent. For example, you might have some natural talent as a piano player. But you will only become a really good pianist by practicing for an hour each day for several years. Every day you must make a choice whether to practice piano for an hour, and this choice is an exercise of free will which results in you improving and becoming a good piano player.

Chapter Eight: The God Experiment

Countless thinkers, philosophers, and scientists have struggled with the question of reason vs. faith. Here I will argue that it is possible, and, if we really want to do this, it is fairly easy, to design an experiment which will specify precisely what we would expect to see in reality if our religion were true, and then observe existence for a period of time, and evaluate whether or not our experiences confirmed or refuted our religious beliefs.

The great challenge in designing a "God experiment" is the human brain's ability to invent post-hoc rationalizations, i.e. after the fact justifications, to justify our doubt-inducing experiences and protect our sacred beliefs from empirical refutation. Two popular post-hoc rationalizations relate to religion. The first post-hoc rationalization addresses the belief in magic, and the second relates to belief in the existence of God. If magic were real, we would expect to see magic and to be able to do magic and make things happen by casting magic spells. When we don't see magic as the engine that drives human life, and instead we see technology like cars, computers and televisions, this shows that magic isn't real and science is real. But the post-hoc rationalization that reconciles the data with the belief in magic is "magic is real, but we can't see it, we can't experience it in physical reality, because it isn't physical, it is spiritual, or because our human brain is too weak and fallible and is unable to see the world of spirit where magic exists."

Let us be intellectually honest and recognize that, if we define a set of expectations before the fact, and we wanted to test whether the hypothesis that "magic is real" is true or not, we would expect to see magic, and we would expect to see that magic has a real, powerful, visible role to play in the world in which we ourselves exist and live our lives. We would also expect to be able to cast magic spells to accomplish real, specific tasks, and we would expect our magical powers to be successful. If we do not see what we would honestly expect to see, then the scientific method refutes the existence of magic, and reason shows that magic does not exist, and can only be believed by means of irrational faith.

Similarly, if we were to list our before the fact expectations about the existence of God, as understood by the Jewish, Christian, and Muslim religions, we would expect to hear God's voice speaking to us in a manner as real and audible as the voices of other people when they talk to us, we would expect to have conversations with God, we would expect God to answer our prayers, and we would expect to see the hand of God in the world and see God controlling everything, and to see God's miracles every day, and all over the place. The stories in the Bible are full of Biblical figures actually hearing God's voice and seeing God's actions and experiencing miracles. If the stories in the Torah or New Testament were factually correct then we would have every reason to expect to see in today's world the very same things that are described in those stories from two thousand years ago. We would also expect all our achievements, and all human achievements, to come from God. And we would expect atheist science to be completely powerless to do or accomplish anything without God's powers, at least to the extent that science disputes religion. For example, we would expect the theory of creationism to be a great and very useful explanation, and we would expect the theory of Darwinian evolution to be foolish and useless.

In sharp contrast to the God expectations, what we experience is the visible, perceivable, physical world, and what we experience is best explained by the scientific interpretation which explains everything in the physical world by reference to physical causes. Science is suggested by everything we see and hear, from the theory of air waves explaining sounds we hear, to the theory of light explaining our eyes and what we see, to science developing every useful and beneficial technology we use every day, from computers to cars to refrigerators to televisions to mobile devices and smart phones. The theory of gravity explains how the Earth formed and why we stand on the ground, and the theory of evolution and natural selection explains the existence of the various species on Earth and has been useful for the progress of human biology and genetics. The scientific understanding of the world, confirmed by our experiences as experienced prior to our mind imposing an interpretation that we want to believe in, leaves no room for God. We see science causing everything, and we do not see God causing anything. If God were the omnipotent power in the world, we would expect to see God's hand and to see miracles every day and all over the place, and we do not. If God existed and was all-powerful as the prophets claimed, then the signs of God's existence would be not only visible, but also quite obvious and unmistakable. We would expect to see the people who believe in science frustrated in failure, not raised in the triumph of the power of technology to enable humans to do things and prosper.

Did you see a miracle today? And yesterday? And the day before, and in most of the experiences you have? I predict that your answer, if you are honest, is no. Did you use electricity recently? And did you do what you were trying to do using your electrical device successfully? If you did, then your experience supports science and refutes God. Instead of seeing God everywhere, we see science and technology every day and everywhere we look. The only way to reconcile our experience with the belief in God is through the post-hoc rationalizations and after-the-fact justifications described above, but post-hoc rationalization is intellectually dishonest, it is manipulation of the scientific method designed to save the ideas that we want to believe in from being refuted by the evidence that we ourselves have seen when we observed the world.

I do not want to tie the proof of science to the existence of evil, but, also, if God existed and a loving all-powerful God mapped out our lives for us then we would not expect to suffer, and to suffer from and be controlled by physical wants such as hunger and thirst. We would expect our faith to empower us to get our prayers answered such that when you make a wish it comes true. And we would expect the evidence to confirm that God created man, and we would not expect fossils and paleontological evidence that evolution created humans.

The idea that it is selfish and petty to expect God to answer all our prayers and grant all our wishes, and we should believe in God even if we don't always hear His voice and He doesn't always answer our prayers, is after-the-fact justification to explain away the fact that we don't experience a God who hears our prayers and answers them. An honest thought, defining expectations based only on the theory of God and the Bible and prior to knowing what we will or won't experience when we run our experiment, would expect God to answer all, or most, of the prayers of the true believers, and to grant the wishes of those who have faith. We would expect God to speak to us regularly as he did to the characters in the Bible, and to regularly make visible, perceivable interventions in our lives like sending plagues to Egypt or handing stone tablets to Moses or removing the body of Christ up to Heaven. The idea that we should believe not for any rational reason, but because of irrational blind faith without any expectation that our prayers will ever be answered, is, as I see it, an admission and confession that the scientific method and the God experiment are capable of proving that God does not exist.

The application of the philosophical scientific method to the question of proof that God does not exist provides the perfect example of what it looks like when we empirically test an idea that is really false but which many people want to be true and wish they could believe. In this situation, people will employ certain standard techniques and tactics to seek to escape from reason and reality. We can picture this as a philosophical game of chess played between an atheist and a theologian, where the chess match plays out as a series of moves in the philosophy game. The atheist designs the philosophical scientific experiment and concludes from the data that the hypothesis was disproved because the observations of experience did not match the expectations. The theologian then makes four counter moves to claim that God was not disproved.

First, the theologian argues that the facts and data really did match the scientific expectations, e.g. by telling an anecdotal story of a near-death experience where someone claims to have gone to Heaven and seen God, or another story where someone says they witnessed a miracle where a terminally sick person was healed by magic. I call this factual manipulation. The atheist replies that he has not had any such occurrences of miracles or magic within his own personal experience, and the stories which assert that miracles exist usually end up being proven to be a hoax or the narrative of a crazy person, so the atheist would rather trust his own observations than doubt his reasoning by having faith in what other people say. Here the theologian uses his knight to take the atheist's pawn and the atheist then captures the knight.

Second, the theologian says that the expectations of the God experiment were incorrect because if God existed we would not really expect him to answer our prayers and we would not expect to see miracles and science would still exist. The argument for this is something irrational and nonsensical such as God is invisible and we cannot know or understand what his hand in our affairs would look like so we would not see plain and obvious evidence of him in our lives despite his supposed omnipotence and all-importance, and we should simply ignore the fact that science is a complete and adequate explanation for everything in the Universe which does not involve God. I call this intellectual dishonesty and post-hoc rationalization. The atheist's counter move is to cling to his intellectual honesty as the guiding star of his thinking and to be honest and sincere in defining what he would expect to see in reality if a specific idea or theory were objectively correct and truthful. Here the theologian attacks with his bishop but the atheist moves his rook and captures the bishop.

Third, the theologian argues that God can't be disproved because religion and belief is of such a nature as to be excluded from the realm of reason and proof. The idea holds that science is one sphere, where reason and logic prevail, and religion is an entirely different sphere, where faith and irrational nonsensical beliefs cannot be disproved or refuted no matter how badly they are contradicted by cold evidence. I call this the exclusion assertion. The atheist replies by asking what it is about religious ideas that would make them special and of a nature such that reasoning from observation and experience would not apply to them, other than the fact that they can be disproved by rational experiments and their advocates want desperately to believe in them. Failing to find a persuasive answer from the theologian as to why the existence of God cannot be proved or disproved, the atheist would be unmoved by this thrust. Here the theologian makes a mistake and the atheist pins his opponent's queen against the king and then captures the queen.

Fourth, when the first three methods of engaging the philosophical method on its own terms have failed, the theological falls back on a sheer emotional assault against the atheist, screaming that reasoning is cold, cruel, and heartless, and human life would be empty without God's love. I call this the pragmatic psychological approach. The atheist, being a decent and upright sort of person, gets confused and says that the two of them were just having a friendly and honest debate, where no insults or ad hominem attacks had been used, and he doesn't understand why the theologian has gotten so angry. The atheist may be slightly shaken up, but he does not alter his beliefs because of emotional manipulation and psychological pressure. Here the theologian sends his two rooks after the atheist's king, but the king steps to the side and the attack is blocked when the rooks are captured by the atheist's queen.

The analysis of the moves in the philosophy game is interesting as an illustration of the philosophical scientific method and the techniques used against it. But it is also worth noting that despite all of the moves the chess match ends in stalemate, because at the end of the game of chess neither the atheist nor the theologian has changed his mind. The theologian will never be persuaded because he has a closed mind and he has chosen not to listen to reason, so reasoned arguments have no effect on him. What is really at stake in the game of chess is whether the atheist will change his mind. The atheist has an open mind and he does listen to reason, so religion's pseudo-scientific arguments could shake his convictions. The philosophical scientific method offers the ideas that enable the atheist to know that his conclusion is correct that God does not exist despite the theologian's techniques used against him.

God is a good example of the moves in the philosophy game, but the same techniques will be used whenever someone tries to persuade you of something which is factually not true but which they want you to believe. For example, say that a used car salesman tries to sell you a car. You see that the bumper and front passenger door are dented, which shows you that the car probably was in an accident previously. The dealer tells you that the car was never in an accident, and shows you a vehicle history report showing that it was never in an accident. This is factual manipulation. The dealer tells you that the bumper and door look that way because the car was designed in a way which makes it look odd, not because it had a crash. That is intellectual dishonestly. He tells you that you are not an expert on cars so you are not qualified to evaluate what a car looks like to determine its condition. That is the assertion that you are excluded from knowledge. And he tries high pressure sales tactics and seeks to coerce you into buying by being aggressive and bullying you. That is psychological pressure. As I said, God and the used car are merely examples of the pattern which tends to happen whenever desire conflicts with fact.

Let me conclude this section by identifying what I am not discussing here. I am not discussing the social need for religion or the psychological need for religion. Religion historically has played a beneficial role in the life of local communities, as a way to meet one's neighbors and socially bond with them. Religion also plays a role in human psychological functioning, especially by making a person feel like they have some control over their situation through prayer when they are really in a situation of total helplessness. It is beyond the scope of this book to discuss whether alternative substitutes could ever be devised for religion as it performs these jobs, or if religion is necessary for human existence. But we have no need to answer such questions. Here I discuss only whether the statement "God exists" is true or false. The God experiment proves that the truth is that God does not exist. Any tangential corollary matters speak only of practical details regarding whether the idea of God helps people, whereas my concern is with the objective truth.

Chapter Nine: The Philosophical Scientific Method as a Critique of Philosophy

One will not always be able to first design an experiment and then run an experiment. Sometimes, especially in the social sciences and the humanities but also occasionally in the hard sciences, one will be faced with a body of observations and/or empirical data which already exists, and one will need to analyze the observations or data to evaluate what theory is true or false. The challenge here, with post-observational analysis, is achieving an honest analysis free from after-the-fact justifications which make the data conform to the conclusion that is desired by the thinker, especially if the thinker's desires are contradicted by what the data actually shows. We will always face some danger of after-the-fact justification in post-observational analysis, because we know how the data turned out, which makes it much easier to force the data to conform to our desired outcome. But nothing makes intellectual honesty impossible in this situation. Our theory will always entail an embodiment or manifestation in reality, so the observations and/or the data will either contain some evidence that proves our theory, or it will contain evidence that disproves our theory, and an honest thinker will usually be able to tell the difference.

An example of post-observational scientific analysis follows. The set of data which I examine is the history of philosophy. My overall hypothesis is that the philosophical scientific method explains a lot about philosophy because philosophy is characterized by the driving motive of the war between science and religion. The religious philosophers have sought to use philosophy to protect religion by arguing that religious ideas are immune to testing via philosophical scientific experimentation, and, generally, by transforming philosophy from a practical skill grounded in empirical reality into mere abstract theory detached from any basis in practical reality, or by corrupting empiricism and the philosophy of science so that it is incapable of refuting religion. If this theory is correct, then we would expect the data to show that most of the popular philosophies are designed, in some way, either openly or secretly, to protect religion from science, or to neuter science and limit it as a

37

general tool for understanding reality and achieving knowledge of truth. What follows is my analysis showing that the data is what we would have honestly expected it to be if we had run a controlled experiment to test the hypothesis.

Chapter Ten: The Philosophy of Science

Many famous philosophers have sought to take the approach and methods of science and translate science into philosophy to create a truly scientific philosophy. These thinkers include Hume, Kant, Wittgenstein, and the experimental philosophers. In addition to the famous philosophers who tried to bring science to philosophy, a thinker also exists who tried to bring philosophy into science to achieve a philosophical understanding of science, namely, Kuhn. I argue that each attempt failed, and some or all of these attempts were thinly veiled disguises for a desire to drag the philosophy of science down a path that would not conflict with religion, for the reasons offered below.

Hume seemed to think that scientific theories are always capable of being disproved by new empirical evidence and that skepticism was the scientific attitude. Hume was a noted atheist who criticized religion, but he equated religion with knowledge as such, and argued that if faith is refuted then all knowledge becomes impossible and every belief should be doubted. He ignored the fact that science, when theories have been proven and experimentally verified, seeks to provide us some degree of knowledge. We can know that the Sun will come up tomorrow, because of the scientific postulates of astronomy, which science has empirically verified. In contrast to science, Hume claimed we cannot know that the Sun will rise tomorrow. Hume's rejection of faith was rational, but his rejection of knowledge based on analysis of the physical world was irrational. Hume was an atheist and an outspoken opponent of religion, but to the extent that he undermined science as knowledge, if science is the alternative to religion then Hume helped religion.

Kant argued that science can achieve certainty and universality only because the mind imposes scientific laws upon the subjective experience of reality. His basic argument was that subjectivism is the justification for scientific knowledge, because the mind imposes the laws of science onto experience in the act of creating what we experience. Scientific laws have no exceptions because everything that is experienced is constituted by the mind. This is, of course, completely backwards. The scientific attitude is that the mind revolves around the physical world. Kant's view that the physical world revolves around the mind is a religious idea because its real world practical application leads inevitably to the conclusion that faith and belief can alter reality, although the real meaning of Kant's philosophy is obscured by the many layers of complexity which he wrapped around his ideas. Kant's actual purpose was to protect religion from the rise of science. Science achieves knowledge that hydrogen and oxygen can combine to form water, for example, from an examination of the molecules and atoms, which are things in themselves in physical reality. The scientific mind learns from reality, it does not impose its subjective beliefs onto sensory experience.

Wittgenstein, particularly in his early "Tractatus" period, sought to apply the principles of mathematics into philosophy, specifically in the form of formal symbolic logic, with the assistance of (the author's namesake) Bertrand Russell and Frege. My favorite argument against symbolic formal logic applies a theory called the Chinese Room, originally developed by Searle in his paper "Minds, Brains, and Programs." Searle's Chinese Room thought experiment was designed as an attack against the theory that computers can think or that it would be possible to design a computer program indistinguishable from a human mind, and by extension it was a refutation of the computer program "functionalist" theory in the philosophy of mind. But I use it as a refutation of symbolic formal logic and the idea that logic is the way that people should think. My use of the Chinese Room is natural given that logic and computer programs are quite similar.

Searle's argument is that if someone in a room is given Chinese words and a computer software program to process them then he could put together Chinese sentences and the sentences would be correct according to the rules of the language. Chinese words would enter the room, he would process the words, and he would hand back Chinese sentences that would leave the room. But if he doesn't speak Chinese then he will have no idea what any of it means, despite having formed sentences that are correct Chinese and which could look like a conversation to a Chinese speaker outside the room. My argument is that logic and Chinese are identical to the extent that they are criticized by the Chinese Room argument. Symbolic formal logic can reach conclusions by processing symbols according to a set of rules, but if you don't know what the symbols mean, if you don't know what the logical notation symbols refer to in reality, i.e. the objective physical objects in reality to which the symbols refer, then logic is merely a meaningless computer program devoid of actual meanings relevant to human experience.

Logic, as the Analytic philosophers such as Wittgenstein posit it, has no theory of induction. It is also worth noting that logic has no theory of truth despite being based on logical symbols being T or F. Indeed, because logical symbols are not based on reference to objects in reality known from induction, it is inherently impossible for logic to achieve knowledge of truth and falsity, because in general a word or sentence is true if the thing that it refers to exists, and it is false if the thing that it refers to does not exist. If P and Q are mere symbols divorced from reference to reality then they can never be true or false. Logic can assume that a symbol is true or false by defining it as such and then calculate what logically follows from a combination of symbols, but it cannot ever actually know that something is true or false because it cannot justify in the first instance that P is true and Q is false. Wittgenstein famously stated that logic could not prove the truth of the statement "a rhinoceros is not in the room," even if one looks around the room and does not see a rhinoceros anywhere, although logic can convert that statement into a symbol and process it using logic. In his later "Philosophical Investigations" period, Wittgenstein said that science does not actually have a principle of induction, and we say that science is true merely as a matter of convenience or popular sentiment. Science, as understood by Wittgenstein and his Analytic heirs, is not something that we would be entitled to call knowledge, at least not knowledge justified by a rigorous intelligent theory of science as having a reasoned basis for belief.

Many Analytic philosophers argue that the philosophical equivalent of scientific experiments, in which theories are experimentally verified using empirical data, is thought experiments, in which a person "tests" a theory by analyzing whether the theory matches the person's "intuitions," which are "teased out" by thinking about the thought experiment. This theory of thought experiments which equates mental intuitions with empirical data has been widely accepted in academic philosophy. This theory has been taken to its most extreme form by the movement known as "experimental philosophy," which seeks to introduce an element of empirical observation into the thought experiment by going out into public and asking groups of people how they feel about a philosopher's proposed thought experiment. The obvious flaw in the methodology of thought experiments as science is that, whereas empirical data comes from objective physical reality, intuitions come from internal subjective feelings, and therefore a thought experiment is nothing like a real scientific experiment. Analytic thought experiments share the Kantian fault, at least in general.

However, a carefully constructed thought experiment based on what is actually physically possible might make a useful or interesting point. We can distinguish good thought experiments from bad thought experiments based on realism. Let me give you three examples of bad thought experiments. For example, a bad, unrealistic thought experiment asks what would happen and who the person would become if one person's memories are removed and transplanted into a brain in someone else's body. Another asks who the person would become if one person steps into a magical machine and two people come out. And a third wonders what the word "water" would refer to if a so-called Twin Earth existed where there was a substance that looked like water and behaved like water and was referred to by the name "water" but had a chemical composition different from H2O. These are all real examples of famous thought experiments. The brain transplant thought experiment was expressed by Bernard Williams. The machine where one person goes in and two people step out, possibly due to a malfunction in a Star Trek teleportation device, was developed by Derek Parfit. And the Twin Earth thought experiment was a famous idea of Hilary Putnam. It did not bother any of these philosophers, and it does not bother philosophy professors in general, that these thought experiments have no relation to physical reality or scientific possibility and they draw conclusions from sheer imagination.

Unrealistic thought experiments accomplish nothing because the scenario which they pose is not physically possible and what would happen in reality if it were physically possible depends upon the details of how it actually works and what made it physically possible. If one person enters a machine and two people come out, we need to know what happened inside that machine to identify which of the two people is the same as the person who went in.

Similarly, if a brain transplant happened, how was it physically possible? How precisely did the old brain and the new body interact? Unrealistic thought experiments are ghostly disembodied thinking with no connection to physical reality. Having debates about ghosts and goblins is a pointless waste of time. And if, for example, one reasons from a thought experiment involving a ghost to the conclusion that consciousness does not physically exist, then the use of thought experiments with premises that are not grounded in physical reality and scientific possibility might actively lead one to irrational beliefs.

Let me give more detailed examples of this principle using the three bad thought experiments that I listed above. If the mind is the brain, as I have argued in this book, then mental continuity and physical continuity are identical, and for a person to be the same person over a period of time consists of both his brain being the same and his mind being the same. Talking about moving one person's memories into another person's brain confuses the analysis of what is a mind by pretending that a person's memories could exist in another person's brain, which is not physically possible. We have no need to decide whether a person's identity comes from his memories or from his brain because a brain, when functioning properly as a mind, will always record the person's memories onto that brain's memory.

Likewise, saying that one person goes into the broken Star Trek machine and two people come out blinds us to what happens inside the machine. Whether one or both of the new people is the same as the old one who stepped into the machine would depend on how those two people were created from the one who entered the machine. And again, we do not need to answer this question because it has no relation to physical reality as it really exists. Brain transplants or body cloning, if they really existed, would be scientific questions of fact which we would need to observe and study in order to learn how they worked, and imagination and intuition cannot tell us anything about them. To discuss Twin Earth, if a substance looks like water and behaves like water and exists on an Earth-like planet and played the role in the evolution of life on that planet which water played on real Earth, then how is it possible for such a substance to exist and yet to not be the chemical H2O? Where in chemistry is there such a substance? If a thought experiment seeks to show that the word "water" could mean something other than water on the basis of the idea that water could be something other than water then the premise of the logic is flawed because water is water. Water behaves as it does because water is H2O, so the thought experiment is nonsense.

For an example of a good thought experiment, my application of Searle's Chinese Room thought experiment as a critique of symbolic formal logic is a proper, rational thought experiment, because it considers logic in the way that real people use it in physical reality, or, at least, in the way that it really can exist, and in the way that logicians and philosophers of language say that it should really be used. A good thought experiment using an example from political philosophy might say "suppose that X earns $10 an hour, and A earns $50 an hour, what follows? Is that fair? Then also assume that X and A are both employed as widget-makers and X builds 10 widgets each day and A builds 50 widgets each day. How does that alter your understanding of whether it is fair?"

Because these thought experiments are physically possible, my realistic scenarios tease out aspects of reality in a way that makes it easier for people to analyze reality, and in doing so the thought experiments show a truth that is useful for real people in the physical world. A thought experiment is also useful if it corresponds to or reduces to something that is physically possible. For example, the omniscient computer model thought experiment presented in the section on free will is not physically possible, but it corresponds to the ability to predict the future which actually exists in real human minds. If I had to categorize the two approaches to thought experiments, I would contrast the philosophical establishment's Analytic Unreality with my Analytic Realism.

But generally, thought experiments should not be viewed as an adequate substitute for philosophical scientific experiments. A scientific experiment goes out into the real world and observes data and experiences. Scientific experiments are not based on intuitions and feelings. The truly scientific approach to philosophy is to take philosophical ideas and actually design real scientific experiments to try to test their truth or falsehood. I call this approach "experimental philosophy." For example, if you think that sensory experience revolves around the mind, and subjectivism and solipsism are true, i.e. perception creates reality and the world is made of your own mind, then you should test your belief. Pick up a piece of warm metal, like the handle of a pan heated on a stove, and see whether your mind can impose a phenomena, the experience of the feeling of ice, upon the noumena, the thing in itself which you are holding in your hand. If the iron feels hot to you and your mind could not control it, this scientifically proves, or at least lends credence and support to the idea, that sensory experience comes from objective physical reality, and not from your mind. On the other hand, if your mind can make you experience a feeling of ice, then your mind is creating your sensory experiences. The Kantian might reply that the structure of the mind could not be controlled by a desire to feel ice, but when we speak of "the mind," we generally mean something that can be influenced by our feelings, desires, thoughts, and beliefs.

Kuhn in his book "The Structure of Scientific Revolutions" sought to explain the history of science, and specifically the history which consists of "paradigm shifts," i.e. periods of time when science undergoes rapid evolution and mutation, where old ideas once widely believed become widely discredited, and brand new ideas emerge and achieve leadership in the world of scientific theory. The best examples are Newtonian physics being replaced by Einstein's Relativity Theory, or the theory of atoms as the building blocks of the universe being replaced by the theory of subatomic particles, although the Copernican Revolution replacing Ptolemaic astronomy is another good example. Kuhn argued that the old paradigm collapses into self-contradiction and the new paradigm emerges from the rubble of the older paradigm's contradictions and failures, but our choice of which paradigm to embrace is based simply upon what new scientists want to believe. Kuhn squarely rejected the idea that a paradigm shift represents a step of objective absolute progress on the path towards achieving knowledge of truth, because Kuhn did not believe that something exists which can be called "the truth" or "objective reality." Instead he thought that only paradigms exist, and that no paradigm can claim to be the objective absolute truth any more than any other paradigm. Thus, Kuhn's belief, although couched in the language of science, is merely philosophical subjectivism and relativism.

Let me offer a different interpretation of paradigm shifts. One thing exists, which we can call "physical objective reality", which all scientists study, and all human beings observe when their senses

45

experience perceptions of the physical world. Theories which match objective reality are true, and theories which do not conform to objective reality are false. Thus, a paradigm which is false can be replaced by a paradigm that is closer to the truth, and when a less true paradigm is replaced by a truer paradigm, this represents actual progress on the path towards science one day achieving perfect knowledge of reality. If reality exists, then one set of theories exist which perfectly describe reality, and the goal of science is to make progress on the steps of scientific theoretical development, and to go step by step until we reach a theory that perfectly describes reality. The best metaphor to explain this is steps on a staircase: none of the steps are the same thing as the door at the top of the stairs, but each step brings you closer to the door. The door exists and will be reached via climbing up the staircase step by step, because the stairs have a specific finite number of steps, so once you have ascended all the stairs you will reach the door.

The Earth really does revolve around the Sun, so Copernican astronomy is not merely a new system which arose out of the self-contradictions of Ptolemaic astronomy and which has truth only in relation to other paradigms, instead, it is the truth. If, in objective reality, space can bend, and time is relative, and subatomic particles actually form the building blocks of atoms, then the Relativity paradigm shift and the subatomic particle paradigm shift brought us closer to understanding the universe as it really exists. It is wrong to say that Newtonian physics and Relativity physics are relatively better or worse than each other but that they cannot be evaluated as true or false in absolute terms relative to "the Truth". If Relativity theory is true, and Newtonian physics is not true, or if Newtonian physics was merely a small part of a bigger picture which also includes Relativity physics and might one day include new theories also, then the paradigm shift brings us closer to the truth in absolute terms of progress towards the end goal of science, which is knowledge of reality.

I theorize that three "holy grails of science" exist, which are the end goals of science. For physics, the Holy Grail is the so-called unified field theory, which would explain and justify every physics equation. For chemistry, the Holy Grail is an explanation of why atoms behave as they do based on the numbers of protons, neutrons and electrons which they contain. For biology, the Holy Grail is how and why DNA defines traits. Once science achieves the Holy Grails, it will know literally everything, and nothing new will be left to learn, and no future evolutions or changes will be possible. Everything about the Universe can be deduced analytically from the unified field theory, and similarly the Holy Grails of biology and chemistry will provide knowledge of every detail of each truth about how life works and how objects exist. Progress will stop, having achieved its desired target, once we climb the stairs and reach the door.

And if we achieve the Holy Grails, then our technological application of this science would enable us to achieve an evolution in our level of civilization. We would use our knowledge of how DNA works to design genes for us to live forever and be super-smart and super-strong. We would use our knowledge of physics to design means of space travel so that we could colonize foreign star systems. If science were to achieve advanced progress, then we could use new chemical reactions to create clean cheap plentiful energy, and genetically engineer food sources to solve world hunger. We would be able to cure every illness and disability, as for example with stem cell research and genetic engineering. This would lift every human being in the world, including the masses of poor, into a standard of living better than the billionaires enjoy right now, such as by creating robotic butlers to serve people and do all manual labor for us. As other thinkers have also said, at some point science and science fiction will be indistinguishable. Science does not move in a circle with one paradigm collapsing into another in an endless rotation that does not move in any direction. Instead scientific progress moves in one straight line, towards the Holy Grails.

Other philosophers of science have offered famous theories, such as W.V.O. Quine, and Karl Popper. I discuss Popper at length elsewhere in this book, and I will give one brief thought on Quine here. To the extent that he posited science as a holistic web of theory which touches empirical reality only at its outer edges, and asserted that objects are merely a useful way to think and do not exist as such in objective reality, Quine is merely a scientific Relativist, with his web being basically the same as Kuhn's paradigm. This becomes clear in Quine's "Speaking of Objects," where he says that the concept of objects comes not from objective reality but merely from cultural modes of thinking and speaking which could be different for a different culture or language.

The ideas at the center of Quine's web of science would never be tested by experiments and therefore Quine's science would actually be unscientific. For example, if Quine had not focused on culture and language and had instead observed whether the contents of his experience of reality were things or non-things then his theory of objects would have been different and scientific, since humans generally observe objects in our experiences. The apple you eat for breakfast is an object that you observe in experience, and this confirms that objects exist in reality. Quine's Relativist theory of science as a web is refuted by the same type of argument which disproves Kuhn's Relativism. Merely saying that as a matter of desire one wants one's philosophy to be based on empirical science, which Quine said, is not the same thing as actually offering specific detailed theories and showing that each idea is in fact verified by experience and empirical data and objective reality, which is what I did in my presentation of the philosophical scientific method.

Chapter Eleven: Mind vs. Body in the History of Philosophy

The history of Western philosophy takes the form of a narrative, a story told to students in countless Philosophy 101 classes in colleges across the United States. By examining this philosophical narrative and the categories that it offers, we can learn more about the historical trends which have made philosophy into the enemy of science and the handmaiden of religion. The story begins with the Ancient era of philosophy, proceeds to the Modern era, and ends in the Postmodern era, which can also be called the Contemporary era. The story basically claims that Western philosophy has always been defined by an opposition between two rival schools of thought. In ancient times philosophy was defined by the conflict between the Platonic and Aristotelian Schools. In the Modern era the conflict was between the Empiricists and the Rationalists. And in the Postmodern era we now have the conflict between the Anglo-American Analytics and the European Continentals, with the latter category including Phenomenology and Existentialism.

The conflict between Plato and Aristotle was simply this: Plato believed that the physical world was governed by a spiritual world, the world of Forms, which are metaphysical ideas that give identity to the physical objects that participate in them. Plato called upon his followers to abandon the physical world and focus on the world of the Forms. He likened the physical world to shadows in a cave, and compared the spiritual world of Forms to the light of day outside the cave. Aristotle, on the other hand, believed that the laws of logic governed the physical world. He called upon his students to focus on, and to appreciate, the physical world. For Aristotle, the physical world was real, and the role of philosophy was to study the physical world, for example by analyzing the causes of material objects. The Plato vs. Aristotle conflict was continued into the medieval period by the Platonic philosophy of St. Augustine and the Aristotelian philosophy of St. Aquinas.

The Modern era abandoned Plato vs. Aristotle as dated, and ushered in the Rationalist vs. Empiricist conflict. The Rationalists began with Descartes, a Frenchman who sought to rebel against

Thomas Aquinas's Scholastic philosophy. Descartes's main idea was that to arrive at belief possessing certainty one must begin by doubting everything and then see which propositions remain as knowledge. He claimed that the physical world changes constantly like melting wax and so no knowledge of the physical world is possible. However, when the mind looks in on itself and relies on reason separated from perception, it can discover necessary truths. Descartes' classic argument is the famous "I think, therefore I am." According to Descartes you can know that you exist and that God exists, but you can't know anything about the physical world.

Spinoza took Descartes' ideas and attempted to apply them in a consistent fashion, arguing for knowledge that was purely from logical deduction from premises; his reason was devoid of any basis in perception. Leibniz, another Rationalist, posited that the physical world is an illusion and reality is made of non-physical spiritual substances called "monads." Rationalists in general are identified by an epistemic method that relies upon ideas derived from a reason freed from the physical world, a reason that provides conclusive demonstrated knowledge, in opposition to the ignorance of sense-perceptions.

The British Empiricist movement, which opposed the Rationalist movement, began with Locke. Locke claimed that one should form conclusions based on the perceptions that were apprehended by the senses. Locke acknowledged that it would be difficult to know objective reality from sensations, but resolved this with his "primary-secondary quality" distinction: some perceivable qualities are objective, some are subjective. Berkeley later argued that if some qualities are subjective then all qualities must be subjective, that "to be" and "to be perceived" are the same, that no objective knowledge can possibly be gained from sense-perception, and that only the perception of God can create objectivity. Hume completed Empiricism with his skepticism, arguing that reality is to be judged solely by the clarity of one's sensations, that on the basis of perceptions it would be impossible to discover cause-and-effect relationships, and that therefore no valid predictions about the future could be made. Empiricists are identifiable as those who base knowledge on perceptions and acknowledge that the world of perceptions does not permit beliefs to possess a conclusive, demonstrative degree of certainty and reasoned knowledge.

50

Empiricists believe in basing ideas on sense-perceptions, but believe that no knowledge of objective reality is possible. Rationalists believe in the use of reason to achieve knowledge, but believe that reason should be independent of sense-perceptions. This is a version of the mind-body split. The body belongs to the physical world of perceptions, a world of contradictions where no knowledge is possible, whereas the mind belongs to the intellectual world, a world of reason and logic where knowledge is possible.

The Rationalists looked to the intellectual world, a world of "pure" reason and mathematical demonstrability, a spiritual world of ideas. The Empiricists looked to the physical world, a crude, vulgar, ugly world where no knowledge is possible, a world where physical matter is imperfect and fails to achieve spiritual ideals, but where one can have some sort of knowledge or belief based on the perceptions that one directly perceives. Rationalists and Empiricists both believe that reason and perception are opposed, but the Rationalists choose reason while the Empiricists choose perception. The Empiricist-Rationalist conflict is really a version of the mind-body fight.

The mind-body split is repellent to a philosophy which seeks to justify and promote science, and it is one of the chief focuses of philosophical scientific criticism. No mind-body split exists. The human individual is a mind in a body, and mind and body should work in harmony together. In terms of epistemology, the mind should operate according to the principles of sound reasoning, and yet reason should be based on sense-perceptions based on the body's sensory experiences, and reason can achieve rational explanations of physical events. Reason does not apply to ideas separate from the physical world. Reason applies to physical objects in the physical world, which obey the knowable laws of science.

The mind-body split is at the root of the two Modern categories, and science must emphatically reject the mind-body opposition. This is why it is impossible to call the philosophical scientific method either an Empiricist or a Rationalist philosophy. This explains why followers of both camps from the Modern era have every reason to scorn and reject a truly scientific philosophy. Because the philosophical scientific method rejects the basis for the Rationalist and Empiricist categories, it is almost impossible to classify it as an Empiricist or a Rationalist approach to understanding the world. It could be called Empiricist in the sense that it looks at experience, but it can also be called Rationalist in the sense that it uses reasoned analysis and claims knowledge, and really neither category fits well as a description of my philosophy.

Now let us look at the background to the conflict which defines the Contemporary era, also known as the Postmodern era, which is the Analytics vs. the Continentals. To examine this conflict we must go back to the mainstream philosophical historical narrative and look at Kant. According to the traditional narrative, Kant is the "hero" of the story because the Kantian philosophy incorporated elements of both Empiricism and Rationalism, giving birth to all future philosophy. Kant claimed that it would be possible to achieve knowledge, such as belongs to the intellectual world, in the physical world by accomplishing a subjectivist "Copernican revolution:" instead of the mind revolving around the world, Kant claimed that the world revolved around the mind.

Kant said that sensations were "phenomena" that resulted from the things-in-themselves, the "noumena," acting upon the mind. He stated that the mind forced the laws of science onto the phenomena in the act of constituting experience, and that this was the reason why such laws are universal and necessary: simply because the mind applies them to every phenomena, the result of which is that there are no exceptions, achieving universality and necessity. But Kant's "accomplishment" consisted merely of applying Rationalist and subjectivist ideas to sense-perception, without doing anything to solve the mind-body problem. Kant articulated the analytic-synthetic division, to distinguish truths which were necessary by definition vs. truths which are contingent and not deductively true, but he said that his a priori "pure reason" worked for both analytic and synthetic truth, such that even contingent truths become necessary because the mind imposes them onto reality.

Kant's ideas have been influential in virtually every area of academic philosophy. He presented the classic mind-body split in a new way, which has developed into the Postmodern categories. In contemporary times the Anglo-American Analytics are on the mind side, seeking knowledge purely from the analysis of concepts (from Kant's analytic), while the Continentals are on the body side, seeking knowledge from history and believing in contradictions (from Kant's synthetic). Analytic methodology focuses on teasing out the implications of concepts and deriving truth from thought-experiments, seeking knowledge entirely from within the mind. Continentals take the opposite approach: they focus on "being" as it relates to human beings, but they rely on a notion of being as containing contradictions, and they avoid logical analysis as a mode of inquiry.

Most academic philosophy is opposed to science, but not all of it functions as an enemy of science. I define myself as an Analytic, and the faults attributed to Analytic philosophy listed below belong to many Analytics, but are not shared by all Analytics. In the next section I will discuss Physicalism and Mind-Brain Identity Theory, two Contemporary schools of thought which are consistent with science.

To add some support to this history of philosophy analysis, we can see in Wittgenstein, Frege, Russell, and the other Analytic philosophers who were the developers of modern logic, a methodology of relying on the truths of symbolic logic, not to analyze the physical world, but to develop tautologies: the basic idea is that the proposition "A rhinoceros is not in this room" cannot be proven by logic, but "If p then q, p, therefore q" or "p and q, therefore p or q" or "p or q, not p, therefore q" can be proven by logic. What is missing from this account of logic is inductive reasoning fueled by perceptual observations. Some advocates of symbolic logic pay lip service to inductive reasoning and sense-perceptual observations, but for the most part they believe in formal logic as something that is "pure" and capable of the same rigorous knowledge as mathematics, as opposed to the confused, muddled physical world of perceptions.

Some Analytics try to solve philosophical problems by clearing up problems in the language that we use, using the "philosophy of language" as their main methodological tool. These attempts fail to see that the philosophical problems are inherent in the things to which our words refer, not in the words themselves. Linguistic philosophers seek to make our problems go away by redefining our words so that our language can no longer say that the problems exist, which is an Analytic approach that seeks to solve problems from within the human mind, specifically from within reason's use of words. From the point of view of objective reality, the philosophy of language is pure subjectivism, which does not engage the physical world and instead hides from philosophical challenges by using altered definitions in order to pretend that the problems have been solved. For example, they think that one can know that trees exist because the word "tree" that names trees is used by humans, rather than knowing that trees exist because we look at forests and see trees.

And with most Analytics we see the widespread use of thought-experiments based on the observation of concepts as a replacement for scientific experiments based on physical observations. Analytics see no problem with basing their beliefs on thought-experiments set up in such a way as to contradict anything physically possible on Earth. For example, Analytics happily use the thought experiment of brain transplants, which are physically impossible, to prove that minds do not reduce to brains. The fact that in actual reality, as it really exists, biology is totally unable to achieve brain organ transplants and move one person's brain into another person's body does not bother the Analytics at all. Instead of basing their thought experiments on data from objective reality, they run the so-called "experiments" entirely in their heads, in their imaginations where brain organ transplants are possible. Thus the Analytic thought experiment shows only what the Analytics already believe within their own minds, and does not collect data from objective external reality. With almost all Analytics we see a method influenced by Kant's pure reason and based on the analysis of concepts without reference to the physical world as it really exists. Analytics use logic to tease out the implications of their intuition, but the source of all their knowledge is intuition. They make no attempt to ground their logic in the world as known from sensory experience.

On the other hand, with Heidegger's Phenomenology we see a quest to replace metaphysics with ontology, by escaping from things as the objects of study by a subject and replacing this with the quest for being, in other words, with the things merely being there and presenting themselves as they are. We also see an attempt to draw conclusions from time and history, which the Analytics would criticize as synthetic and contingent and not logically necessary. Marx engaged in a deeply historical analysis, and Foucault also looked to history for answers, something no self-respecting Analytic would stoop to. Analytics apply their analytic ideal by seeking knowledge purely from concepts. Continentals, on the other hand, actually look at the world around them.

However, the world that the Continentals see when they look around is not one governed by the rules of reasoning. Heidegger sought to escape from the knowledge of a subject knowing an object. He wanted being to reveal itself after reason and analysis had been removed, but if you remove reason and analysis from philosophy, then merely nothingness and irrationality are left behind, and this is precisely what one finds in Heidegger's philosophy. Kierkegaard, the founder of Christian Existentialism, based his metaphysics on a contradiction: Christ is both human and divine. And, for Kierkegaard, devotion and faith can never earn for a believer the right to go to Heaven, so one must throw oneself on God's mercy for no rational reason whatsoever, and be religious out of pure faith, with a void of nothingness replacing any rationalization or reasoning for why one should choose to be faithful. In Kierkegaard we see a stark contrast between the two sides of the battle of reason vs. faith, and we see what it looks like to make a principled decision to choose faith and reject reason.

Sartre, taking his turn at Existentialism, based his beliefs on what he saw as a set of contradictions inherent in human existence, such as the contradiction between the in-itself (reality) and the for-itself (consciousness). Continentals care about existence and being as it is for human beings, but they see this as inherently relying on contradictions. When a Continental attempts to explain his or her experience of reality, as a consequence of not using reasoned analysis, the account is unclear, vague, and pseudo-mystical. Failing to find anything comprehensible in the absence of reasoned logical analysis, some Continentals come to believe that there isn't anything to find, and turn to Nihilism. At its heart, Existentialism, in Sartre and others, is really a new word for Nihilism, the philosophy that believes that everything is nothingness. As an inherently negative philosophy, Nihilism attacks all positive ideas and says that they are meaningless and empty. Then, instead of offering an alternative to other philosophies, Nihilism says that nothing exists and nothing can be known, and therefore one must either live a life that is a meaningless lie that pretends to have meaning or else one can accept the darkness of nothingness and live without meaning. As the ultimate conclusion of a life devoid of meaning one can commit suicide, which explains the role of suicide in Continental thought.

Nietzsche has been claimed by the Continentals, and that is an accurate historical classification to the extent that Nietzsche rejected disembodied abstract reason as a means of knowing timeless universal truth, and he asserted that knowledge was impossible. But to the extent that Nietzsche put forward a substantive body of ideas, a philosophy focused upon achieving joy and a zest for life on Earth for physical human beings, and particularly an ethics with new values, Nietzsche was not precisely a Nihilist, because he believed in something (the Overman). He does not fit comfortably with Existentialism and Phenomenology.

In a very real sense, there are two Nietzsches. First is the Nietzsche whose commitment to radically new ethical values and the physical world gave rise to Ayn Rand as the heir of Nietzsche. Secondly is Heidegger's Nietzsche, who was merely a traditional Continental who opposed Analytic reason in favor of the Existential/Phenomenological rejection of timeless truth and absolute knowledge. Rand's Nietzsche rejected Christian anti-physicality in favor of moral values based around survival in physical life on Earth, whereas Heidegger's Nietzsche opposed the idols of reason and knowledge in favor of a wine-soaked frenzy of irrational emotions. Having mentioned Rand, let me point out that, although she is rejected by the academic philosophical establishment and is perceived as the rebel of philosophy, Ayn Rand's philosophy is simply Aristotle's metaphysics and epistemology paired with Nietzsche's ethics. Rand's original contribution to philosophy was to apply her unconventional Aristotle-Nietzsche combination to political philosophy in order to deduce capitalism as a logical politics suited for the Overman as productive genius businessman.

The Analytics seek to proceed from disembodied logic, while the Continentals seek to directly experience an illogical physical existence. The Analytic-Continental conflict is merely the new version of an old classic, the mind-body fight. The philosophical scientific method, which does not accept the mind-body split, the analytic-synthetic division, or the reason-sensation contradiction, does not fit into these philosophical categories, and rejects the basis for those categories. The philosophy of science would condemn the Analytics for seeking knowledge from the mind without reference to the physical world, and it would scorn the Continentals for looking at the world without any rational analysis.

A fundamental attitude about the intellectual world, the world of math and concepts, being opposed to the physical world, the world of flesh and blood, in other words a mind-body split, is at the heart of the Analytics and Continentals. Science must not accept that fundamental idea. Science should reject the mind-body split as detrimental to reason and happiness. Reality is not two worlds, an intellectual world and a physical world opposed to each other. Reality is one world, which is both intellectual and physical, a world where physical objects obey intellectual rules.

Chapter Twelve: Physicalism and Mind-Brain Type Identity Theory

In contemporary philosophy at the academic level, two popular theories support science, contrary to the overall theme of philosophy as a tool to protect religion from science. These two theories are Physicalism and Mind-Brain Type Identity Theory. This section offers two interesting ideas that might be useful for defending the doctrines of Mind-Brain Type Identity Theory and Physicalism against two of the most damaging assaults against those theories, namely Hilary Putnam's Multiple Realizability argument and Frank Jackson's Blind Mary Knowledge argument.

At the outset, let me offer a set of definitions so that we know what we are talking about. By "Mind-Brain Type Identity Theory" I refer to the claim that the mind and the brain are identical, i.e. they are both one and the same thing, but talked about using two different sets of terminology, an internal terminology of introspection used by a person to describe what his brain looks like from the inside, and an external terminology of neuroscience used by scientists and biologists to describe what someone else's brain looks like when dissected under a microscope. By "Physicalism" I refer to the claim that reality exists physically and that only physical material objects exist, and no non-physical or spiritual substance exists or is capable of existing. The insight offered here is to employ Mind-Brain Type Identity to assist in the defense of Physicalism, and to apply Physicalism to assist in the defense of Mind-Brain Type Identity, such that Physicalism and Mind-Brain Type Identity reinforce one another.

Frank Jackson claimed that if a blind person (named Mary) knows everything scientifically true about light wavelengths then her missing experience of what colors look like is a non-physical component of experience and consciousness. In detail, Jackson said that Mary could see but was trapped in a black and white room and also that she was a brilliant scientist, although these details are largely immaterial. To summarize my argument up front, in reply to Jackson I argue that the thing "red" is a wavelength of light frequency, which exists physically, and that the experience of "red" when seen by the eye does not entail any non-physical aspect, but that different information about "red" is entailed in either seeing it with the eye or learning about it conceptually by a scientist talking to you about light waves. What red looks like is one aspect of red and the frequency of red light wavelengths is another aspect of red, in the same sense that having four sides is one aspect of a square and having four right angles is another aspect of a square. They are the same thing, but as two difference aspects of it. A square is one thing, but it has multiple aspects which are distinguishable as different aspects. Color, as well as other physical objects such as an apple, can similarly be one and the same thing but with multiple different aspects. For example, an apple can be red and round, and the same thing is both red and round, but its being red is its color aspect and its being round is its shape aspect, so the one thing has two different aspects.

Mind-Brain Type Identity Theory assists in the Physicalist counter to Jackson by enabling us to argue that types of thinking reduce to sections of the brain, so that our experience of mental events can be understood as such. The collection of sensory rod and cone cells in the eye's retina, connected to the occipital lobe in the brain, is one type of information processing device, which analyzes light waves, while the brain's frontal lobe of the cerebral cortex is another type of information analysis tool, which processes a different set of information, namely ideas and concepts. It is unreasonable to expect every aspect of one physical thing to be the same or look the same as every other aspect of the object, especially when the two aspects are processed by different brain structures. It is coherent for one physical thing, e.g. the color red, to have both a visible aspect and a conceptual aspect, which are distinguishable from each other. Red as seen and red as thought about are one thing, they are the same thing, but it is the same thing analyzed using two different methods of analysis based in two different regions of the brain, one region of which processes data received by the eye's retina to know perceived properties, and the other region using reasoning and concepts to process properties that are thought about. So we can understand what knowledge of redness Mary possesses and what knowledge she lacks without saying that the color which we experience through sight is a non-physical consciousness. Mine is an approach which takes the principle of Mind-Brain Type Identity and uses it for the benefit of Physicalism against Physicalism's critics.

The Frank Jackson argument says that Mary knows everything scientific about the physical characteristics of the color red although she has never seen a red object. Jackson implies that an abstract theoretical knowledge of color is all that there is to knowing color as a physically existing thing. But if color physically exists then it exists as physical objects, and the direct knowledge of color comes from the physical eyes seeing the physical light reflected off of those physical objects. When you look at a red apple, the apple reflects light which enters your eye so that your brain then has a perception of redness which your brain then processes to see the red apple in reality. If your eyes don't see the red apple then there is no physical redness which you know. If someone describes an apple to you or if you see measurements from scientific equipment which shows you numbers then you know something, but not redness itself directly from the physically red thing.

If Mary has never seen red then she does not in fact know everything about the physical properties of redness, despite the possibility that she understands the scientific statement "redness is the quality of reflecting light wavelengths of frequency R." She has a filtered secondhand knowledge of redness but no direct knowledge of the physical existence of red objects, since she has never seen a red apple or anything else that is red. Mary's brain might have abstract theoretical knowledge of how eyes work, but that is not the same thing as her eyes actually working, and her eyes have never seen a red thing. There are no physically existing red objects in the black and white room that are present for Mary to see them and experience their color. Mary's eyes physically exist, and the absence of red objects in the room is an objective physical fact, so her lack of knowledge can be attributed to her physically existing eyes not seeing colorful things. We do not need to appeal to non-physical spiritual consciousness to explain the experience of color which she lacks. Mary lacks the experience of redness because no red objects exist in her room and the experience of redness is identical to red physical objects.

If you remove the mistaken hidden premise that abstract theoretical knowledge, e.g. a scientific explanation of color, is the equivalent of, and identical to, direct knowledge of a physical object, e.g. seeing a red apple, then you collapse the Blind Mary argument against the experience of color being explained purely in physical terms. If the color red consists of all the physically existing red apples and all other red objects in physical reality then only the physical means of knowing redness, e.g. sight and color vision by the eyes, would give direct knowledge of red color.

Similar to the use of Mind-Brain Type Identity for defending Physicalism, Physicalism can also support Mind-Brain Type Identity. Putnam argued that types of mental states such as pain cannot reduce to types of physical states such as the excitement of pain neurons in the human brain because a non-human mind with a different neural substrate, e.g. an alien's mind, could also feel pain but have a completely different neurological or chemical process. To summarize my reply to Putnam, I begin by arguing that, if we believe that only physical substance exists, then the mind must reduce to the brain, because we can see that we have minds in our heads and nothing else is there for the mind to be, so any arguments to the contrary cannot be true. Having established that, I would then note that Putnam's argument takes advantage of a sloppy and imprecise definition of what we mean by a mental state. If the experience of pain consists of a harmful or dangerous object acting upon an organism's sensory nerves and sending signals to the organism's neural decision-making center to inform the entity that it has been harmed, e.g. a person's hand gets burned by fire and he feels pain, then pain would be the same for all living things, both human and alien.

For example, if an alien was vulnerable to fire and it was burned and the fire acted upon its sensory organs and this sent an idea of danger to its organ that makes decisions, then this would be the alien's mental state of pain reducing to a physical state, in the same manner as could be described for a human in the abstract sense of how we defined pain. In other words, if we define "pain" as "the biological reaction of a living organism to objects or stimuli that cause harm or danger or are recognized by the organism's sensory, neural, or decision-making organs as causing harm or danger to it", then only one possible realization could exist to pain according to our precise but broad definition of pain, because every human and alien pain would satisfy our definition. The different various alien and human physical realizations of the mental state called "pain"

61

would all be identical in the sense of meeting the criteria to be called "pain", which is what matters if our analysis deploys that definition of pain.

On the other hand, if we define "pain" as "that which happens when a human has an experience that hurts, in other words an experience that is usually described as painful, for example being burned by fire or cut by a knife," then our narrow definition limits us to only consider human brains in order for the argument to produce a consistent analysis, which then enables us to say that "pain" refers to a type of thing in human brains. If this is so, then pain has one realization, the excitement of pain neurons in the somatosensory regions in the human brain.

Putnam obfuscates the fact that, if we are to analyze the reduction of minds to brains, then we must keep our concepts and definitions distinct in terms of distinguishing when we speak specifically of humans or when we speak of minds that could exist in any evolved life form. Because humans are the only animals with minds that we currently know of, have access to, or can study, it is easy to blur the two and assume that all brains would be like human brains, but the best approach is probably to limit our analysis to humans, because we do not know what alien or non-human brains would be like if they actually existed, and science cannot offer deep meaningful assistance to philosophy because science has no mind-possessing alien brains to dissect and study. Because philosophy done by human beings is done for human beings, i.e. with the purpose of being useful for real people, I think that the biggest prize that Mind-Brain Type Identity Theory can win would be from proving that human minds reduce to human brains, without worrying too much about non-human alien minds. For a philosophical analysis undertaken by human beings, Putnam's objection is largely irrelevant.

And, if we use precise but broad definitions to analyze minds as such, and if we define "brain" as "the physical structure or structures within an entity which react to external sensory stimuli in reality by means of thinking and thought," then we can say that the mind is a type of thing which reduces to and is identical to the brain as a type of thing, even if the thing in an alien which fits the definition of "brain" is six different organs which communicate via a flow of green slime between them, while the "brain" in a human is one organ that communicates via neural synaptic connections. The "brain" will be the "mind" in any possible physical realization, in humans, aliens, or artificially intelligent computers.

Chapter Thirteen: Science vs. Religion

We have now seen what the philosophical scientific method consists of, and tested the hypothesis that philosophy is defined by the conflict between philosophers who seek to use philosophy to protect religion from science, and those who seek to justify and defend science from anti-physical attacks. We will now test another hypothesis, namely, the idea that science can replace religion as an overall explanation for the human condition. The next three sections, from here up to the end of Part One, will explore this in detail.

Here I will discuss technology. First, let me explain that science enables technology because science is the system for examining and understanding objective reality. Reality exists objectively, and science learns how reality works, so it follows naturally that science would produce technology, which consists of the practical ideas and knowledge of how to do things which are achieved as a result of the scientific method. Technology works because science succeeds as a method, and science succeeds as a method because reality exists and science looks at reality.

Second, we can ask, and must answer, this question: does technology drive civilization and progress, or does philosophy drive civilization and control the pace of progress? My answer is that technology drives civilization in the sense that technology controls what humans are able to do and defines our material wealth and our standard of living, but philosophy either allows technology to exist or else philosophy (in the form of religion) destroys technology. Scientists are the true heroes of progress and humanity, because technology adds to our material wealth and raises the standard of living, and technology is the engine which drives longer lifespans, more prosperity, and happiness in life on Earth for physically existing human beings. The scientists are the ones who actually explore the world and see how reality works, and who use complicated mathematical formulas, clever experiments, and exceptional achievements of reasoning in order to translate theoretical science into technological applications to improve the lives of you and me. However, science has an enemy, which is religion. The philosophers are the ones who determine whether science or religion triumphs in the court of popular opinion and intellectual culture.

It is worth noting the "philosophy of history" analysis of Ayn Rand. My interpretation of the Randian argument is that in ancient Greece and Rome, some degree of technological and social achievement was accomplished because the Homeric Greco-Roman religion was one in which humans were seen as good, noble, virtuous, and heroic, and humans were not viewed as pathetic insects meant to cower before the glory of the gods and reject the physical world in favor of spiritual revelation. The poems of Homer taught the ancient Greeks not to be afraid of the gods, and the Romans borrowed this attitude.

The rise of Christianity reversed the Greco-Roman trend. The Christian hatred of the physical world, introduced into the religion by the Platonic elements of the Gospel of John, destroyed ancient science and collapsed humanity into the Dark Ages. The Enlightenment saw a new awakening of science, but this was due to the Renaissance in Europe, which meant literally the "rebirth" of the Homeric Greco-Roman spirit. Philosophically, the work of Thomas Aquinas made Aristotle and Aristotle's pro-science view compatible with Christianity, such that religion was forced to give some breathing space to science. All modern science, from Newton to Einstein to today, is the descendent of the Renaissance spirit, which shifted from anti-science religion to a compatibility of science and religion. However, the fundamental principle of science, which is belief in the physical world, is inherently incompatible with the basic idea of religion, which is belief in the spiritual world. Religion and science are always in tension, and ultimately either one or the other will win out and triumph over the other.

Most people in the world are members of a religion who make use of technology in their daily lives. As such, most people believe that science and religion are compatible. However, when faced with a choice between idealism or practicality, most people will choose to do what they feel is the right thing to do, so the usefulness and pragmatism of technology will not save science if religion leads people to reject and abandon science. Although many religious people watch televisions and use computers and wash their laundry in washing machines, the religious believe in the principle of religion. That religious principle will lead the world to reject science

65

and destroy technology, although this historical trend will happen gradually, one step at a time, and so the extreme result of collapse seems unlikely to we who stand at the top of the slope looking down.

But a time will come when technology will offer immortality, by changing the genes which cause aging so that people won't get old and by finding medical cures for all diseases, and technology will also create an end to suffering, by raising our standard of living to the point where, for example, even the poorest person can buy cheap robots to work for him and clean his house and do his chores and can buy cheap abundant genetically engineered foods with delicious tastes and great nutritional value. At that point religion, which sells the afterlife and is the psychological opium for sorrow, at that point religion will either destroy science or be destroyed by science. The principle of science will undermine, marginalize and refute religion, as was understood when teachers first started teaching evolution in science classes in schools. The dueling ideals of science and religion will eventually play out to their logical conclusion, a war with a winner and loser.

The war has already begun. Science vs. religion has seen some famous battles, many of them in courts of law. One battle was the trial of Galileo by the Catholic Church for the heresy of advocating Copernicus's idea that the Earth revolves around the Sun. Another battle was the Scopes Monkey Trial where a school teacher was tried in court in the United States for teaching evolution to students. The religious authorities correctly feared that progress in astronomy (the heliocentric model) and biology (evolution and natural selection) would be the harbinger and herald of science replacing religion as the public's tool for understanding reality. In recent times religion has struck back against science, with extremism among Muslims who hate the West and view the West as the embodiment of science and progress, and also with fundamentalist Christians in America who reject science and advocate extremist faith. The world now faces a "culture war" of science vs. religion.

Obviously if science wins then technology will one day enable humans to live forever and travel among the stars. But if religion wins then science and technology will be wiped out and the human race would quite possibly go extinct. Our grandchildren and great-grandchildren will be condemned to a life of want and impoverishment and horror, with cars and washing machines and computers and genetically modified food gone, because electricity and technology will be wiped out for standing proudly against the Dark Ages mentality.

The path of a civilization is determined by the principle which guides the people, and the principle of science leads in one direction while the principle of religion goes in the opposite direction. It is philosophy, as the handmaiden of intellectual culture, which will determine the winner and the loser in the war of science vs. religion. Hopefully this book will provide some useful intellectual ammunition for the battles that lie ahead.

Chapter Fourteen: Religion as Myth and Mental Illness

If science is correct and religion is wrong, then we would expect science to have a plausible explanation for why so many people believe in religion. My theory of religion is that religion is a mixture of (1) mythology and (2) psychological defects arising from "design flaws" in the human brain. I can explain the myth part of religion very simply. Ancient humans circa 10,000 BC to 1500 AD did not have science, but they needed theories to explain the natural phenomena they observed, so they dreamed up gods and supernatural forces to explain lightning, fire, death, winter, etc. They saw lightning and dreamed up Zeus, the Greek god who threw thunder bolts at people in the ancient Greek mythology. The mental illness side of religion is the more interesting aspect, which I will discuss here.

What is a "mental illness"? Let me begin with two foundational premises. First, humans are animals. Second, the mind is the brain. If the brain's purpose as an organ in the body is to think, then we can define "mental illness" as a physical malfunction of the brain which interferes with the brain's ability to think properly.

My basic idea is that religion is a type of mental illness, not as extreme as insanity or delusions or schizophrenia, but lacking in sanity and based upon brain malfunctions. It should be clear that religion, in itself, would be defined as a "mental illness" by most psychiatrists if not for the fact that billions of people believe in religion and that makes it socially acceptable and immune from criticism. If someone came up to you on the street and said to you that God has spoken to them through a burning bush, you would think they were hallucinating. If some stranger told you that his wife was a virgin who had given birth to a child who was the son of God, your natural reaction would be to think that this person is completely insane. But for some reason, when a book written 2000 years ago says that God spoke to people and that the Virgin Mary gave birth to Christ, it is regarded as truth and not as evidence that the Bible's authors had mental health problems.

It is perfectly understandable that so many people believe in God. I believe that the human brain has evolved several "design flaws," problems with how the brain works which make it easy for human beings to believe in religion. A list of these mental illness vulnerabilities follows:

1. Confirmation bias. Brains have "confirmation bias" of thinking that something might happen, and then when it happens they think it is supernatural because there was a chance it wouldn't happen, or else they notice when what they expect happens and don't notice when what they expect doesn't happen so they believe in the supernatural. Thus when a person has a religious belief they tend to feel that their experiences confirm their beliefs, but when they have an experience which would refute their belief they simply ignore it. For example, when a believer wins the lottery they think God gave them good luck, but if the very next day this same person's best friend gets sick and dies they won't notice it as it reflects upon the existence of a supposedly omnipotent loving God. Another example of confirmation bias: you have a feeling that something will happen and then it does happen, you think you predicted the future, but you do not notice the ten other times today that you felt something was going to happen and it did not happen.

2. Prayer as a reaction to helplessness. Ayn Rand had a theory which she called "the primacy of consciousness," which is really subjectivism, the feeling that people have that thoughts, beliefs, and feelings, can alter reality supernaturally without any physical activity or physical causation. The brain knows that what it perceives is real, so some brains incorrectly infer that reality is coming from perception, when in fact perceptions and beliefs are coming from physical reality. This explains the belief in the power of prayer. When the brain is totally helpless it turns to prayer as the only thing it can do, and then when the person who prays is somehow saved they credit the prayer and believe in the supernatural. The people who survive disasters and who prayed to God and were saved then become fanatical believers who tell everyone else to pray and to believe in miracles. But nobody ever hears from all the people who had emergencies and prayed and their prayers were not answered and they died. When people think that subjectivism is true they think their beliefs and feelings can change reality, so when they are helpless they pray. I think the brain actually has a design flaw which makes the mind try to get control of the situation in situations where

the person is totally helpless, instead of simply accepting one's helplessness as a result of a reasoned evaluation of one's limitations.

3. The blank slate problem. The brain is born empty and is essentially programmed like a computer by what it is taught from ages 4 to 15. Young children have no knowledge or experience and most often they simply believe what they are taught. And as adults the human brain tends to simply go with what it already believes instead of questioning beliefs or using critical thinking directed in upon itself. This is why most people believe in the religion they were raised on and were taught by their parents. If there was one true religion then you would expect everyone to believe in the same religion, and if people reached religion as a result of thought and reasoned analysis then one would expect various religions to overlap geographically. The fact that each region has one majority religion (e.g. Christianity in Europe and the USA, Islam in the Middle East, Hinduism in India) is what you would expect if religious beliefs come from what parents teach to their children. It is rare for a belief to come as the result of what a person reasons independently, and it is unusual, although not impossible, for a person to doubt what he or she was taught by authority figures as a child. This tendency towards belief is caused by a design flaw in the human brain that causes brains to soak up beliefs that they are fed as young children instead of thinking critically from a young age.

4. Conformity and obedience to authority. Psychology at the academic level has established by scientific experiments that the human brain has two tendencies: (1) the obedience to authority tendency, and (2) the conformity tendency, both of which are wired into the human brain. Humans will take orders from a perceived authority figure that they would not take from normal people. Religion creates authority figures by declaring that the religious leaders are chosen by God, and the human brain has a tendency to obey these authority figures. Religious figures of authority like priests and reverends issue commands that people tend to obey, and the prophets of scripture are obeyed as authority figures. Also, an individual will tend to say that he believes things which parallel the beliefs of the group that he or she is a member of, even in defiance of his or her own independent perceptions. So religions get people to say they believe in the religion because everyone else also believes, and people tend to conform to the beliefs of the group. If atheism

was popular then most people would be atheists, but because religion is popular most people are believers.

5. The stress reaction. It is well established that the emotion centers in the human brain react to danger by shutting down the thinking centers in the brain and giving energy to the "fight or flight" biological response, which is accompanied by fear and panic in the emotional parts of the brain. This design flaw in the human brain makes a person behave stupidly and irrationally in crisis situations. Emotional panic might also make the person seek help from the supernatural to save him or her instead of calmly, rationally figuring out how to solve the problem.

6. Habits. The subconscious mind parts of the brain tend to develop habits, and the argument can be made that the human brain can be conditioned, like a dog or a bird, to repeat the same behavior if it has been given a reward repeatedly for performing that behavior in the past. This conditioning and subconscious repetition tends to embed traditions in human behavior, which would not exist if decisions were always made as a result of reasoning.

7. Hallucinations. Most people erroneously believe that they see with their eyes. That is not how the brain works. The eyes send neural signals to the vision centers in the brain, and the vision centers in the brain are what actually form visual perceptions which are "seen" by the mind, really, by the conscious mind which I believe is the frontal cerebral cortex or perhaps the entire frontal lobe. When the brain's neurotransmitter chemicals malfunction or the brain's neurons malfunction, it is very possible for the brain to believe that it is "seeing" something which simply does not exist. And these hallucinations are influenced by the ideas which are already in the mind, so if your brain malfunctions you might think that you are hearing God speak to you or seeing spirits which look like your religious beliefs. The brain plays tricks on the mind, and this is probably what happened to ancient "prophets" who thought God was speaking to them. I do not believe that the possibility of hallucination makes it impossible to achieve certainty and knowledge, and hallucination is not an excuse for philosophical doubt and skepticism. But we humans do need to be aware that our brains are vulnerable to design flaws, and also that the people we trust, and the groups we follow and go along with, are all of them vulnerable to human brain design flaws.

8. Problems in the different parts of the brain communicating with each other, i.e. a flaw in the evolution of interneurons and the neural synaptic connections that connect the different brain regions. This takes many forms:

A. Confusion regarding who hears your thoughts: When you pray, you hear your own thoughts and think God is listening to you. This is a brain malfunction. You also talk to yourself in your thoughts and think you are talking to God, which is part of the origin of prayer. Prayer is talking to yourself in your thoughts and hearing yourself think, and a brain malfunction and the lack of an evolved set of neurons within the brain for the brain to see its own activity permits the mind to think that it is talking to God and that God is listening. The mind sees someone listening but the neurons don't make the connection that the thing which is listening is also the thing that is thinking. As a matter of neuroscience it is well established that the center in the brain which speaks and forms words is not in precisely the same location in the brain as the center which hears and interprets words. I believe that religious mental illness can arise when these different hearing and speaking centers in the brain do not communicate properly with each other and when the brain does not pay attention to its own thinking and fails to pay attention to its listening to its own thoughts and is not properly self-aware of its own thinking, in other words if the brain does not know that it thinks to itself then it believes that it is speaking to God.

B. The visual parts of the brain when they hallucinate send signals to the frontal cortex/conscious mind, and you see the hallucination but your brain does not tell you that it is a creation of the imagination and not from the sensory receptors in the eyes. This is a design flaw in communication between the eyes and the brain and between the vision and imagination parts of the brain as they connect to the conscious thinking part of the brain. This would not exist if the human brain's neurons had evolved better. This would also apply to auditory hallucinations, waking dreams, etc. This problem relates to something that I call Oracle Syndrome. Oracle Syndrome happens when the brain's subconscious is behaving in an insane, irrational manner, and sends ideas to the conscious mind, but the conscious mind has no direct neural perception of where these crazy ideas are coming from, so the conscious mind believes that these ideas have a mystical source, and come from God, or the gods,

72

or from the supernatural. This explains why the ancient Greek and Roman Oracles who inhaled drug vapor fumes were regarded as capable of making prophesies which came from the Greek and Roman gods. The ancient world believed that these ideas came from the supernatural, because they were blind to the origin of these ideas in the malfunctioning subconscious brain of the drug-hallucinating Oracles.

C. The ability to "will" or wish, of the conscious mind to send signals to the brain stem or motor cortex (which control the muscles and move the body) desiring things to happen, is a design flaw because the brain perceives no difference between sending a signal to the muscles to move the hand, which it can do, and sending out signals to cast a magic spell to make it rain, i.e. to "wish/will" it to rain, which it cannot do. If the frontal lobe of the cerebral cortex, i.e. the conscious mind, received accurate signals from the brain stem and motor cortex to see where its signals went and what they did then this would simply not exist.

9. The Assassin's method: During the Middle Ages, a cult in the Middle East existed which was known as the Assassins. Founded by a man known as Hasan (not to be confused with the author), who was also called "the old man of the mountains", the Assassins was a religious cult which indoctrinated people and then planted them as sleeper agents in the courts of various lords and kings. When Hasan sent the signal to his servant, the assassin would shed his or her disguise and murder the political target. This is where the English word "Assassin" comes from.

Interestingly, Hasan had a technique for making people believe in his religious cult. He would stage events that actually physically existed but which looked like magic or a religious miracle. After seeing a real event by means of sensory perception that looked like a religious event, his followers would then have no ability to tell the difference between physical reality and religious doctrine. For example, Hasan would stage a fake public execution made to look like someone's head had been cut off. He would then set up a table where the person would hide underneath the table with his head poking up through a hole in the table, and the head would talk to Hasan's followers, telling them about the afterlife. Later, Hasan would really cut the person's head off, and then let his followers examine the severed head and corpse to empirically verify that the execution had happened. When the brain misinterprets a real experience as a spiritual experience, it loses the ability to distinguish fiction from reality.

10. Obsessive-Compulsive Disorder-like symptoms: People who are afflicted by the mental illness known as "OCD" tend to perform ritual motions over and over again in a specific sequence or order. Such people feel a strong sense of purity and being right and clean when their order or habits are strictly observed, and they suffer from a feeling of wrongness and corruption and contamination whenever their pattern is disrupted. Religious practices look quite similar, especially in the area of rituals and rites, such as organized prayer, where following a set pattern is felt to be holy and any deviation from the orthodox is perceived as unclean and unholy.

In conclusion, if religion comes from brain malfunctions then it is highly debatable whether people are "to blame" for mistaken religious beliefs, since the physical matter of the brain is to blame, at the level of neural structures and brain segments defined by DNA, and conscious volitional choice plays a minimal role. People are not weak and stupid. People are merely human, all too human. Perhaps religion will remain in the human species until the human brain takes its next steps of evolutionary progress. Perhaps wars between Christians and Muslims will lead to nuclear war and our total extinction before that evolution can happen. These are merely idle speculations. However, from a biological, scientific point of view, religion is adequately explained purely from physical causes and with no reference to the actual real existence of God or anything supernatural. A popular phrase to explain religion is "if God did not exist then He would have to be invented." A more accurate aphorism is "God does not exist, but the human brain has a psychological tendency to invent Him."

Chapter Fifteen: Post-Genetics vs. Genetics

If the philosophical scientific method is a valid way to organize our thinking, then we would expect science to explain a lot about human reality. Science does explain human history and culture, and the specific idea which explains humanity is evolution. The theory of "post-genetics", which applies evolution to the analysis of human civilization, states that most human behavior can be understood as an expression of the urge of living organisms to cause their genetic material to continue to exist. In my Liberty Magazine essay "Playing the Race Card" I described how this theory explains racism. The members of each race tend to share more genetic material with each other than with members of other races (from interbreeding due to the isolation of the various races from each other in ancient times), and therefore the members of a race are urged by their genetics to promote that race over rival races. That is the genetics theory.

The post-genetics theory represents my idea that human beings have evolved minds which now demand different behavior from mindless DNA, and the evolution of minds shifts the dynamics of human civilization away from a pure drive to perpetuate genetic material and towards a paradigm of individual ethical achievements. For example, racism is wrong because what matters are people's individual identities and personal choices and not what race they are a member of. Defining someone by his or her racial DNA ignores that a person's individual mind and choices are the only valid criteria for an evaluation of a person's ethical character. If people are mere expressions of DNA then a race can be good or evil, but if a person is an individual mind which makes right or wrong decisions then only a person, and not a race, can be good or evil.

In this essay I want to elaborate my theory of post-genetics and show how it can be used to interpret various phenomena of human civilization, including marriage and the abortion debate. First I will present a general overview of genetics theory. The genetics theory holds that each individual exists only to procreate and ensure the survival of the species, and each human exists only to compete with others in order to spread their genetic material so that the best genes disseminate and the species is made stronger. Human beings realize their purpose only through the survival of the species.

Genetics is a cynical but scientific explanation for several human behaviors. Why do parents love their children? So that the young of the species will be protected until they reach sexual maturity and can reproduce. Why are children cute? So that they will be protected. Why do people watch sports? Because it encourages physical fitness, which increases the health of the species. Why have women historically been politically and socially dominated by men? Because the female of the species is designed to produce and nurture the offspring, in the womb and through the production of breast milk, which requires that the male protect the female during the gestation in the womb and during the rearing of the young children, and this puts the male in a position to patronize the female. Why do men and women marry? Because it is the most efficient way to create and nurture offspring. Why are people obsessed with having sex? Because more sex produces more offspring, increasing our chances of survival. Why is sex pleasurable? To motivate people to procreate. Why do the most attractive and successful people find lovers more easily? Because mating with the members of the species with the most desirable traits encourages those traits to be more widely expressed among future generations. Why do men like women with large breasts, and why do women like athletic or wealthy men? Because they want those qualities for their offspring, and breasts produce breast milk, and men use their strength or money to provide the food while the woman bears the child and lactates.

One of the main principles of genetics theory is that the purpose of sex is procreation for the survival of the species to produce as many young as possible, and young with the best DNA possible, and love and dating and marriage are merely social institutions designed to promote this genetic purpose. For example, the behavior of teenagers, listening to "cool" music and dressing in "cool" clothes etc., is behavior designed to attract a mate, like a bird preening its feathers and chirping a love song.

Now let's consider homosexuality. Why are gays hated and persecuted? It must be because gay sex does not produce children, which results in fewer human young, which decreases the species' chances of survival from an evolutionary perspective. Therefore genetics urges the species to destroy homosexuality. This can be confirmed by looking at the social institutions of marriage and contraception. The Christian Church, which denies the truth of biological evolution, is, ironically, the main enforcer of genetics and its mandates. If the purpose of sex is merely procreation and the production of children, then condoms violate this purpose and should be outlawed.

The Catholic Church dictates that marriage must be between a man and a woman. Why? Obviously so that marriage will produce children. Marriage evolved so that a woman could only have sex with one man so that the man could know for certain that the woman's babies came from the husband's DNA, so that protecting his children would promote his DNA. In ancient times, say from 20,000 BC to 5000 BC, the woman needed a man to protect her while she was burdened by a fetus in the womb, so the males evolved to be stronger than the females, and this is the origin of the social custom of male domination of women. Later, somewhere around 5000 BC (the precise years are inexact, and best left to historians to debate), marriage became a way to tie families together, when family and clan were the central organization of society. Why was the joining together of two family clans by marriage such a big deal? Because the marriage of a son of one family and a daughter of another family united their DNA in the children of the marriage, so their DNA was tied together and therefore the purpose of the two families was united in their purpose of promoting their DNA.

If this is genetics, then what is post-genetics? Post-genetics says that, while the purpose of sex from an evolutionary perspective was the creation of babies for the survival of the species, the purpose of sex for minds is not the same thing as the purpose of sex for genetics. Therefore the evolution of the mind, which is probably the frontal lobe of the human brain and which has evolved relatively recently, has drastically transformed human behavior. The purpose of sex for minds is firstly for pleasure and enjoyment, and secondly as an expression of love and affection. This represents a huge change in human evolutionary purpose, and a lot of the "culture war" between social liberals and social conservatives is explained by the war between humans as minds fighting for sex as pleasure on the one side, and humans as DNA fighting for sex as procreation on the other side. If the purpose of sex is pleasure, then condoms to prevent fertilization make perfect sense, and homosexuality also makes sense, and premarital sex makes sense, and a lot that is socially liberal makes sense. And if those things are okay, and the perpetuation of DNA is no longer paramount, then gay marriage also makes a lot more sense, since the married couple would not have an imperative to procreate their DNA.

Post-genetics views the purpose of sex as not only pleasure but also as an expression of love. In the post-genetic model you have sex with someone and then you fall in love with them, until you reach the point at which you want to express your love for them by marriage and having their children, and love then becomes the glue which holds the family together. One individual's love for another individual is paramount, and DNA is secondary. The focus of post-genetic love is on your lover's personality and mind, which come from their choices and not from their DNA. But it is also worth noting that, if your lover's self is identical to his or her body, which follows from mind-brain identity, then your lust for their body is an expression of your desire for their self. Marriage for love is a modern invention. History had marriage for family connections, not for love. We are still coping with this evolution.

Generally, religion is almost always the genetics view because religion is the embodiment of tradition, and tradition dates back to the era when only genetics existed and post-genetics did not exist because the conscious mind had not yet evolved. So genetics vs. post-genetics is tradition vs. evolution, entirely as a result of the structure of the history of human evolution.

Let me then briefly apply this interpretation to the pro-choice vs. pro-life debate. Obviously the people who are pro-life are not advocates of their position for any of the reasons that they say. If killing all life were evil then Christians would not eat meat which comes from killed animals. The pro-life position is completely irrational, as proven by the classic Rothbard argument that a baby which is forced upon a mother against her will has no right to exploit the mother's body even if the fetus really is a human life. In other words, if a fetus has all the rights of adult human beings, but no more, then abortion would be okay, because an adult human cannot enslave the body of another human being. So we must look to genetics to understand the motives of pro-lifers. And the motive is obvious. Genetics views the role of women as to produce children,

78

so anything that frees the female womb from the production of children is an enemy. Abortion enables women to shrug off their purpose of birthing children, which then frees their minds up for mental pursuits, like pursuing a career. So genetics wants women to be merely wombs for childbirth, and post-genetics views the female purpose as individual personal achievement, such as by pursuing a career. This explains the pro-choice vs. pro-life fight, and the participants in the battle are where the theory expects them to be, with social liberals fighting for post-genetics and social conservatives fighting for genetics.

Part Two: Pure Empirical Essential Reasoning

Chapter Sixteen: Overview

This Part begins with a short paper, called "Rand's Axiom Problem", which summarizes the theory of empirical essentialism, followed by a longer book-length treatise titled "An Essay on Reason and Perception" which presents the theory of pure empirical essential reasoning in complete detail. If you want instant gratification, then read the summary, and if you desire a deeper exploration of the ideas, then first read the summary and then read the treatise. "Rand's Axiom Problem" expresses my unique, original philosophy as an alternative to Ayn Rand's philosophy of Objectivism as a basis for reason, objectivity, and scientific knowledge. No prior familiarity with Objectivism is required, and my positions are summarized neatly in the paper. Then, in "An Essay on Reason and Perception," I show how my philosophy is useful for refuting the arguments of philosophical doubt and ignorance which assert that science, based on perception of the physical world combined with the human brain's reasoning, is incapable of achieving something that can be properly called "knowledge" and which we can rely upon as possessing one hundred percent perfect certainty.

Part Two-A: Rand's Axiom Problem

Chapter Seventeen: Introduction to Rand's Axiom Problem

The purpose of this paper is to analyze and solve problems in Ayn Rand's epistemology, particularly Rand's theory of axioms. Her main problem is that she fails to offer proof of the axiomatic propositions, which I believe can be debated and denied and are therefore in need of proof, so that Rand's epistemological foundation has a hole in it where a proof of the axioms should be. I will challenge Rand's intrinsic-objective-subjective analysis and posit that a theory of essences intrinsic in reality is necessary to prove identity and existence and to support a theory of reality as objective and concepts as non-arbitrary. I will also seek to offer a new understanding of proof and demonstration as capable of being grounded in empirical observations rather than logical or mathematical deductions. Such an understanding benefits a philosophy of empirical reasoning such as Rand's.

My approach in this paper will employ two different tactics. I will argue that Rand's epistemology, particularly her theory of axioms, is guilty of an internal contradiction. The internal contradiction will be explored in depth below, but can be summarized by saying that the axioms assert that they cannot be proved by reason, yet they form the foundation for all reasoning, hence all reasoning depends upon things that cannot be known by reason, and if a thing cannot be known by reason then it must be accepted on faith, therefore all reason is ultimately founded on unproven beliefs that must be accepted on faith, therefore Rand's epistemology reduces to the statement that reason is based on faith.

But I will also argue that Rand's epistemology contains what I call an "external contradiction". By "external contradiction" I mean that reality exists objectively and Rand's philosophy is not a complete and truthful account of reality as it really exists, hence one may say that Rand's philosophy is contradicted by reality. An internal contradiction happens when someone says that a thing is and also says that it is not, while an external contradiction happens when reality says that a thing is and someone says that it is not. To show Rand's external contradiction, I will offer my own unique, original theory of epistemology, based on the theory of essence. My theory describes reality as it really exists, and Rand's philosophy fails to match it. Note that my argument is not that Rand is wrong because she disagreed with my view, which is of no importance. Rather my argument is that Rand is wrong because her theory does not match reality, which can be known because my theory is a better description of reality than hers. Also, here at the beginning of my argument, I will concede that this paper is merely an introduction to my theory, and many details to my ideas exist that would require a longer discussion to be explained, such that the account given here is incomplete. I fully explain my theory elsewhere, but this paper contains enough of the gist of my ideas for the reader to undertake an evaluation of my claims regarding Rand's epistemology.

In this paper I am going to criticize Rand's position regarding axioms, by which I mean her claim that the propositions "existence exists" and "A is A," which she also referred to as "existence" and "identity," are axiomatic (Rand 1990, 55). She also considered "I exist," in other words consciousness, to be axiomatic, and although I won't deal specifically with that axiom, my argument applies to it also. Once I have established that her position on axioms is contradictory I am going to offer a different theory to replace hers. My purpose is not to show that axioms are false and therefore Objectivism falls apart. On the contrary, my purpose is to show that there are holes in Objectivist epistemology where a proof of the axioms should be, but the axioms can be proved, so that Rand's philosophy can be made much stronger and put on a more stable foundation if the holes are filled.

I do not dispute the truth of existence and identity. I merely dispute whether they are axioms. My main argument is that Rand, by claiming that existence and identity are axioms, claims that these propositions cannot be proven, but that they must be believed, and the only logical result of this is that they must be taken on faith. Nothing should be taken on faith, and I believe that those two propositions should be proved and can be proved. The problem I will seek to solve is how to prove the truth of "existence exists" and "A is A." I also believe that Rand's mistake regarding axioms trickles down and causes problems with her account of objectivity, necessity and universality.

I will begin by offering Rand's view of axioms in her own words. I will proceed to explore what Rand's axioms mean and discuss the problems with axioms and axiomatic concepts. I will look at Rand's arguments about the impossibility of rejecting the axioms without also accepting them, the non-existence of non-existence, and the stolen concept. I will then examine the difference between premises and implications. I will explain my approach to proving existence and identity by using perceptions as premises combined with essential reasoning. I will conclude by critiquing Rand's intrinsic-objective-subjective analysis and her theories of objectivity, necessity and universality.

Chapter Eighteen: Rand's Axioms in Her Own Words

Rand's position regarding axioms is evident from a passage in John Galt's speech:

"'You cannot *prove* that you exist or that you're conscious,' they chatter, blanking out the fact that *proof* presupposes existence, consciousness and a complex chain of knowledge: the existence of something to know, of a consciousness able to know it, and of a knowledge that has learned to distinguish between such concepts as the proved and the unproved.

'When a savage who has not learned to speak declares that existence must be proved, he is asking you to prove it by means of non-existence—when he declares that your consciousness must be proved, he is asking you to prove it by means of unconsciousness—he is asking you to step into a void outside of existence and consciousness to give him proof of both—he is asking you to become a zero gaining knowledge about a zero.

'When he declares that an axiom is a matter of arbitrary choice and he doesn't choose to accept the axiom that he exists, he blanks out the fact that he has accepted it by uttering that sentence, that the only way to reject it is to shut one's mouth, expound no theories and die.

'An axiom is a statement that identifies the base of knowledge and of any further statement pertaining to that knowledge, a statement necessarily contained in all others, whether any particular speaker chooses to identify it or not. An axiom is a proposition that defeats its opponents by the fact that they have to accept it and use it in the process of any attempt to deny it'" (Rand 1957, 956).

Rand says much the same thing in her "Introduction to Objectivist Epistemology":

"Since axiomatic concepts refer to facts of reality and are not a matter of 'faith' or of man's arbitrary choice, there is a way to ascertain whether a given concept is axiomatic or not: one ascertains it by observing the fact that an axiomatic concept cannot be escaped, that it is implicit in all knowledge, that it has to be accepted and used even in the process of any attempt to deny it" (Rand 1990, 59).

We can summarize Rand's argument regarding axioms. First, she claims that the axioms cannot be proved, because proof presupposes or assumes knowledge already contained in the axioms or in an attempt at proof. Second, she attacks anyone who asks her to prove the axioms by claiming that they seek knowledge of existence from non-existence. Third, she argues that the only way to reject the axioms is to shut up and die, and fourth, she argues that one must believe the axioms even though they cannot be proved, because it is impossible to reject them without also accepting them at the same time. We can reduce this to three basic ideas: first, the impossibility of denying the axioms, second, the non-existence of non-existence, and third, the stolen concept. I will address each of these ideas later, but first I would like to look at the theory of axioms more generally.

Chapter Nineteen: What Do Rand's Axiomatic Propositions Mean?

Rand believed that "existence exists" and "A is A" are axioms, but to see whether this belief is problematic we must inquire into what Rand's axiomatic propositions mean, and what it means to be an axiom (Rand 1957, 933-34). One dictionary defines "axiom" as "established or accepted principle" or "self-evident truth," and defines "self-evident" as "obvious; without the need of proof or further explanation" (Thompson 1998, 52, 826). Rand refers to Aristotle (not by name, but as "the greatest of your philosophers") when first explaining her belief in axioms, so it may be useful to examine Aristotle's understanding of axioms (Rand 1957, 934). Aristotle, in describing his theory of logic as demonstration from premises in "Posterior Analytics" Book I Chapter 2, defines an axiom as a "basic truth," i.e. a proposition in a syllogistic demonstration that has no proposition prior to it, which a pupil must know in order to learn anything even though it cannot be proved by the teacher (Aristotle 1947, 12-13). In Book I Chapter 10 he says of basic truths that they cannot be proved, and that axioms express necessary self-grounded fact and must necessarily be believed (28-29).

I would define axioms as propositions that are self-evident and form the basis of all reasoning. Rand says that "axioms are usually considered to be… fundamental, self-evident truth," and axioms are made of axiomatic concepts (Rand 1990, 55). Rand's axiomatic concepts are "primary," meaning that they are irreducible (55). She doesn't say that axioms are the beginning of all cognitive activity, but her description indicates that the purpose of the axioms is to act as "guidelines" for and to "underscore" and "delimit" all conceptual thought (59). The function of the axioms is "delimiting (knowledge) from non-existence, imagination, falsehood, etc." (261). She claims that axioms are the "base" of knowledge (Rand 1957, 956). From this it seems that Rand's axioms are the foundation for all conceptual thought in the stage of adulthood, and that Rand's axioms really are axioms in the Aristotelian sense.

What did Rand mean by the phrases "existence exists" and "A is A"? The closest Rand comes to explaining what "existence exists" means is her explanation of existence as an "axiomatic concept" as found in "Introduction to Objectivist Epistemology" (Rand 1990, 55-61). Rand claims that existence is a quality that applies to everything (56). If existence were a concept that applied to everything, then it would have no real meaning, at least no more so than the concept "everything." The phrase "existence exists" should mean more than simply "everything is everything," which is an empty statement. "Everything is everything" is an empty statement with no real meaning because the proposition "everything is everything" does not reveal new information about the term "everything."

What, then, is Rand's definition of "everything," or of "existence"? Rand doesn't believe that "existence" needs to be defined at all: "Since axiomatic concepts are identifications of irreducible primaries, the only way to define one is by means of an ostensive definition—e.g., to define 'existence,' one would have to sweep one's arm around and say: 'I mean *this*'" (41). The only problem is that "*this*" has no meaning, or if it has a meaning it is so vague and imprecise as to be useless for rational thought. Rand offers a way to refer to existence, not a definition of existence, not a description of what existence is. If it is impossible to give an intelligible definition for some idea, a definition capable of explanation, then that idea's usefulness for rational thought is highly dubious. Concretes are perceived, and so they can be defined by pointing to them; for example, I could point to an apple on my desk and say "I mean this apple." But existence and identity are abstractions, not concretes, and as such they are not directly perceived and should not be defined that way.

Of course, Rand believed that existence and identity are concepts that are perceived directly (55). My objection to this is the observation that I can see a chair, or a table, or a book, or an apple, but I do not see the concept of "existence" as such, and Rand fails to explain how one arrives at the universal abstraction "existence" via induction from perceptions of specific concretes.

"Existence exists" has a meaning. One dictionary defines "exist" as "have a place in objective reality," and defines "existence" as "all that exists" (Thompson 1998, 303). I argue that "existence" means not just being, but also having a physical, objective existence at a

89

specific place and time, and "physical" and "objective" are not axiomatic concepts and can be debated. "Existence exists" means "reality is real," which means "everything in this world is physical and objective," and that is a statement which has content. I would argue based upon my understanding of the English language that the definition of the concept of "exist" reduces to three distinguishing characteristics: that the existing thing has the quality of being at a specific place and time, that it is made of physical substance, and that it is objective.

Rand rejected this view, arguing that "existence exists" does not reduce to "there is a physical world" (Rand 1990, 245-51). But Rand fails to adequately address two problems: first, that if "existence exists" does not mean that, then it has no meaning other than "everything is everything," which has no content, and secondly, that based on the definitions of words in the English language, the best interpretation of "existence exists" is "reality is physical and objective." Rand claims that even a primitive savage who believes in supernatural spirits has grasped that "existence exists" (248). But if that is true then we must ask: what does the savage really know? And why doesn't this knowledge prevent him from believing in spirits? If "existence" is so general, then it really is no different from saying "everything is everything," which has very little meaning.

It seems that Rand really intended "existence exists" to mean "everything is everything." She says "…what's the difference between saying 'existence exists' and 'the physical world exists'? 'Existence exists' does not specify *what* exists" (247). She also says "the concept 'existence' does not indicate what existents it subsumes: it merely underscores the primary fact that they *exist*" (59). If it is really true that "existence exists" does not specify what kinds of things exist, and is not the same as "existence is physical and objective," then Rand's axiom would not contradict the proposition "existence is made of Platonic Forms." If Rand is right then "existence exists" has virtually no content whatsoever.

Perhaps looking at the purpose of the axioms will shed light on their meaning. Rand says that axiomatic concepts enable one to know that what is true at one point in time is true at every point in time (56-57, 260-61). But how do axiomatic concepts accomplish that purpose? Her answer to that question is vague and imprecise (261). She says that the axioms' purpose is to distinguish object from subject (57, 261). But she is unclear about precisely how they do so. And she says that axioms underscore primary facts and confine knowledge to reality (261). But if "existence exists" does not say anything specific about what reality is, then how does it limit knowledge about reality? I found no intelligible explanation of how axioms accomplish these purposes in "Introduction to Objectivist Epistemology". I don't consider it a defense on Rand's part that she believed that axiomatic concepts could not be analyzed, such that the answers to my questions should be self-evident (55). If she had consistently applied the belief that axioms are incapable of analysis then she would not have been able to fill a whole section of her book on epistemology with an analysis of axiomatic concepts.

I will concede that Rand attempted to add content to her axioms in Galt's speech. She says: "existence exists—and the act of grasping that statement implies two corollary axioms: that something exists which one perceives and that one exists possessing consciousness…. To exist is to be something, as distinguished from the nothing of non-existence, it is to be an entity of a specific nature made of specific attributes. … Reality is that which exists; the unreal does not exist…" (Rand 1957, 933-34).

She says that existence exists and that to exist is to be something, in which case "existence exists" reduces to "everything is something." I argue that there are three problems with that reduction. First, "everything is something" might reduce to "A is A," making the first axiom redundant. Second, Rand does not give a significant analysis of what it means to be something or to have a specific nature, such that "everything is something" has little content.

My third and central problem is I believe that non-existing things are things with identities, so that "existence exists" cannot reduce to "everything is something." Regarding this analysis, I argue that "to exist" and "to be something" are different concepts. I call this analysis the Japanese Distinction, because in the Japanese language there are two completely different words for "to be" and "to exist". Japanese uses the word "desu" for "to be something" and the word "arimasu" for "to exist." (Japanese also has a third word, "imasu" which means "to be alive, to exist as a living thing".) For Rand's understanding of "existence exists" to be correct there must be no difference between the terms "to be something" and "to exist". In the English language the word "to be" has both meanings, but I argue that this is a coincidence and not a philosophical insight. This ties into Rand's idea of the "reification of the zero," which is that, according to her, only existing things are things, and nonexistent things are not really things at all, nothingness as such does not exist and is not a thing, and every truthful description of a non-existent actually reduces to a description of existents because non-existents can be known only in relation to existents (Rand 1990, 58, 60-61, 149-50).

My critique of Rand's reification of the zero argument has several parts. First, the terms "exist" and "be something" are conceptually distinguishable. One can say "a nonexistent thing" and "a non-identity" and mean two different things. One can also say of a gigantic bronze statue of Ayn Rand inside the main concourse of Grand Central Terminal in New York City that it does not exist, however it is something, and it has an identity. It has attributes of substance, appearance, and location, so one cannot say that it is not a thing. Second, it seems to me that it should be impossible to know anything about something that has no identity, so that if non-existents are not things, they would be unknowable. But one can possess some degree of knowledge about Platonic Forms, pink elephants, fictional statues of Ayn Rand, what one experiences in one's dreams, or for that matter Howard Roark, the hero of Rand's novel, who is fictional and therefore does not exist but who has a knowable personality. Rand's idea, if followed to its logical conclusion, prevents us from describing the things that don't exist. I concede that nonexistent things cannot be known except in relation to that which exists, but that doesn't mean that they have no identity.

Third, it seems to me that when something is destroyed and ceases to exist, what happens is not that it ceases to have an identity, but that it ceases to have an identity with attributes of being at that place and time in physical reality. And fourth, if Rand were right then the number "zero" would not be a thing, yet it seems to me that zero is a number and has a mathematical identity. I agree that zero does not exist, but I do not agree that it has no identity. Regarding this point Rand might have argued that zero is a concept of method, a claim that she made regarding imaginary numbers (35-36, 304-6). My dispute with numbers as concepts of method is that while the numbers may be used as a means of discovery, the numbers in themselves, e.g. the square root of negative one, must either be something or not be something, because anything else would be a contradiction. Rand's concepts of method are ambiguous as to whether they refer to things or not.

I must clarify that I believe that everything that exists is a thing, but not that everything which is a thing exists. It is difficult to make a distinction between "to exist" and "to be" because every existing thing is something. However the distinction can be done on the basis of things that don't exist, e.g. zero, and it is equally difficult, or impossible, to distinguish "to exist" from "to be objectively physical at a specific place and time," because every existing thing also has that attribute. If Platonic Forms and pink elephants are non-existing things, as I argue, then there must be something more to "exists" than merely "is something." And I posit that the something more is being physical and objective at a specific place and time.

Chapter Twenty: Can Rand's Axioms Be Debated or Denied?

If it is truly impossible to debate or deny existence and identity then they are not in need of proof and Rand's axioms do not have a problem. I believe that Rand's axioms can be denied, even if they mean exactly what Rand intended them to mean and do not mean what I claimed they should mean in the previous section. In order to argue that Rand's axioms can be debated, first I will argue that linguistically it is possible to frame propositions that contradict the axioms in a manner capable of being understood and discussed, second I will argue that if the axioms cannot be proved then logic is circular reasoning, and third I will briefly discuss two famous philosophers, Plato and Derrida, whom I think would not have agreed with Rand's axioms as Rand meant them.

To argue that one can state contradictions in a way capable of being understood and debated it may be useful to analyze the term "coherent." That term has two meanings, a common English meaning and a technical philosophical meaning, and it helps clarify the use of the word to distinguish the two meanings. The technical meaning of "coherent" is "not entailing any contradictions," or in other words, non-contradictory and logical (personal correspondence, 3 May 2009). The common meaning of "coherent," found in a dictionary, is "intelligible and articulate," and this dictionary defines "intelligible" as "able to be understood" (Thompson 1998, 158, 460). If one were to fail to carefully distinguish the two meanings one might think that incoherent propositions cannot be understood because they are contradictory, but I argue that it is possible to understand ideas that are contradictory and illogical. For example, the statement "hgteagleajrg bjergaoejighl gikhalighi" is unintelligible, whereas the statement "if you achieve knowledge of spiritual reality then it will be revealed to you that the physical world is really not the physical world" contains a contradiction, but has a meaning that is capable of being understood and debated.

If you can state contradictory propositions in a way that is intelligible then you can debate whether they are true or not, and it becomes necessary to prove that A is A in order to refute the truth of illogical propositions. Contradictory propositions that are intelligible and debatable can be formulated. It might be hard to imagine someone believing "a cat is not a cat" or "a triangle is a square circle." However, a more easily imaginable example is that a person can say "humans do not need to think in order to survive, but only need to believe in God," which Rand might have thought violated the principle "Man is Man," which violates the axiom "A is A," so Rand might have considered such a statement to be contradictory (Rand 1957, 934). Similarly one can imagine a person believing the economic theory that "inflation creates jobs," which someone else might believe to be a contradiction.

One might believe that existence and identity cannot be denied except by an insane person. It is difficult to imagine a sane person believing that A is non-A, and it is difficult to find a philosopher who openly claims that logic is useless. But a quote from Rand's Toohey speech sheds light on this: "Men have a weapon against you. Reason. So you must be very sure to take it away from them. Cut the props from under it. But be careful. Don't deny outright.... Don't say reason is evil—though some have gone that far and with astonishing success. Just say that reason is limited. That there's something above it. What? You don't have to be too clear about it either. The field's inexhaustible. 'Instinct'—'Feeling'—'Revelation'—'Divine Intuition'—'Dialectic Materialism.'... Suspend reason and you play it deuces wild. Anything goes in any manner you wish whenever you need it" (Rand 1943, 637). Rand seems to be saying that illogical thinking is widespread, and that a person can deny logic not by denying identity openly, but by believing in something besides logic. Based on this quote Rand seems to agree that logic can be denied, in fact she seems to be saying that religion and Marxism both reject reason, and if reason can be denied then it can be debated, and if it can be debated then it is in need of proof.

If "A is A" is not demonstrated then all logic is mere circular reasoning: if logic is valid because "A is A," and "A is A" is true because it is logical, then logic is a circle. It might be illogical to disbelieve that "A is A," but if logical belief in a proposition requires proof by demonstration, and "A is A" cannot be proved by demonstration, and all logic is based on belief in the proposition "A is A," then logic is not based on logic, which means, logic is not logical. To say that logic needs special rules for axioms is to concede that axioms are not logical. Either obedience to logic has an intellectual origin, in which case we have the problem of showing what that origin is, or else there is no reason to believe that one should be logical, and logic is merely arbitrary and without a foundation. My problem is more than the mere question of why one should believe in logic, although it includes that question. My inquiry is the question of where logic comes from, because if logic's foundation comes from intuition and is incapable of rational proof then it is no better than any other intellectual strategy that comes from intuition, particularly faith in God. If you cannot demonstrate that "A is A" then faith is the only basis for claiming "A is A," and if you choose to believe in unproven, undemonstrated axioms then it is mere preference whether you prefer the axiom "A is A" or the axiom "God exists," since both would be equally unproven.

I accept that it is impossible to dispute identity, the principle of non-contradiction, without being illogical, but if a person does not begin with the belief that thinking should obey the rules of logic then one has no way of knowing that it is wrong to think illogically. Rand's argument that you can only put the negation of the axioms into practice by abandoning language and committing suicide would be true only if one had to put the negations into practice in a way that was strictly consistent. But if you negate identity you no longer bind yourself to be strictly consistent, and the axioms can be denied in a way that is irrational and contradictory, but not impossible, e.g. you can believe in the physical world and in a spiritual world at the same time.

I must also note that I can think of two philosophers who might dispute Rand's axioms; I cannot be sure about what either of them would have said about Rand's axioms but I can make educated guesses. I would guess that Plato would not agree that "existence exists." He would say that existence, the perceivable world, is a mere shadow of the world of Forms, as for example in his famous cave metaphor in "Republic", or his discussion of Forms in "Phaedo" (Plato 1997, 57-73, 1132-35). And based upon what I have read about Derrida I would guess that instead of agreeing with "A is A" he would try to deconstruct that statement. Insofar as "A is A" claims to be a perfectly rational statement that perfectly represents the real world, Derrida might critique it as logocentric presence, and would dispute that any language can capture objective universal truth (Appignanesi 2007, 77-81). Because the axioms can be debated and denied we need a demonstration of them to show to doubters in order to prove to them that the axioms are true.

Chapter Twenty One: The Non-Existence of Non-Existence and the Stolen Concept

Before I show how it is possible to prove that existence exists and that A is A by means of essential reasoning from sensory experience, let us consider two of Rand's ideas that she offers in support of her axioms: the non-existence of non-existence and the stolen concept. Rand's non-existence of non-existence idea contains two parts: first, that the negation of existence, i.e. non-existence, doesn't exist, and therefore you have to accept existence; and second, to obtain proof of existence one would need to go outside of existence, into the realm of non-existence (Rand 1957, 956). This argument can be seen as a form of proof. But this argument is a negative proof, not a positive proof: she proves that non-existence does not exist, not that existence exists. To justify Objectivism we need positive proof. If a negative is not a positive, as Rand claims (for example, "light is not 'the absence of darkness,'.... Existence is not a negation of negatives,") then disproving the negative of "existence exists" is not the same as proving the positive "existence exists" (941-42). Rand's argument defines reality by what it is not rather than by what it is, yet her sharpest criticism of the concept of God is that it is defined only by what it is not (951-52).

Rand argues that the person who wants proof of existence wants it to come from outside of existence (956). Her general idea is that existence is all around us, so the people who doubt it want knowledge from outside of existence. This may be true for her anti-reason critic, but it is not true for the honest inquirer. Do we need to not have bodies in order to prove the truths of biology? Do we need to not contain chemicals in our bodies in order to prove the truths of chemistry? Do we need to live in a world with no geometric shapes in order to prove the truths of geometry? Does proving something within existence require you to have a point of view from outside of existence? I do not believe that it does. Why would we need non-existence in order to prove that existence exists? Critics of objectivity might claim that you would need a point of view outside of experience to claim objective knowledge, but that view is not consistent with a belief that objectivity is possible.

The stolen concept argument is that when you reject the axioms you also accept the axioms, or when you deny the axioms you use the axioms in the process, and therefore they must be accepted and can't be rejected (956). But to make this argument two things are necessary. First, you have to prove that the concepts are stolen. When this is done you establish a contradiction: you both accept and reject the axioms. Proving the theft in a persuasive manner is a big problem. It is hard to argue that the proposition "existence does not exist" contains or relies upon the proposition "existence exists," or that "A is not A" borrows the idea "A is A." The Socratic claim "I know that I know nothing" does not seem like a claim to knowledge. It seems like a claim to ignorance, with a meaning that is identical to "I know nothing" or "I do not know anything." This refutes the Randian argument that the statement "I know that I know nothing" both claims and denies knowledge, such that it entails a contradiction and accepts that knowledge is possible. I will flesh out my objection to this aspect of Rand's argument further when I discuss premises and implications in the next section.

Second, having established that a theory or idea contains a contradiction, you have to also prove that contradictions don't exist for the contradiction to be a problem. Even if you could prove that the axioms are "stolen," which would be difficult, you would still need to prove that the laws of logic should be obeyed in order for the theft to be problematic, assuming that you don't begin with an unproven law of non-contradiction. The argument that "contradictions exist" is contradictory is a circular argument: if you don't already have a proof that contradictions don't exist, then proving that "contradictions exist" is contradictory does not invalidate it, because something could exist despite being contradictory. The negative proof results in circular reasoning, and the only solution is a positive proof of "A is A."

Chapter Twenty Two: Premises and Implications

Nietzsche in "Twilight of the Idols" claimed that philosophers confuse the last with the first, the most abstract with the most basic, and the ultimate goal of thought with the intellectual point of origin (Nietzsche 1990, 47). This is precisely Rand's mistake. "Existence exists" and "A is A" are perhaps the most abstract propositions possible, so it makes little sense to say that thought begins with those, or that they form the base of all reasoning. It would be better to say that thought begins with induction based on the observation of the objects of perception, the most specific or "concrete," and ends with axiomatic propositions, the most general or "abstract."

I argue that Rand's theory of axioms is itself a contradiction. But I believe that her mistake is probably not random. I would conjecture that Rand made her mistake for two reasons: first, she herself did not fully understand how she had arrived at her knowledge of existence and identity, and because she did not know how to reason them, she claimed that they could not be proven. When you are sure about something but you don't know the reasons behind your knowledge, the temptation is to say that it is self-evident, which translates into the statement "I know it but I don't know why." Second, she saw that it seems as though you would have to know existence and identity in order to do any reasoning, including any reasoning that would prove them: after all, doesn't existence have to exist in order for you to be able to reason that existence exists, so wouldn't any proof be circular? (Rand 1957, 956). Regarding the first problem, I have something that Rand lacked, a proof that existence exists. But regarding the second problem we must ask: does all reasoning use those two axioms as premises?

How can you possibly reason that A is A without already knowing it? How can you engage in any reasoning or any thought without the premise that A is A and the premise that existence exists? If that were true then these ideas would have to be in the mind at birth, and they would be intuitions accepted on faith, which is precisely the problem with Rand's belief in axioms. The problem here was that Rand was confused about the difference between implications and premises. Something that is implied by an act of reason is different than something that is a premise in an act of reason. I would define an "implication" of reasoning as something that makes the reasoning true and has to be true in order for the reasoning to take place, and a "premise" of reasoning as that belief, proposition, or source of knowledge that is actually what the conclusion of the logical argument is deduced from. Identity and existence are implied by all reason, in that obviously there would be no reasoning taking place if existence did not exist in order to our brains to exist and think, and there would be no logical thought possible if things were not themselves.

But the principles of existence and identity are not what knowledge, in its origin, is actually deduced from. They are implied in all reason, but they are not the premises of all reason. Sensory experience is what all rational conclusions are ultimately reasoned from. Sense-perceived objects are the premises of all reason, since all reason begins with specific sense-perceived objects and proceeds via induction to general abstractions. Rational knowledge begins with the objects of sensory perception, and then proceeds by deriving general principles from specific perceived objects. In other words, reason integrates concepts from sense-perceived concretes. The foundation of all thought is not axioms. The foundation of thought is the specific sensory perceptions from which our concepts are formed. Rand's belief about axioms is chronologically backwards. She has us beginning with what comes only at the end. Identity and existence are not our most fundamental premises. In fact, they are the two most abstract and general concepts of all our conceptual abstract generalizations, which means they are the pinnacle of thought, not the start of thought. They are implied by all reason, but they are not the first premises of all reason.

Rand did put forward a theory of "implicit" concepts (Rand 1990, 159-62). But she seems to suggest that an implicit concept is one that is implicitly present because the material from which to integrate it is present and the mind is in the stage or process of integrating it (162). Thus, Rand's "implicit" is something in the process of being integrated, rather than something that must be true in order for reasoning to take place, and so her analysis is not precise enough to mesh with my argument. For example, I would disagree with Rand and say that the laws of physics are implicit in all reasoning (162). I would also say that an infant or young child has no awareness that existence exists or A is A, because sophisticated cognition is necessary to grasp those concepts (162). Because she was imprecise about whether an infant is actually aware of an implicit concept or not, simply saying that it is "not yet conceptualized, but it is available," it is difficult to know to what degree Rand would have disagreed with me about that (161).

I have a problem with Rand's belief that axioms are the foundation of all thought. You do not begin with the laws of logic inherent in the mind and then apply logic to the perceivable world. That is a Kantian approach to epistemology (Kant 1977, 38-49, 63-64). I believe that in the human mind as it develops for real human beings, in babies and young children, you begin with an empty mind, you experience sense-perceived objects, and then you reason from the specific perceived objects to general abstractions. You begin with sense-experience and then derive the laws of logic from that experience, using inductive reasoning. You do not begin with the laws of logic and then use them as the foundation for all cognition. Rand might have agreed with me about intellectual development because she believed that children integrate abstractions from concretes (Rand 1990, 11-13, 19-21). But if she did agree she would be contradicting herself, because beginning with the laws of logic and applying them to the world is no different from beginning with unjustified axioms and applying them to the world. Both approaches lead inexorably towards Kantian epistemology in which the laws of logic come from pure reason prior to empirical experience. The origin of reason can be either self-evident axioms or inductive reasoning from sensory experience, but not both. If axioms could be derived by inductive reason from experience then they would not be self-evident, because the observations from which they were reasoned would prove them and be the evidence supporting them.

Chapter Twenty Three: Proof, Demonstration, and Perceptions as Premises

To solve Rand's axiom problem we must discover a way to prove existence and identity. Before I offer my solution to Rand's problem, I must address the question of what "proof" is and what it means to be "demonstrated," and also what "self-evident" means. It is possible for someone to believe that nothing is logically demonstrated or proved unless the demonstration begins with premises and demonstrates that the premises prove the conclusion by means of demonstrated mathematical computations or the rules of symbolic formal logic. Such a person might believe that logic enables deductions, but that sensory experiences only enable mere inferences, but never deductions. Aristotle may have believed this: in "Posterior Analytics" Book I Chapter 31 he claims that perceptions do not provide the scientific knowledge of demonstration, because perceptions are of particulars and demonstrations are of universals (Aristotle 1947, 66-68). But in Book II Chapter 19, where Aristotle grappled with the problem of the origins of axiomatic knowledge, he states that knowledge of basic truths comes from induction from sense-perception, but he then contradicts himself by saying that it comes from intuition (106-9).

The belief that demonstration comes from logic and mathematics but not from perception is a flawed, one-sided account of reason: it accepts deductive reason as proof, but rejects inductive reason, which takes sensory experience and derives generalized abstractions and premises from that experience. If inductive reason is incapable of proof, then how can you prove the premises that logic begins with? The argument "If P then Q, P, therefore Q", or "If P then Q and R, not Q, therefore not P", and all other similar syllogisms, are mere abstract theory detached from any basis in practical reality if we lack a principle of induction capable of telling us whether P is true or false. Similarly, the logical argument "all money is made by doing work, it is good to make money, therefore it is good to let people be free to do more work," depends upon proof that making money is caused by doing work, which can only be proved via induction from sensory observations, and cannot be derived entirely from deductive logic and axioms. Unlike Rand and Aristotle, I have solved the problem of induction, by means of my theory of essence, and through empirical essential reasoning I can achieve proof and demonstrations of universals by reference to specific concrete observations.

In order to understand the theory of essential reasoning as inductive proof, we must be precise in what we mean by proof and demonstration. Something has been demonstrated, and you have proof of something, if you can point to a source of knowledge and show how you can reason a conclusion from it that is true if the source of knowledge is real. In other words, something is proven if you have evidence that proves the conclusion. The claim "a cat is in the room" can be proven if you see a cat in the room: if your perception of the cat is accurate, in other words if the cat that you see is real, then the cat itself, which you perceive, is the proof of the truth of the proposition, and you can prove it to someone else by pointing out the cat to them. The problem of induction is how we can point to one cat in the room and infer "all cats are mammals" from this one cat which we perceive, and essence is the tool that enables inductive reasoning to infer from specific to general.

It seems to me that an induction-deduction dichotomy, which could also be called a perception-logic split, is appropriate for philosophers like Kant and Descartes who want all proof to come from the analysis of concepts and who believe that logic and mathematics are pure but that the temporal physical world is muddled and illusory. But for a philosophy that believes that reason should properly be focused on the physical world and which thinks that truth can be known from empirical observations, we need to accept that induction is necessary for logic and that perceptions can be premises in deductions. I do not believe that we should draw a sharp distinction between inductive and deductive reasoning, because the two terms merely describe different phases of the same process. It is beyond the scope of this essay to prove that sense-perception provides accurate knowledge of reality which admits of absolute certainty (although I will discuss objectivity below), but it seems as though Rand accepted that it does (Rand 1957, 934).

Regarding the term "self-evident," I would argue that perceptions are the only premises that can be self-evident. I believe that a "self-evident" premise is not one that needs no proof, because something that cannot be proved would require faith in order to be known. A self-evident premise is one that contains the proof of itself within itself, in other words it proves itself just by being perceived. When I see a red apple, the apple itself is the proof that it is red, and that it exists. It contains the proof of itself within itself, therefore it is self-evident. If someone argues that the apple is not red, you simply show him the apple: the apple is the proof of its redness. One might say along these lines that "the proof is in the pudding," or as the English say, "the proof in the pudding is to be found in the eating." The proof of the thing's existence is in the thing itself, as perceived by the senses. Self-evidence is therefore a property that belongs only to specific sense-perceived objects, and not to ideas. To rephrase this in Randian terms, concretes can be self-evident, but abstractions, such as existence and identity, cannot be self-evident.

Chapter Twenty Four: The Solution of Essential Reasoning

My solution to Rand's axiom problem is to claim that perceived things, the objects of sensory perception, are self-evident, and universal laws such as existence and non-contradiction can be reasoned from sense experience by means of inductive reasoning. If you can reason from the experience of an apple that the apple cannot be ripe and rotten at the same time, then you can deduce a universal idea of non-contradiction from a specific instance of non-contradiction. Once derived from specific cases, universal principles can then be applied by deductive reasoning to all future instances. The propositions "existence exists" and "A is A" are best defended, not by any of Rand's arguments, but by arguing that the knowledge of existence and identity can be grasped by inductive reasoning applied to sensory experience, and that this constitutes proof.

But there is a big problem here: how do we go from specific experiences to universal principles? How does inductive reasoning work? We can go from a specific instance to a universal law by a process that I call "the necessary consequence of the ontological essence." Some Objectivist readers might react with horror to my use of the concept of "essence," but this issue will be addressed in the following section. My solution to Rand's axiom problem depends on the idea of "essence," so I will proceed to argue that reason is best understood as relying on essences. In order to argue this I must explain my view of how essential reasoning works. It took me 120,000 words in a philosophical treatise to fully explain my theory of essences, essential reasoning and essential things, but let me try to do so now in five paragraphs. First I will define "ontological essence" as whatever quality or attribute a thing has that makes it be a particular kind of thing, and then define "the consequence of an essence" as whatever being or attributes a thing has that is caused by it having an essential attribute. In other words, a thing's essence is what makes it be that thing, and the consequence of the essence is whatever results from having that essence.

There are three ways in which an essence of a thing may result in the consequence of the essence in that thing: first, by causation;

second, by requirement; and third, by containment. For example, the essence of iron will cause it to sink in a pool of water, because being iron causes a thing to be heavier than water, so that it will sink in water is a consequence of the essence of iron. Wood will require something that cuts it to be harder than wood as a means of cutting it, so the essence of being a thing that can cut wood will require the thing to be harder than wood, so that if iron cuts wood then a consequence of the essence of iron is being harder than wood. And being a metal is part of the act of being iron, because to be iron is to be a metal, so the act of being iron contains the act of being metal as a part of it, hence being metal is a consequence of the essence of being iron. From this essential reasoning we can conclude that iron is a metal, that it sinks in water, and that it is harder than wood. This will be true of all iron everywhere, such that if a thing is iron then it will do those things, and if something does not do those things then it cannot be iron.

The best example of essential reasoning comes from geometry. Reason begins with inductive reasoning: for example, you see the cover of a box, which is a red cardboard square. Your eye sends information to your brain, and this information in your brain is the perception of the red cardboard square. You analyze this information and separate the different aspects of the object. In its aspect as a red object, it is red. In its aspect as a shape with four equal sides, it is a square. By thinking this through you create the concept of a square in your brain: the processed perception becomes a concept. This is inductive reasoning. Reason can then take the essences known from perception and abstract more essential qualities from them to build knowledge of more abstract essences, such as going from the concept "square" to the concept "shape."

What makes a thing a square, in this case having four equal sides, is the essence of the square, and the consequence of that essence is whatever qualities a thing has that are caused by having that essence. You can reason the consequence of the essence of a square by thinking about the essential square, an object that has four equal sides and has no other qualities. The essential square has four right angles, because the consequence of the essence of having four equal sides is having four right angles. Having four equal sides causes a thing to have four right angles. This is deductive essential reasoning.

What is true of the essential square must be true of it because of the essence of square. Therefore what is true of the essential square must be true of all squares in the real world, because to be a square all those objects must have the square essence. If a thing did not have the essence of a square, then it would not be a square, so for a thing to be a square requires it to have the square essence, and the essence causes the consequence, so we can infer that everything that is a square has four right angles, because being a square will cause it to have four right angles. Therefore it is impossible for something to be a square and to not have four right angles. It is necessary for every square to have the consequence of the square essence, because everything that is a square must have the square essence in order to be a square, and having the square essence causes a thing to have the square consequence. By isolating the essence of a thing from a specific perceived thing and then employing essential reasoning, a person can achieve inductive reasoning and can infer general propositions from specific instances. This also shows that universal laws are obeyed because of things in themselves, not because of the structure of the mind, contrary to Kant.

Every essence is an attribute (and every attribute is an essence relative to the other non-essential attributes of that thing), and the attribute of having four equal sides is the square essence. The square essence is an aspect of squares in the real world, but the essential square does not exist: it has no specific place or time, and existing things always exist at a specific place and time. You can take essential reasoning from the essential square and apply it to new squares that you encounter. You see something square which is new to you and which you have never encountered before and yet you know instantly that it has four right angles because it is square. This is essential identification. You can take one thing as the representative of all of its kind, for example geometric reasoning from one square applied to all squares, only if you reason from the square essence of your representative square, which everything of that kind must have in order to be that kind of thing. This is why a geometric demonstration using one specific square as an example to show that every square in the universe has four right angles is effective, even though the example is just one square and there are other different squares with different measurements of the lengths of its sides. A demonstration is universal because the demonstrator deduces his proof from the square essence of the example square.

The three central mistakes that people make in their thinking are all mistakes in essential reasoning. First, some people think that what happened repeatedly in the past will always happen again in the future, confusing coincidence with necessity because they don't look for the consequence of the essence. For example, if someone sees one hundred wooden square boxes they may infer incorrectly that all squares are made of wood. This is why correlation is not causation. This is also the origin of belief in magic, e.g. someone does a dance and it rains so they conclude that doing a dance causes the sky to rain. Second, some people believe that something necessary is merely a coincidence, again because they don't grasp the consequence of the essence. This is the origin of skepticism. And third, some people believe that essential things are real, leading to a belief in a non-physical world of spiritual ideas. I posit this last mistake as the origin of Platonic Forms. This is a concise summary of my theory of essences.

If we can reason that an apple cannot be both ripe and rotten because it is an apple, then we can say that if a thing is both ripe and rotten, it is not an apple. In this way we can go from one apple to every apple, and the ontology of the thing itself, in itself, requires reality to universally obey what we believe. If the essence requires the consequence then everything of that kind must have that consequence because to be a thing of that kind requires having that essence. It follows that if a thing did not have the consequence of the essence it could not be that kind of thing because having that essence causes that consequence and the absence of the consequence would show the absence of the essence, and so being what it is requires a thing to obey the demonstrated conclusions of essential reasoning.

The line of essential reasoning that proves "A is A" is: a contradiction is both A and not-A, there are some pairs of essences for which being one kind of thing causes a thing to not be the other kind of thing (this would be reasoned by induction from specific instances, e.g. being a dog causes a dog to not be a cat because a dog must have different organs and DNA than a cat in order to be a dog, being ripe requires an apple to have a different appearance and chemical molecular substance than a rotten apple), A and not-A are such a pair because a thing is not-A by not having the essence of A and a thing is A by having the essence of A, therefore the consequence of the essence of being A is not being non-A, therefore the essence of being A will cause the thing that is both A and non-A to cease to be non-A if it continues to be A as a consequence of being A, therefore a contradiction cannot exist.

The line of reasoning proving that "existence exists," if it means "everything (that exists) is something," is: to exist is an attribute, to have an attribute is to be a something (which is deduced from the essence of "thing," which is abstracted from perceived things,) therefore everything that exists is something.

If "existence exists" means "reality is physical and objective," it is reasoned as follows: one forms a concept of the perceivable world as physical and objective by observing perceived objects, e.g. by touching a book you learn that it is physical, by seeing a book both before and after closing your eyes you learn that it exists separately from your perception of it (because it continued to exist while not seen and therefore your vision did not create it), which means that it is objective. Because the essence of existing is being physical and objective, it follows that everything that exists, in other words existence, must be physical and objective. Another line of reasoning proves this: a thing to be perceived must be at a specific place and time, to be at a place it must exist in space, to exist in space it must be physical, to be physical it must be made of physical substance, and to be made of physical substance it must exist outside of consciousness and perception, which proves that the perceivable world exists physically and objectively.

When you reason something from some source, you can point to your source and your reasoning and say that this constitutes proof, that you have not asserted your conclusion, instead you have demonstrated your conclusion. You need only point out an apple to someone and show how you can reason that it cannot be contradictory and that it is physical and objective, and then show how to go from one apple to everything, and you have proved identity and existence to that person.

Chapter Twenty Five: The Flaw in Rand's Intrinsic-Objective-Subjective Analysis

Any reader of mine who is well-versed in Rand's epistemology will now exclaim about my theory: "But that's intrinsic, not objective!" That objection is the fundamental source of Objectivism's rejection of essences (Rand 1990, 52-54). So let's deal with that objection, and see how my theory can improve Rand's account of the objective.

When Rand rejects the theory that essences are in objects rather than in the mind, she is thinking of Plato as the representative of the "intrinsic" position; her problem with Aristotle is that he is too much like Plato (141). The differences between my theory of essences and Plato's Forms come in two parts. The first difference is that Plato's Forms can never be known by sense perception, and my essences can only be known by sense-perceiving objects that have those essences, or by conceptually combining or analyzing the different essences that you have reasoned from perceived objects. The second is that Plato's Forms are in a world apart from the physical world, they are spiritual, and my essences are in the physical world, they are in physical objects. Really the essences are the physical objects, or, to phrase it more precisely, they are physical objects when they are thought about in a certain way, by focusing thought on the essential aspect of the objects. For example, the essence of a dog is every real dog thought of as dog, focusing only on what makes it a dog and not on any of its other characteristics. Also, as I mentioned, it is useful for essential reasoning to think about essential things, but an essential thing does not exist, it is merely a conceptual took for isolating and analyzing the essences of real physical things.

Thus, my essentialism does not reduce to Platonic Forms, and is fully compatible with Objectivism. Rand's objection to intrinsic essences centers on her claim that they can only be known by revelation, and that they exclude consciousness from reality (53, 141). She doesn't consider the possibility that essences could be known by perception, or that they might be in things in themselves but be known though an intellectual process such as essential reasoning which abstracts essences from perceived objects, perhaps

111

because this might damage her intrinsic-objective-subjective idea, which is itself a contradiction. She claimed that a theory such as mine reduces thought to perception (53). But that is not true: the essence is perceived, for example you can see that a dog is a dog, but the act of abstracting the essence from a specific perceived object is an act of thought, for example it is an act of thought to go from a perceived real dog to the essential dog. In fact, her criticism here could more accurately be directed back upon herself, in the sense that she reduces thought about "existence exists" and "A is A" to perception, as if the abstract axioms were seen directly instead of being integrated from inductive reason.

Being, what makes a thing be what it is, what can be called essence, is either in things in themselves, outside of the mind, which Rand rejected as "intrinsic," or else it is in the mind, in which case it is subjective. There is nothing in between. For being to be both in things and in the mind is a contradiction. And if being is in the mind, but is based on reality, then what precisely is it in reality that it is based on? Rand writes that *"A concept is a mental integration of two or more units possessing the same distinguishing characteristic(s), with their particular measurements omitted"* (13). Either the "distinguishing characteristic" is in the thing, which Rand rejects as "intrinsic," or it is in the mind, in which case Rand's epistemology reduces to subjectivism. Rand also says that the definition of a concept must be by its most essential characteristic, the one that causes and explains all or most other defining characteristics (45, 71, 230-31, 238-39, 304). If this quality of being essential is merely in our minds and not in reality, if it is epistemological rather than metaphysical, then how is our concept not subjective?

Rand's "objective" is a contradictory compromise between "intrinsic" and "subjective," an attempt to claim that being is somehow in between reality and the mind. The truth is that being is in reality but is known by the mind, but Rand's "objective" does not accomplish this.

She says that essence is "epistemological," not "metaphysical," but essence is another word for being or what makes a thing be what it is, therefore Rand's statement reduces to the claim that being is in the mind, not in reality (52). If we are going to be rational then we must conclude that the "intrinsic" position is what makes objectivity possible. This also reflects what is visible, for example, what makes

112

a square be a square is its four equal sides, and this is in the square, not in our minds: it is metaphysical, but it is known by epistemology. Rand's intrinsic-objective-subjective account contradicts what she says elsewhere, namely that the attributes of entities are metaphysical, not epistemological (277-79). A square's essence, as square, is its attribute of having four equal sides, and every being's essence is merely whichever attribute it has that is essential as that type of thing.

When Rand claims that the objective is between the intrinsic and the subjective, the implication is that the objective is in the mind rather than intrinsic to external objects, but yet it is still somehow related to something outside the mind. That is a contradiction. If essence is in the mind but is related to reality, then what in reality is it related to? If definitions refer to essential characteristics, what are these characteristics if not essences? Either the defining characteristic of a thing is in the thing itself, which Rand rejected as "intrinsic," or it is in the mind, in which case it is subjective. Rand asks "to what precisely do concepts refer in reality? Do they refer to something real, something that exists—or are they merely inventions of man's mind, arbitrary constructs or loose approximations that cannot claim to represent knowledge?" (1). Yet this, the question of what concepts refer to in reality, is what Rand fails to answer, and what she cannot answer because she rejects the idea of essences. An essence is a thing's identity as a specific kind of thing which is in the thing itself, so if there are no essences then concepts have nothing to refer to.

If we reject Rand's intrinsic-objective-subjective analysis, then what becomes of objectivity? A good theory of objectivity was put forward by Aristotelian philosopher Mortimer J. Adler in his "Ten Philosophical Mistakes" (Adler 1996, 5-82). Adler argued that the problem of objectivity is an interrelated conceptual and linguistic mistake. The word that is missing from the discussion is "of," which denotes the object of a subject. When I see an apple, the apple is not a perception. The apple is the object of a perception. The perception is in my mind, whereas the apple that I see is out in the external world. When I think about a dog, the concept of "dog" is in my head, but the dog itself is out in the world, if it exists, or if the dog doesn't exist, then it isn't anywhere, and I am thinking about something that doesn't exist.

Perceived things and thought-about essences are the objects of consciousness, whereas the perceptions and concepts are the means by which I perceive or think about things. Perceptions and concepts are in my mind, whereas what I see and think about is in reality. If I am thinking about a square, the concept of "square" is in my mind, but the essential square, the object of thought, does not exist, and I am really thinking about all the squares that exist as such, as squares. Adler argued that since perceptions and concepts are in the mind, if we see perceptions and we think about concepts, we become trapped in subjectivism and we can only see and know what is within the mind. I extend that argument by claiming that what we perceive is objective reality precisely because what we perceive is that which we perceive, not the means by which we perceive it, and this fact is itself perceptually verifiable, e.g. when you look at a book you see the book, not your eyes.

If we use "of," if we are aware that what is in the mind is only the means by which we perceive and think, rather than the objects of perception and the contents of thought, we can escape from this problem. Rand, who advocated a belief in an objective reality, should have said that we perceive an objective reality, a world outside of our mind, a world separate from the mind and independent of the mind. But she didn't understand how to put this belief on a solid logical footing. Because Rand claimed that the things that we see are perceptions made of sensations, and that things such as existence and identity and everything else abstract are concepts, her epistemology reduces to subjectivism (Rand 1990, 5, 10). But the problem is easily fixed: simply know that existence is not a concept, it is a thing to which a certain concept refers, and the things that you see are not your perceptions, they are the things in objective reality which the perceptions in your mind are your means of perceiving. Being itself is in things in themselves, but our concepts of beings are in our minds.

This enables us to say things that Rand rejected, but should have said: that concepts and perceptions physically exist in our brains, but refer to objects outside of our brains, that the mind reduces to the brain, that the self reduces to the body, and that the mind-body relationship is a question whose answer lies in philosophy, not science (154, 290). This also clears up Rand's confusion and imprecision over whether concepts physically exist as concretes or

114

not (154-58). I would argue, contrary to Rand, that a mind cannot understand itself or the world around it in any meaningful way until it grasps its relationship to its body, the relationship of consciousness to self. And once you deduce that perceptions come from the impact of external objects on your body's sense organs, you can reason that everything that you perceive exists objectively.

Chapter Twenty Six: Other Problems with Rand's Epistemology

Rand might have challenged me to show in what way the essence is not arbitrarily defined, but I don't consider that to be a problem. The essences are in things in themselves, but which essence you are thinking about is a matter of your choice. A thing is a set of attributes, and essential reasoning chooses one attribute and then reasons the consequence of the essence which then applies to everything which has that attribute. For example, if you have a red cardboard square, thinking about it as a square rather than as a red thing is a choice, but the four sides producing four right angles is not a choice. Rand's disciple Peikoff might have objected that if non-essential qualities are contingent rather than necessary then we enter the problem of the "analytic-synthetic dichotomy," but I disagree with that as well (115). A thing's non-essential qualities are contingent relative to the essential quality, but are absolute when the thing is considered as a whole. Rand seems to think that an Aristotelian essence is a physical object within a thing that gives it identity, as if there were an organ inside of humans that made them human, and each organ is identical (139). I dispute this. An essence is an attribute of an object, such that the details of what I do and what you do may differ, but our act of being human, as such, is identical, in that we are both beings which think and have human DNA and generally human-shaped bodies. If there was nothing identical between us then we would not both be humans.

Where are the universal laws of logic, such as identity? Where are metaphysical facts? If they exist in this world, where are they so that we can point to them and look at them and touch them, or if they exist in a different world, how is that not a non-physical world of spirit, which contradicts Rand's metaphysics? Why are the laws of logic necessary, and why are they true everywhere and at every time? Rand says that things act a certain way because a thing "must act in accordance with its nature," but where precisely is a thing's "nature"? (287-88). Is it in the thing itself, or in some other plane of existence? What exactly is a thing's "nature"? And if one goes from concretes to universals by means of measurement-omission, precisely how does the mind go from a concrete measurement to a universal measurement-omission? (11-12). Specifically, how does the mind know that any measurement may be substituted for the omitted measurements? (18). What are the "characteristics" that determines whether a concrete is included in a concept or not, which Rand seems to admit are the basis of conceptual categorization, if not an essence in the things in themselves? (17).

Rand's theory has no answers for these questions, but my theory does. Metaphysical facts and the laws of logic do not exist separately from the physical world. They are the essences in physical objects. The essence of "thing," the essence of a thing as a thing, in other words a thing as such, requires it to be itself and to not be not itself. A thing must have the essence of a thing in order to be a thing, so the essence requires obedience to the laws of logic, and the metaphysical fact of identity is the essence of a thing. This is why propositions stated as metaphysical facts are necessary for all things of that kind. For example, a consequence of the essence of water is that it is a liquid when it is at room temperature, so all water, not just some water, is a liquid at room temperature. A thing's essence as that thing is its "nature," that is why it behaves in a certain way, because it has that being. The essence of water is being H2O, so if a scientist reasons that water is liquid at room temperature because of its chemical composition, then this is proved as a universal truth for all water. For a thing to be water it must be H2O, and being H2O causes a thing to be a liquid at room temperature, so the essence of water will require all water to be liquid at room temperature.

The essences are in the things in this world. Essences are the things of this world thought about using essential reasoning, so they exist physically in our world, and not in a separate Platonic world of spirit. But it is the essence, not the mind's essential reasoning, which requires that things in reality obey the metaphysical facts and act according to their nature. Necessity comes from the essence itself. The essence requires the consequence of the essence, and having that essence is being that thing, so being the thing will cause the consequence. Essential reasoning enables the mind to go from specific things to universals, because induction derives a concept of the essence of a type of thing from one, a few, or a group of some real things with that essence, and deduction then reveals the consequence of that essence which must be true for all things with that essence.

Rand admits that concepts must match "essential" characteristics, and that "essential" characteristics exist, but claims that this is the case because cognitive efficiency and practicality dictate it (52, 65, 70-72). It would make more sense and be less arbitrary if the justification was that the essential characteristics actually exist, in other words if essences exist. Rand acknowledges that definitions should be based on "fundamental" characteristics which cause or explain most other characteristics (42, 44-45). If this is the case, how is the fundamental characteristic not an intrinsic essence if it is in things in themselves, and if it is not in things in themselves, then how are definitions based on objective reality? Rand posited rules for the formation of concepts, but the same critique applies to her rules as well: if a black swan is a swan because of attributes of the black swan itself, then those attributes are an essence of "swan," and if it is a swan merely because it is mentally useful for humans to classify it as one, then the concept of "swan" is hopelessly subjective (70-73).

Chapter Twenty Seven: Rethinking Randian Necessity and Universality

My theory also corrects a problem in Rand's account of necessity, which is based on her theory of the metaphysical vs. the man-made (299). For her the metaphysical could not be different than it is because God did not create the Universe, but the man-made, because man has free will, could have been different. My problem with her analysis is, first, that she says that metaphysical facts were not created by God, but she does not explain where they come from, and second, that while she explains why metaphysical facts are necessary, she does not explain why it is necessary for the objects in reality to obey those metaphysical facts, she offers no detailed mechanism for explaining how necessity functions. My theory explains where metaphysical facts come from, e.g. "squares have four right angles" comes from the essence of squares. My theory of essences provides a clear, understandable mechanism for how necessity is created: because a thing must have the essence in order to be that kind of thing, and the essence causes the consequence as a result of having that essence, everything of that kind must have the consequence because it is what it is, e.g. every square must necessarily have four right angles because it is a square, and if it did not have four right angles then it could not have the square essence and it would not be a square.

My analysis extends to solve another contradiction in Rand's epistemology, her belief that all knowledge is contextual (42-43). Rand's thoughts on necessity as it relates to her metaphysical vs. man-made distinction and universality as it relates to contextual knowledge are captured in a bit of dialogue about water in "Introduction to Objectivist Epistemology" (295-301). A "professor" asks her how we know that water boils at a certain temperature, and why we know that all water boils at this temperature rather than merely all the water that we have known in previous experience. She begins by saying: "By whether you can or cannot establish a causal connection between what you have determined to be the essential characteristic of water and the fact that it boils at a certain temperature" (295). I fully agree with this statement, and this quote supports my theory of essences rather than Rand's own intrinsic-objective-subjective idea. But when asked how to establish the connection, Rand ducks the question by saying that the answer lies in science, not philosophy, because it involves molecular chemistry. When pressed on why molecular knowledge matters, she says that it is simply because we have gained more knowledge. This I disagree with: it is not merely more knowledge, rather it is knowledge of cause and effect, of the consequence of the essence of water, and this is where necessity and universality come from.

Rand, without this idea, then posits that water boiling at that temperature is necessary because it is metaphysical and not man-made, and says that this knowledge about water applies only to specific cases, and could be proven wrong if we were to learn more knowledge, because all knowledge is contextual (296, 298). It is true that there are certain specifics to the case of water boiling at a certain temperature, but these all must be considered in essential reasoning, to arrive at the essence whose consequence you want to reason. In other words, water is an essence, water at a high altitude is another more specific essence, water at a high altitude with salt in it is an even more specific essence, and each essence has its own consequence. Thus, when you have reasoned the consequence of an essence, you can say that your knowledge is necessary and universal, without any possible exceptions, regardless of any new "context." But when Rand claims that not just some, but all knowledge is contextual, she paves the road to skepticism or relativism. For her position to be applied logically, all knowledge is merely relative to one's context, and everything that one knows, literally everything, could be disproved by a new context. This is a form of skepticism since it makes absolute knowledge impossible, and it is relativism because it makes all knowledge relative to one's context.

Rand claims that a new context will never contradict previous knowledge, but she never proves why this will be the case (43-45). If your context changes and forces you to give up on knowledge or definitions that you had previously held to be true, it would seem that your previous knowledge has been disproved. She claims that mathematical knowledge is not contextual, but this leads to an analytic-synthetic split: math would be analytical, empirical science would be synthetic (202-3).

Rand's contextual knowledge is a contradictory concession to those who claim that empirical reason cannot provide absolute certainty. If all knowledge is contextual then no knowledge can claim to be truly universal. The solution is to see the difference between more knowledge and consequence of the essence knowledge: the former can be overturned by still more knowledge, whereas the latter is necessary and universal, in all cases. Essential knowledge is not just more knowledge. It is a special kind of knowledge.

Our theory of necessity and universality is of the utmost importance. In my opinion the great appeal of Kant is that he offers a comprehensive theory of necessity and universality. Kant claimed that necessity is created because the mind applies the laws of science to everything that it experiences with no exceptions (Kant 1977, 63-64). Thus Kantian subjectivism creates universality. If we are to challenge Kant's supremacy we will need an account of necessity and universality that comes from things in themselves, which is impossible if we accept Rand's intrinsic-objective-subjective analysis.

Chapter Twenty Eight: Conclusion to Rand's Axiom Problem

For Rand, because concepts are "merely... our way of organizing concretes," without reference to essence in things in themselves, it is impossible for Rand to justify universal knowledge, or to claim knowledge of things in the future (Rand 1990, 307). Nowhere in "Introduction to Objectivist Epistemology" does Rand explain how, if concepts are not based on essences, they are not arbitrary, nor how, if they are merely groupings of concretes from previous experience, do they offer knowledge of the concretes that one is going to experience in the future. If concepts are merely a way to mentally organize concretes, then how is conceptual reason capable of proof? Rand's concepts, just because they are "open-ended," do not enable her to prove truths about the future, because her concepts do not refer to essences in concretes and so she has no basis for knowing that concretes in the future will resemble concretes from the past (17, 27, 66). She claims to have solved the "problem of universals," but her answers are problematic (1-3).

Because things in themselves are required to have the essence in order to be that kind of thing, since they would not be that kind of thing if they did not have the essence, and the essence causes the consequence, my essences offer universal knowledge. Only essential reasoning enables your knowledge of the present and the past to produce conclusive, demonstrated knowledge, in other words proof, about the future and the things that you have not yet experienced. My theory also enables reasoning that is purely empirical, in other words, which is purely the result of deductions from sense perceptions. Essential reasoning solves the problem of induction. Rand's axioms are assumptions dressed up in a fancy name, and as soon as you base your reasoning on unreasoned assumptions instead of perceptions, you can no longer claim to base your knowledge on empirical observation of the perceivable world. Indeed, reliance on the axioms is based on faith in the axioms, and Rand's theory of axioms contradicts her commitment to reason.

My theory will give an epistemological foundation to Rand's metaphysical, ethical and political ideas that is stronger and more logical than the theory of axioms. My theory will also introduce a more logical conception of objectivity, necessity and universality than is found in Rand's writings. I agree with Rand's vision of reality as objective and as something that can be understood and mastered by the reasoning human mind, but the details of her epistemology betray and contradict that vision. Fortunately we have a solution to Rand's axiom problem.

Let me add a concluding postscript to this paper. When I first submitted it to "The Journal of Ayn Rand Studies," it was rejected for two reasons: first, because the reviewer asserted that I was accusing Rand not of an internal contradiction, but merely of disagreeing with my personal epistemology, and second, because I did not fully explicate my epistemic theory of essences alluded to in the paper. The conception of this paper as incomplete is a straw man. The rest of this book explains the details of my philosophical beliefs, but the above presentation gives enough of the gist of my ideas for my critique of Rand to be thoroughly understood. Secondly, my argument is undeniably one based on Rand's own internal self-contradiction, as well as her external contradictions.

Rand believed in reason and rejected faith, but she accepted the axioms without reasoned proof that the axioms are true. If the reader of this paper considers the natural, reasonable proposition that something which cannot be proven by reason must be accepted on faith, since the origin of any idea is either reason or faith, then it follows that Rand's premises collapse into a contradiction, in that reason is based on axioms and axioms must be accepted on faith, hence reason is based on faith. Merely disagreeing with my argument, in the absence of a well-reasoned critique, does not refute my argument. Disagreement does make clear whether the reader of this paper possesses blind faith in the ideas of Ayn Rand, or whether he or she is willing to check Rand's premises and question the Randian philosophy using a serious philosophical methodology.

REFERENCES:
Correspondence
Chris Matthew Sciabarra, 3 May 2009
Publications
Adler, Mortimer J. 1996. Ten Philosophical Mistakes. New York: Touchstone.

Appignanesi, Richard and Chris Garratt. 2007. Introducing Postmodernism. Cambridge: Icon.

Aristotle. 1947. Introduction to Aristotle. Edited by Richard McKeon. New York: Random House.

Kant, Immanuel. 1977. Prolegomena to Any Future Metaphysics That Will Be Able to Come Forward as Science. Translated by Paul Carus, revised by James W. Ellington. Indianapolis: Hackett.

Nietzsche, Friedrich. 1990. Twilight of the Idols/The Anti-Christ. Translated by R.J. Hollingdale. New York: Penguin.

Plato. 1997. Complete Works. Edited by John M. Cooper. Indianapolis: Hackett.

Rand, Ayn. 1957. Atlas Shrugged. New York: Signet.

___. 1990. Introduction to Objectivist Epistemology. 2nd ed. Edited by Harry Binswanger and Leonard Peikoff. New York: Meridian.

___. 1943. The Fountainhead. New York: Signet.

Thompson, Della, ed. 1998. The Oxford Dictionary of Current English. New York: Oxford University Press.

Part Two-B: An Essay on Reason and Perception

Chapter Twenty Nine: The Problem of Knowledge

In philosophy it has been taken for granted for a very long time that deducing something that can properly be called "knowledge" from objects perceived by the body's five senses (sight, sound, touch, taste, and smell) is impossible. The idea that appearance and reality are different and opposed, and that reason helps free the mind from the deception of the senses by helping the mind come to know true ideas, dates back to the philosophers of ancient Greece, such as Plato. It was popularized by Descartes and Kant during the Enlightenment, and it remains the dominant trend in both Analytic and Continental philosophy today. The illusion of the physical world is a dominant idea in Western philosophical thinking. The desire to go beyond appearance to deeper truth has been called the hallmark of the philosophical mind.

However, just because everyone believes that something is true should not be taken as evidence that it is true, at least not by an inquiring mind. I challenge the notion that it is impossible to reason knowledge of reality from sensory experience. To advance my position I have written this essay. In this essay I describe the four most powerful arguments against the possibility of empirical reason achieving knowledge, and I present my general strategy for refuting them. In the main body of the essay I explain how reasoning works. Then I prove that empirical essential reason can produce true knowledge of existence. In the conclusion I apply my proof to refute those four arguments.

I call the first argument against the senses the "deception" argument. This argument has different versions, but its main idea is that the senses are capable of deception and therefore cannot be trusted. One good version of this argument was put forth by Descartes. He said that because everything that you see might be part of one gigantic dream, or else might have been put before your eyes by a deceitful spirit, that the reality of what you see cannot be taken for granted. If the senses cannot be trusted then a source of certainty outside of the senses is required, because all perception could be wrong. Descartes, like many other philosophers, proposed God as the extra-sensory source of knowledge. To extend upon that argument, if what one sees might be an illusion, then one could not trust any principles one might reason from perceived objects. Furthermore, it would be impossible to use the senses to prove that the objects of perception exist, since those very objects in the physical world from which one might try to deduce that reality can be sense-perceived might themselves have been deceptive. Thus, certainty in the senses is logically impossible. For example, if you see a pizza, and smell it, and taste it, each of these sensations might be deceptive, being a total illusion or a distortion, and therefore your senses cannot provide you with knowledge about the pizza, nor can you prove that the pizza exists merely on the basis of the pizza itself.

I call the second argument against the senses the "subjectivity" argument. This argument also has many variants, but its main idea is that the human senses are subjective and therefore are incapable of providing any knowledge of objective reality. As an example of this, Hobbes argued that, given the many cases of our senses deceiving us which we have all experienced, it is impossible for reality to be directly perceived by the senses. He proposed that the objects of human perception are representations of external objects created by the brain in response to sensory stimuli. Later philosophers including Berkeley, Hume, and Kant have all pointed out that if the objects of perception are mere images produced by the effect of objects upon the human mind, then the senses never come into direct contact with existence external to the mind, and therefore it is impossible to reason anything about objective existence from sensory perception, which is purely subjective. For example, since your experience of tasting a slice of pizza is subjective, and the sight of the pizza is subjective because it comes from your eyes, your senses cannot be relied upon for any objective knowledge of the pizza in itself.

I call the third argument against reasoning from the senses the "coincidence" argument. The idea behind this argument is that, given the possibility that everything that one sees could be one gigantic, long-running coincidence, that it is impossible to reason anything about identity or cause-and-effect from sense-perceived objects. I can give two historical versions of this argument. First, the ancient Greek philosopher Heraclitus believed that because what everyone sees is constantly changing it is impossible to reason any stable knowledge about anything that one sees, since it will change and one will no longer know what it is. His famous saying was that you cannot step into the same river twice. Descartes adapted that argument when he compared all the objects of perception to wax from a melting candle, which constantly changes and ceases to be what it was.

Another version of the coincidence argument was presented by Hume, the famous skeptic and doubter. He said that, because the fact that something causes something else is reasoned by the repeated observance of the cause followed by the effect, that the conclusion of cause and effect was mere conjecture, since for all that one knows the cause could be perceived in the future without the effect, and that therefore there was no way to reason knowledge of necessary cause and effect from perceived objects. For example, one believes that the Sun will rise tomorrow, but how can one be certain that this will happen? One believes that the Sun will rise tomorrow because the Sun rose today, rose yesterday, and rose every day since one can remember. But all of those things could have been a coincidence. Causation is not evident in the things in themselves, and the past is no basis for any claim to certainty in the future. We cannot know for sure that the Sun will rise tomorrow. Empirical observation is not the basis for any certainty about future events. Hume's position was that no principle of induction exists. This means that we cannot infer a general principle from observing a group of many specific instances. For example, even if we saw the Sun rise in the morning one thousand times, we cannot infer from our experience by means of induction to a reasoned conclusion that the Sun will always rise in the morning.

I call the fourth argument against reasoning from the senses the "assumption" argument. This argument says that reasoning can never be certain, because if one were to use reason as a means of certainty in the ability of reason to provide true, certain knowledge of existence, say for example by reasoning the nature of existence and of the mind from sense-perceived objects, and then reasoning that they are such that reason provides true, certain knowledge of existence, then the very act of reasoning that conclusion requires beginning with the assumption that reason provides true certain knowledge of existence and the natures of things. Only with that assumption can you have a basis for being certain of the reasoning that tells you that reason is certain, and without it you have no basis for certainty, because your only basis for certainty in the reasoning telling you that reason provides knowledge would be the conclusion.

In other words, using reason to prove reason would be a circular argument, requiring that one begin with an assumed premise that reason is a valid method to achieve knowledge. If the conclusion were false and reason was deceitful by nature, or merely capable of error, then your reasoning could be lying when it tells you that reasoning is certain, and you would not be able to distinguish truth from lies using reason.

For example, if one reasons that reasoning as a means of discovery provides true, certain knowledge of existence, i.e. reason is right, valid, truthful, etc., and reasoning did not provide true knowledge, then, if one used reason to analyze the nature of reason, one might be wrong and not even know it. One must assume that reasoning is right in order to believe that the reasoning which showed reason's validity is itself valid. And if reason is founded upon an assumption, which by definition is not reasoned, then one cannot rationally be 100% certain of reason. One could only be 100% certain of reason through faith or intuition grounding the assumption that reasoning works, which would have to be discovered by some other non-rational method. Because any attempt to reason the validity of reasoning relies upon reason, reason needs something irrational for its foundation.

Each of the above arguments seems irrefutable, but they are not. Each of these arguments makes use of a very clever trick which philosophers have been employing for thousands of years. This trick is the introduction of an unreasoned, unproven assumption at the beginning of the argument, and then using that assumption to reason a conclusion, a conclusion which in turn is used to justify the initial assumption. I call this kind of technique a loop argument. By identifying and refuting the hidden key assumption, the loop argument is refuted. My general strategy for refuting the above arguments is to use reason from the senses to refute the assumptions behind the arguments.

One could argue against me here, along the lines of the "assumption" argument above, that what I would be doing would be no better, since I too would be beginning with assumptions in the use of reason to dissect those arguments. But this is mistaken. There is a classical misunderstanding about how reason works which makes that objection possible. One of the goals of this essay is to show that the classical understanding of how reason works is flawed, and to explain how reason actually works. Reasoning does not begin with assumptions, and the historical idea that reason begins with self-evident propositions called "axioms" is deeply flawed, and boils down an attempt to give reason a foundation in unreasoned knowledge, which is the popular answer to the "assumption" argument. Reason does not begin with assumptions. It begins with sense-perception, and then proceeds to reason principles from the information it receives from sense-perception. Reasoning from an assumption and reasoning purely from sense-perception without the use of any knowledge besides what is perceived, i.e. reasoning from the senses without assumptions, are two fundamentally different actions with two fundamentally different kinds of result, as I shall explain.

Chapter Thirty: Metaphysics: Being

We must now ask and answer the question of how reasoning works. Reason thinks about things and deduces knowledge about things. In other words, reason is a means of human knowledge, and the objects of human knowledge are things. We must begin our inquiry, then, with a pursuit of the answer to this question: what is a thing? It may seem like a tangent to explore this question, and the question may seem to be too broad to be answerable, but we will see later that any inquiry into epistemology can benefit from, and ultimately necessarily depends upon, some exploration of the question of what it is to be something.

To begin I must introduce an idea that I call the Japanese Distinction. Unlike English, the Japanese language has two different words for the two different uses of the English verb "is," or "to be". The English language has two related but conceptually distinct uses for this one word, and English speakers typically confuse the two and think of them as one thing. The first use is as the verb of identity, as in "X is something". The second use is as the verb of existence, as in "X exists." I draw a distinction between being something, e.g. "X is Y," and existing, e.g. "X is." The Japanese language proves that it is possible to form a coherent linguistic system built on this distinction. The Japanese use the word "desu" when they wants to say that a thing is something, which means that it possesses a certain identity or has a certain quality or property. The Japanese use an entirely different word, "arimasu," when they want to say that a thing exists. (They use a third word, "imasu", to say that a living organism exists, i.e. that a thing is alive.) Although most Western languages use the same word for these two different meaning, (1) being something and (2) existing, this does not have to be done. The two meanings are not identical. I employ the Japanese Distinction for my analysis in this essay. Any reference to "being" in the following text refers entirely to the first usage, to being something. In contrast to being, I use the word "existing" whenever I want to say that something exists. What I have to say will not make sense unless the Japanese Distinction is understood.

We should base our consideration of being upon concrete empirical observations. To that end, we shall use an object as an example to observe what it means to be a thing. Consider, for example, a rock. Please, feel free to go out into the street, find a small rock or pebble, and examine it. Rocks are not very expensive, and this example is intended to be based on my reader's direct experiences of reality.

Your rock has certain properties. It has a property of color, which you can see. For example, the rock that I am holding is mottled white, brown, and gray. It has a property of strength: squeeze it and experience how hard it is. It has the property of texture: run your fingers across it, and you can feel how rough or smooth it is. It is easy to perceive that this rock has properties.

This rock can also do actions. If I flick it, it will roll a little. If I drop it, it will drop to the floor and roll. And if I throw it, it will fly through the air and land a few feet away. When the rock sits on my table, I cannot see it move or change in any way, so I may think that the rock's act of sitting there on the desk is a passive, inactive action, since the rock just sits there and doesn't really do anything. When the rock rolls, drops, or sails through the air, on the other hand, it is easy to see the rock moving and thereby performing an action, these actions being rolling, dropping, and flying respectively. It is fair to say that these are active actions, passivity being characterized by the absence of motion, and activity being characterized by the presence of motion. Even though I am the source of the motion, it is the rock that flies through the air. This is not a passive act, since it consists of motion. Thus it can be clearly perceived that the rock does actions.

Now let's go back and consider our rock's properties. Using my rock as an example, I can say that it is mottled white, brown, and gray, that it is hard, that it is rough, and that it is light-weight, and that these are some of its properties. But do we really know what this rock's properties actually consist of? What does it mean for my rock to possess these properties? I propose that the property of color of this rock, including all the different white, brown, and gray areas of its surface, consists of the surface of the rock affecting the light that hits it in a certain way. Similarly, the hardness of the rock consists of it pushing back against my fingers when I squeeze it, and the property of texture which the rock possesses consists of its surface affecting the surface of my fingers in a certain way as I touch it. It seems to me that each property consists of an action done by the stone, whether it is light or my fingers which the stone acts upon. What the properties actually consist of is the thing performing the action that it does, for example, by reflecting and absorbing light as it does, by exerting force in response to force as it does, and by scratching as it does.

Now let's turn our attention back towards the actions that the stone does. One can form an idea of what the rock's actions of rolling, falling, and flying consist of by causing it to do these actions and observing the actions. However, one aspect of what these actions consist of might not occur to the reader. To draw your attention to this I must note the classical distinction between being-type actions and doing-type actions. This distinction has been made as follows: that for a thing to be, or for a thing to have a property, is a being-type action and is inactive, and that actions which consist of visible motion are active and are doing-type actions. For example, for a person to be a nice person is a being-type action, and for a person to kick a soccer ball is a doing-type action.

But now examine your rock in this light. You should be able to drop it and see its doing-type action, falling. Now while the rock is falling, it is a falling rock. In terms of the above distinction, the rock's property, the being-type action of being a thing that is falling, is an inactive, immobile "predicate" that is "predicated" of the rock's "subject," or an "attribute" that is "attributed" to a "substance". Now I ask this: what is the difference between the property of being a falling thing, and the actual visible motion of descent that the rock is seen doing? What is the difference between being a falling thing, and falling? No matter how the rock is examined while it falls, no possible difference is observable. For that stone to be a thing that is falling, which is for that stone to be something, which is a being-type action, is for that stone to fall, a doing-type action. It is precisely because it falls that the stone is a falling thing. In the other examples I mentioned, empirical observation will reveal that the soccer player is a soccer player because he kicks the ball, and that a nice person is nice because he is nice to other people. The act of kicking and the act of being nice are equally visible, active actions.

In light of the above, my reader can understand the following proposition, that there is no difference between a thing being something and a thing doing an action. I propose that properties and actions are completely identical, and are in fact the same thing discussed using two different sets of terminology resulting from a mistaken distinction made by philosophers and linguists. If I am right, then each property that a thing has is actually a physical, continuous, active action done by that thing, and the thing "possessing" the property consists of the thing doing that action. In other words, to be something is to do a certain action, and a thing is something by doing the action which that certain something does. In support of this proposition, I would like to draw your attention to the above example, in which only a few of the actions and properties of a rock are shown to be consistent with this idea, and argue that what was demonstrated regarding those properties and actions can be noticed in any action or property that you might care to examine.

Furthermore, in order to disprove this proposition, the reader would have to find a property which did not consist of any action, or find an action done by a thing which is not a thing doing that action, both of which I believe that my reader, if he or she searches for it, will concede that it is impossible to find. If my reader points to something red and says that it is not yellow and that this is a property which does not consist of any action, I will reply that while being something consists of doing, not being something consists of not doing, as not-being is the absence of being, and showing that an absence of identity consists of an absence of action only supports my theory. This proposition is best supported by my reader undertaking the experiments using a rock that I have described and seeing what he or she sees, while asking the questions "what is it doing?" and "what is it being?" (My answer to the objection that size, shape, mass, etc. are inactive properties will follow shortly.)

Now let us consider the example of a symphony. A symphony does an action. It causes you to feel the emotions that result from you listening to the symphony. For example, perhaps listening to the Marche Slave symphony by Tchaikovsky causes you to feel cheerful and excited. The symphony acts upon you. That is an action. Now consider the individual notes that you heard. What does each note do? It seems to me that, if one asks this question and examines how the symphony accomplishes what it does, that one can see that the individual action of each individual note, by being what it is and relating to the notes around it as it does, contributes to accomplishing the effect of the action done by the symphony as a whole, and that the actions of the individual notes form the entirety of the action of the symphony as a whole. The action done by the notes is what the action done by the symphony consists of. The action of the symphony is that action which the actions of the notes form as a whole. Thus it seems evident that the notes are the parts of the symphony, and that the symphony itself consists of the notes acting together as whole. You could pluck any individual note out of a symphony and listen to it by itself, and it would not have the same effect, but if you listen to it in the symphony, if you hear all the notes in the symphony, then they work together and have a single cumulative effect. And what could this be the effect of, if it is not the effect of the action done by the notes together, as a whole?

This seemingly simple observation, when considered with respect to the being of the symphony and its notes, indicates that a thing is its parts acting as a whole, and that a thing's actions, the single actions done by that thing, are formed by the interactive actions of its parts. This will answer many of the questions regarding the proposed theory of properties which may have occurred to my reader. Take the question of size. In what way is the property of size an action done by the thing which has that property? To answer this, take a piece of standard nine-by-twelve inch paper. We might agree that it being nine inches wide and twelve inches tall are its size properties. Now cut it into four equal pieces, and stack them up. You will still have all the matter of which the paper consisted, but it will no longer be a nine by twelve piece of paper. How did its property of size change? My answer is that the paper is no longer doing the act of filling an area of space as a whole that is nine inches wide and twelve inches tall. Furthermore, the four pieces of paper are no longer a single piece of paper because they are no longer acting as a unit. They no longer move as a whole, nor do the parts hold each other together, although this is true of each of the four new pieces, which is why each of those four new pieces is itself one piece of paper. They sit on top of each other without doing any action in common, which is why they are a pile of four wholes rather than one whole.

If my theory is right, then for a thing to be a whole is caused by its parts acting as a whole, by their actions forming the single action which the thing as a whole does. Shape can be seen as an action in the same way as size in the example above. A thing's parts, its matter, does the act of filling an area as a whole, and thus it fills the outline of that area, which is its shape, as a whole. Thus a thing's shape is an action done by it as a whole.

I believe that the most potent objection to my theory, which is that some properties are totally inactive and do not involve any action, i.e. some properties are still and motionless, would be disproved by examining these properties and asking what it is they consist of. One can examine the property of mass in this way. It would seem that the property of mass, which is the amount of matter in an object, is totally inactive, as it consists of nothing more than an object's mass just sitting there and being counted without doing anything.

Let us return to the example of my rock, which weighs one gram for argument's sake. I could break it into two pieces. One of these pieces might contain .25 grams of mass, and the other .75 grams of mass. All the matter which was just sitting there is still just sitting there, in fact I have still have one gram of mass in total, and yet I no longer have an object with the property of being one gram of mass, instead I have two objects with the properties of being .25 and .75 grams of mass. Why don't I have a one gram object anymore? Because the act of being one gram was done by the parts of the rock acting as a whole, as a single thing, as that rock. Without this action, there is no thing being one gram, and the property of being one gram as a whole actually consisted of the act of being a thing as a whole done by the particles which make up that rock.

Along these same lines, say that one has a bar of iron, and one considers one atom of iron inside the bar, and one atom of iron a few feet away from the bar. On what basis is one atom a part of the bar and the other atom not? The atom inside the bar is a part of the bar because it acts as a whole with all the other atoms to form the bar, because it functions as a part of the bar, and the other atom isn't a part of the bar because it does not act as a part. Parts of something are parts of it because they do the action which a part of something does. Contrast this with the case of a parasite inside an animal. The parasite is within the animal, yet it is not a part of it, because the parasite acts as a whole and does not act as a part of the animal. If it did then it would become a part of the animal; this is how some of the parts of human cells evolved from cellular parasites.

Another way to look at this is in terms of form and substance. Consider the example of a wooden square. One might say that the square is the form, and the wood is the substance. Classical philosophy has no reasonable way to understand the relationship between the two. But the relationship can be understood through my theory. Being square is an act done by the wood. Form is action and substance is doer. The wood particles fill space in the shape of a square, the particles which act as a whole filling the area.

Another example might clarify this. You can think of your own body that you are an organism formed of organs, which are formed of cells, which are formed of molecules, which are formed of atoms, and that each of these things is the substance of the form, e.g. the organs are substance and the body is form. But doesn't it also make sense to say that atoms acting together are molecules, that molecules acting together are cells, that cells acting together are organs, and that organs acting together are the body? That is what I mean by the theory that substance is doer and form is action. The substance of a thing is what the parts are and the form of a thing is what the whole is.

For another example, consider a golden square. The substance of the golden square is gold because each atom is gold individually independently of the rest, while the parts are square as a whole and not square separately from each other. The square is gold because each part of the square is gold, but each gold atom is gold as a whole, and so all the gold atoms are the square as a whole. Similarly, regarding a molecule of water, the hydrogen and oxygen are the substance and the ratio of hydrogen to oxygen, or more precisely the interaction of the hydrogen and oxygen atoms, produces the whole molecule, which is the form.

Now consider this example. Say that you take ten dominoes and set them in a row, and place a tennis ball behind the domino at the end of the row. I urge you to actually set up this demonstration. If you lack dominoes, thick books stood vertically will do just as well, and you can replace the tennis ball with anything. When I push the first domino over, it hits the one behind it, which hits the next one, which hits the next one, etc., until the last one is hit and then hits the tennis ball. One can ask, what causes the ball to be hit? Is it the final domino, or the hand that pushed over the first domino? Observing the experiment, one can see that the final domino hits the ball. But would it have done so, if domino number five had not hit domino number six, or if the hand had not pushed the first domino? Try removing the fifth domino, or not pushing the first one, and see. It seems to me that the action of the hand and the action of each individual domino contribute to, are a part of, the action that causes the ball to be hit. Given what I have said above, if the action of all the parts as a whole causes the ball to be hit, then the thing which is the thing that hit the ball is the hand plus the ten dominoes, and they are a whole, a single thing, in the act of hitting the ball. Hitting the ball is this thing's total act of being, and the act of each part which contributes to the whole's effect is a part of the whole's total being act. If you consider the experiment in this way, it is a perfect example of a thing and its parts.

One can also see from these things that one kind of cause consists of the transference of motion. The transference of motion and the act of one thing altering another are both visible, and so the existence of that kind of cause can easily be perceived. The hand causes the first domino to move, the first domino causes the second domino to move, that domino causes the next one to move, etc., and the last domino hits the ball. I might also note that the motion of the ball, however small, when it is hit, is the action the ball does which makes it be a hit thing.

Considering the above, it is reasonable to say that the last domino which hits the ball is the immediate cause of the ball being hit, and that the hand that pushed the first domino is the ultimate cause of the ball being hit. The last domino is the thing that transfers motion directly to the ball, while the hand is the original moving object that transfers motion through the dominoes to the ball. One can also see in the domino example that for a thing to cause something is for a thing doing an action to cause a thing to do an action. Thus the hand and all the dominoes are the entire cause of the ball being hit, they as a whole are the thing that causes the ball to be hit, and each individual part of that cause is a part of the thing that hits the ball. The hand is the first or ultimate cause of the ball being hit, and the last domino is the last or immediate cause of the ball being hit. The hand is why the ball is hit, and the dominoes are how the ball is hit. The dominoes are the means by which the hand touches the ball, in that the dominoes transfer the hand's motion from the hand to the ball. It is in this sense that the hand hits the ball.

This point will become important later in my section on perception, when I discuss rays of light hitting human eyeballs.

The domino example shows something else also. The act of hitting the ball consists of a set of acts, including the act of the hand upon the first domino, and the act of each domino hitting the next. There are other examples of this. The act of going from point one to point five of a line consists of the act of going from point one to point two, from point two to point three, from point three to point four, and from point four to point five. The act of two chemicals combining and exploding consists of the action of each individual molecule on those around it. The act of a bird flying consists of each flap of its wings and the act of each individual air particle upon it. The act of filling a shape consists of filling each section of that shape. These actions are the actions of which the whole act consists. In order to fully understand an action, the different parts of the whole action must be understood, for the whole action can be known only by knowing how the different parts' action forms a whole, since the total act is the act which the parts do as a whole.

So a thing is something by doing certain actions as a whole. But what about being something, not in the sense of having a property, but in the sense of having an identity: a thing being this rock, with all of the properties that it has? Being the rock itself, the whole thing? That is what I mean by a thing's identity. The former being something, like rough or hard, is a property, the latter being something, like being this rock, is an identity. I propose that my above theory can be extended to say that the rock's identity is the rock as the doer of its entire being action. If what I have said is true, then a thing's identity consists entirely in being what it is, and its total being action is a set of actions, the set of all the actions which it does. Each action that a thing does is an act of being. The being action of a thing itself, the being action of a thing as a whole, its entire being action, is composed of all the being actions it does. A thing's entire being action, the act of being that thing, is formed of all the acts of being which it does.

For two things to be identical, for them to have the same identity, and for those two things to be one thing, is for them to do all of the same actions. A thing is itself because it does all of the actions which it does. A thing's identity is that thing as the doer of every action which it does. The being of that thing is the set of every action which it is doing. If you were asked "what is this?" and you answered, you would find yourself listing the parts of its total being action, e.g., for the rock you would say that it is rough, hard, round, etc. In this sense, a thing is a set of properties, and the entire thing itself is the set of every property which the thing is. Also, two properties connect to form one thing because one property does the act of being another property, e.g. for a thing which is a red sphere, the red color does the act of filling a sphere-shaped space. You might say that the parts of a chair are its legs, seat, arms, and back, but, from the above examples of the iron atom, the organism, and the dominoes, we can see that being a leg, being a seat, being a back, etc., are all parts of the total act of being a chair. Just as a thing is the doer of every action it does, so it is the doer of each action it does, and each part of the total being act which becomes known adds to the knowledge of the total being act.

Some philosophers assume that the attributes of a thing are infinite, but the argument of Locke, who noted that the existence of an infinite thing is impossible because infinity is the quality of a series being permanently incomplete, and if the series of the parts of the being action was incomplete then there would be no total act of being and the thing would not do an act of being something, is sufficient to disprove that possibility. No difference exists between "a thing" and "what a thing is," there being merely a difference in whether emphasis is placed on the total being action or on the sum of each of the individual being actions which are properties. "The thing" is the thing as the doer of the total act of being, whereas "what it is" is the thing as the doer of each individual act that it does.

Few people in the past have ever asked and tried to answer the question of what a thing is and what makes a thing be a thing, and those who did accomplished more harm than good by creating terminology with which things are examined that serves to obscure rather than reveal what a thing consists of. The most dangerous such term is "property." Let us see why. My rock is a thing. It is a thing that is a certain shape, a certain size, a certain mass, a certain roughness, a certain hardness, a certain color, etc. Classically, each of these things is termed a property that the thing has. Now let us consider what it implies for that thing which is the rock to possess those properties which are its shape, color, etc. A thing having a property means that a thing owns that property, and the clear implication is that the thing itself and its properties are separate, and that the thing is the area of reality where all the properties exist together, a kind of framework or "substratum" underneath the properties which they can hang onto and which connects them to each other.

But if you consider a rock, or any other thing, anything at all, what else is there besides its "properties"? What is a thing besides what it is? Any reasonable person who asks these questions will realize that a thing is a set of properties and a thing is not anything else besides a set of properties. The physical substance of the thing is no different from those properties, as an examination of what the properties consist of reveals. But then why has the study of what a thing is been dominated by the idea that a thing is different from its properties? Two reasons explain this mistake. First, people have not understood how a thing could both change and remain the same thing, even though they see this happen all the time. It happens because a thing remains the same thing by doing the same actions, and even after it begins or ceases to do one action, a thing remains the same as the doer of the other actions it does by continuing to do them. People always assume that action involves things changing and is the opposite of identity, but the reverse can be seen in the perceivable world, for example by being continually nice to people I make my identity as a nice person continuous. For another example, the rotation of the Earth, ever so far from changing it, keeps things on the Earth the same even as its parts move in relation to each other. The question of why things act as they do is answered by looking at the dominoes example: the dominoes act as they do because motion has been transferred to them, and they respond to action as they do because of the actions which they are already doing.

Secondly, no one has been able to figure out how a thing goes about being what it is, or how different properties connect to each other. The theory of a substratum and its properties, which is also called substance and attributes, or subject and predicates, is the most convenient solution. The theory of substance and attributes (or subject and predicates, etc.) asserts that a substance is different from its attributes, and the attributes somehow hook onto or rest upon or exist within the substance, although no philosopher has ever explained precisely how this works. I have shown that a thing is what it is by doing the action that it does, and that the properties connect because they do not in fact have any existence separate from each other but are rather a single thing as the doer of each action that it does. Each action that a thing does is a part of the single total act of being that it does. As a whole, each thing only does one action, its whole act of being, of which the individual actions that it does are parts.

For example, a wooden square is one thing doing two property actions, it is a thing which is wooden and is square. Being made of wood and being square are two acts, but they are both done by one thing. The wood does the act of being square, and the square does the act of being made of wood. This can be restated by saying that the thing does the act of being a wooden square. It is the wood that does the act of being square, and it is the physical matter which occupies the space within the square that does the act of being made of wood. Three acts of being can be distinguished: (1) being a wooden square, (2) wood being a square shape, and (3) a square being made of wood. These three acts are all actually one act. They are the same thing, which is the total act of being a wooden square.

To summarize what has been said, just as doing the act of falling causes a thing to be a falling thing, so every action that a thing does makes it be everything that it is, and acting as a whole makes parts be a whole. A thing's identity is that thing as the doer of every act it does, and a thing is nothing besides the doer of every action it does. Its action as a whole is the total act of being of a thing, in other words is the act of being that thing. This can all be empirically verified, as things doing actions are the objects of perception and experience, and it can be seen that a thing is what it is on the basis of the action it does.

What difference does any of this make? It makes a difference, because, from the conclusions I have drawn above, one can deduce something very important, which is that if you know what a thing is doing, then you know what a thing is. It is a valid question to ask why metaphysics matters at all with respect to the topic of this essay, which is how it is possible to have rational certainty in empirical reason. Every system of metaphysics, which is a theory of the nature of things, implies a system of epistemology, which is a theory of how one learns about things, how knowledge is possible. The implications of my metaphysics are clear. Since the identity and every possible aspect of real things consists of being the doer of a number of specific actions, then when one knows the actions of a thing, one knows the thing, and when one knows all the actions which a thing is doing, one knows the entire thing. If reasoning is the exercise of reason in order to know things, and if things are the objects of human knowledge, then reason must be the means by which a person discovers the actions done by things. Once the actions of real things are reasoned, then the things in reality can be known. I shall proceed to explain how one reasons the actions of things from sensory perceptions and then employs this knowledge in higher reasoning. But first I must address one other question: what is the essence of a thing?

Chapter Thirty One: Metaphysics: Essence

If a thing is a set of properties, then to say that something is a specific type of thing means that it is a thing which has a specific property. For example, to say that an apple is a red thing means that it has the property of being red. This explains the meaning of the word "as," such that a thing as something is that thing picked out and pointed out by reference to a specific property which it has. A red apple as an apple is the red apple with special emphasis on its property of being an apple. A red apple as a sphere is the apple with a focus on its spherical shape property. And a red apple as a red thing is the apple with emphasis on its color and with all other properties ignored. This is important because, as I will explain, if a thing is a type of thing as the doer of an act of being, and we learn something about an act of being, then we can apply our knowledge to all things which have that being, i.e. which are that type of thing. For example, if we learn about the property of being an apple, from considering an apple as an apple, e.g. that an apple is a fruit, then we can apply our knowledge to everything which has the property of being an apple, and deduce that all apples are fruit. Or if we learn about an apple as a red thing, e.g. by scientifically describing the light wavelength which signals redness, then this would apply to all things which have that property, i.e. all red things. I will spend the next two sections elaborating upon this insight.

In this section I will be asking an important question: what is the essence of a thing? Aristotle once said that the essence of a thing is that thing which when removed does not leave behind that thing of which it was the essence. The essence of a thing is that which is necessary for it to be, that which is necessary for its identity, that which is essential to the very core of its being. If you remove the essence, then the thing is gone. But surely, if that is the case, then the essence, whatever it might be, was whatever was causing the thing to be what it was. When you eliminate a cause, its effect ceases, and when you eliminate the essence, the thing ceases. The essence of a thing is the source of its identity. The essence of a thing, then, is what causes that thing to be what it is.

As I have explained above, it is the actions done by a thing that cause a thing to be what it is. I could consider the pen currently in my hand, for example, and note that it does many, many actions, for example, it absorbs and reflects light in many directions, it rubs smoothly against my fingers as I write, it leaves a trail of ink upon this piece of paper I write on. Furthermore, it acts in this room, at this time, and the various being actions of its parts, like being made of plastic or consisting of ink, are also parts of its total being action. By the way, the act of being at a certain time and place consists of acting at that time and place. The pen does all these actions, and many others, and every action it does goes to make it be exactly that specific thing which it is.

However, I believe that one single action which it does makes this thing be a pen. This act is obviously a part of its total being action. And I believe that this act which makes this thing be a pen is the act of being able to leave a trail of ink from its point upon objects which the point touches. This act consists of its various parts interacting with each other, for example the ink being ink, the shaft being a shaft, the tip being a ball, and holding each other together in a certain proportion. When it writes, that act of being a pen and the act of leaving ink makes it be a writing pen. Given that this act is the essence, anything which does this act is a pen by doing this act. If I had a chair which was capable of being used to write in ink on paper, it would be both a chair and a pen. Similarly, a feather which writes is a quill, and the essence of being a quill pen is the act of being a feather and the act of being a pen, which together constitute the act of being a feather-pen, which act is the essence of being a quill pen. As I noted, there are single acts which consist of multiple acts. For example, the essence of a mobile device, i.e. a cell phone which is also a music player and an internet-connected computer, is a composite essence. The act of being a mobile device consists of the composite of four acts, (1) the act of being a phone, which consists of giving you the ability to talk to other people, and (2) the act of being a music player, which consists of the ability to process music file data and emit their correspondent sounds out of speakers or headphones, and (3) the act of being a computer which is connected to the internet and can surf the web, and (4) the act of being small and portable and convenient to carry everywhere.

This pen in my hand is a real thing, so it does not only the act of being a pen, but many other acts as well. But it would not be a pen if it ceased to do the essential act of being a pen, which is the essence of the pen. This pen is a pen only because it does that one particular action, the act of being a pen. With the essential action identified, it is possible to consider this pen, to consider all the many acts which it does, and then to cease to consider and be aware of every action which this pen does, except for one, which is the act of being able to write, i.e. to deposit ink upon objects. If one performs this simple mental exercise, one will have identified the essence of this pen, namely, the pen's essential action. All one needs to identify a thing as that kind of thing, is that thing doing the essential action, because doing the essential act causes a thing doing it to be that something, and therefore a thing doing that action must be that kind of thing.

The essence of a thing is its essential action. If I consider my pen, and keep in mind its other actions but concentrate only on its essential action, then I am considering the thing as a pen, or in other words I am considering the pen aspect of this thing, because this thing as the doer of that essential action is this thing as a pen. If I consider this pen as the doer of that essential action, and totally and completely cease to consider, to be aware of, or to reason from any of the other actions that it does, including its act of being a certain length, width, at a certain location, etc., i.e. simply all actions which are not the essential action, then what I am aware of, a thing doing only that essential act, is an essential thing, in this case the essential pen.

Let me describe my theory in more general terms. Say that one has a real thing named X. X does many actions. One can identify and understand the different actions which X does separately. Say that one considers one action done by X, let us call this the X action, and one then stops thinking about all the other actions done by X. If one thinks about a thing doing only the X action, and no other actions, one will be thinking about a thing that is being X only, and nothing else. One will be thinking about the essential X, which I abbreviate as "X(e)" or "E(X)". It is easy to see that the real X one is looking at is an X(e), since it does the X action, and doing the X action causes a thing to be an X. The real X one looks at is an X(e) and more, because it does the X action and also all the other actions which it does at the same time.

You can see how it is possible to get an idea of an essential thing, e.g. a car, a tree, a rock, etc. Each of these things has an essence, a certain action, e.g. the car act, which is the act of being a car. The car's essential action consists of certain specific sub-acts, acts which are part of the total being act, done by the parts of the car in relation to each other, and the action of transportation via engine and wheels which these acts form as a whole. Similarly, one can consider the tree act, the rock act, etc. Whenever a real thing does that action it is a thing of that type. If a thing does the act of transporting via engine and four wheels, with various other details, then that thing is a car, because doing that essential act makes a thing be a car.

If you consider just that action, and imagine a thing doing that action and no others, you can think about the essential thing itself. If you consider a thing doing the act of being a car, and no other actions, then you are thinking about the essential car. Another way of saying this is that you make every aspect of a thing's identity except for its essential action into a variable. A thing's identity is merely that thing as the doer of every action it does. Every action it does is a specific action done by it. But by considering one specific action, and no others, all the other actions become variable. The variables could be anything, within the scope of those other actions that a thing with that essence can do, e.g. a car can be painted red, or can be a sports car, or can be an SUV, or can be located in New York City, etc.

Now I will explain an important concept, which I call "the consequence of the essence". The nature of an action is such that when a thing does an action, doing this action can cause the thing to do other actions also, as a consequence of doing that action, either by requiring the thing to do the consequential action in order to be able to do the essential action, or by the essential action either producing or consisting of the consequential action in some way. Thus if one can reason what actions are a consequence of the essential action, the consequence of the essence, as I call it, then one can reason that the consequence must be true of every real thing which does the essential action. I will abbreviate the consequence of the essence as $X(c)$, or as $C(E(X))$.

Essential reasoning works by examining the essential X and deducing the consequence of the essence from the essential action.

Essential reasoning then concludes that every X in reality must have the consequence of the essence. This is true because the real X is an X only because it does the X act, and it was reasoned that doing the X action causes a thing doing it to do the consequential actions, therefore every real X must be as essential reasoning discovers the essential X to be. The act of being X causes a thing to be C, and each thing which is an X is an X because it does the act of being an X, therefore the act of being X will cause every X in reality to also be a C. In other words, every line of essential reasoning follows one basic path, with a conclusion deduced from two premises: (1) $X=E(X)$. (2) $E(X)=C(E(X))$. Therefore (3) $X=C(E(X))$.

The specifics of every real X are variable in the essential reasoning, so they cannot prevent a real X from being $C(E(X))$. Since the X action causes the consequence because of what the X act is, the specifics would have to change what the X act was, or prevent the thing from doing the X act, in order to stop or alter the consequence. If the specifics did either then the thing in question would no longer be an X, since it would no longer be doing that precise X action.

Note that the objection which most academic philosophers would make to this argument is merely a linguistic trick; they would point to something that is not X, say "this is X," and then conclude that their example is an X which is not $C(E(X))$, supposedly disproving my theory. For example, if I say that X=a swan and $C(E(X))$=having white feathers, they will then point to a black swan, but their critique would be based upon confusion in the definitions of the words which are used, for example, did X mean that particular cluster of DNA which biologists call a "swan", or did it refer to birds with beaks and bodies of that shape which live in such habitats and eat such food, etc.? If a swan is a bird of a certain shape and body, with color as a variable, then it can have black feathers, but if a swan is an animal with a specific set of DNA, and that DNA defines white feathers, then a black "swan" would not be a swan with the essence as defined, and if by "swan" we mean water fouls with white feathers, then a black swan would not be a swan.

Essential reasoning is based on acts of being in the things in themselves contained in objective physical reality, and the necessity and universality of essence comes from a thing in itself, not from language or a definition of the word which refers to it, e.g. a swan is a swan because of its essence, not by definition of words in a dictionary. Nonetheless, in order to discuss or analyze essences, it is necessary to be careful and precise in defining "X" and "E(X)" and maintaining a consistent definition, so that one can be accurate in one's conclusion of C(E(X)), which would be distorted if the meanings of the words changed in the middle of the line of reasoning.

Take as an example of essences the scenario in which I am holding in my hand the lid of a box, which is red and square. The lid is a red, square thing. It does the red act: the act of reflecting the red frequency of light and absorbing all others, which is the essence of a red thing. Doing this act makes it be a red thing. It does the square act, which is the act it does of filling an area with four equal sides. That act is the essence of a square thing. This thing does the act of being a red square, which consists of those two actions combined.

But now imagine a red square. What you are imagining is simply a thing doing the act of being a red square, and no other acts. That is a red square (e), the essential red square. If you understand what I said in the section on metaphysics, then this should easy to grasp. A pen, a red thing, a red square, a tree, a rock: all of these things are what they are because they do that essential action which causes their identity, and the essential red thing (i.e. "redness"), etc., is a thing doing the essential action with all other actions not considered, making the specific actions which all real things do into variables. When any real thing has the essence, i.e., does the essential action, then that thing is that something, as for example this red square box lid is red and square.

The act of being red is a certain act of reflecting and absorbing light with a particular wavelength. The essential red thing does that act. It only does that act. A real red thing does that act. It also does other acts, like the act of being when and where it is and having a shape. But, just as a thing is the doer of what actions it does, a real red thing is a red thing in the exact same way that an essential red thing is a red thing, by doing the red action. An essential thing can be thought about by thinking of a thing as the doer of the essential action only.

I can state this a different way also. Say that a red thing is in New York and another red thing is in Los Angeles. They are both red things because they do the red act. Being in New York is the red thing in New York's location action, and for that thing as a red thing that is its specific action of location, a specific of the thing. For a thing which does a certain action, every other action that it simultaneously does which is incidental to that action is a specific of that thing with respect to that action. The specific actions of a thing are all the non-essential actions of that thing, those which are not part of the essence of the thing. For example, considering the red square as a red square, the acts of being a lid, being a specific length, width, location, and every other act it does, are specific acts. The act of being a square requires a thing to have a length, but does not require the specific length. Thus being a length is a consequential act, but being six inches long, for example, is a specific act.

Precisely speaking, it isn't true to say that the essential thing has no specific actions, that it has every specific action done by that kind of thing, or that it has any specific actions. Its specific actions are not considered. That is precisely what makes them variable, that is why the products of essential reasoning are true of any specifics with the same essence. To be precise one cannot say that an essential thing exists without the qualification that thinking about essential things is a tool by which essential reasoning thinks about essential actions done by real things. Essential things, as such, do not exist. They are the tools of essential reasoning when it analyzes essences in real things.

It is important to understand something which can be deduced from my theory of identity. A real thing is a group of physical particles, i.e. atoms, each doing certain motions. These motions may be the parts of many different actions. One may say that a lid is red, square, and wooden. The lid is a group of moving particles, and being red, square, and wooden, as a whole, are each parts of the total action of the motion of the particles. Each of these acts is a part of the total being act which consists of all of the actions of all of the particles, and each of these actions can be considered separately. The acts of being square, red, and wooden are different acts, even though the same thing does all of them at the same time, and they are all part of the lid's total action. They can be distinguished because they are different actions, as can be seen in this case. If one looks at the lid and distinguishes between these three actions, the mind has the ability to consider one of these actions and to cease to consider the other two. It does this by reasoning things about the box from one of the actions it does and not reasoning anything about it from the other two. To not be reasoned from is to not be considered.

One turns an action into a variable by ceasing to consider that action, and considering only the essential action. One turns a thing into a variable by turning its actions into variables. In this way it is possible to conceptually "remove" the specific actions of a thing, removed not from the real thing itself, but from that thing which is the object of your awareness. The reason why essential reasoning chooses not to consider the specific actions of a thing is that if you can reason that Y is true of something that does X, because it does the X action, then everything which does the X action has Y true of it, and must have Y true of it, because the essential X act caused the Y act to be done, and if the thing did not do X it would cease to be an X. Every action of the essential X besides the essential action is a variable in the line of reasoning. One could plug any specific action into that place, or more precisely put, Y must be true of every X no matter what specific action it does.

It is possible to make the essential red thing an object of thought by reasoning what a thing is because it does the red act (that particular act of absorbing other light and reflecting red light) and no other acts. This is how it is possible to have the idea of an abstract "red thing," or of "redness," by thinking of a thing whose only act is being red. That is what a red thing(e) is, and it is no different from what one means when one says "a red thing" in English, when one is not talking about a specific red thing but about any red thing. "Red" and "a red thing" is the same thing.

Essential reasoning is reasoning that makes use of essences. The next thing to explain about essential reasoning is how an action can be consequential of doing the essential action. The nature of action is such that when a thing does an action it is possible for that action to require a thing doing it to do another action in one of three ways. First, by the first action being such that doing it causes a thing to do the second action. Second, by the first action being such that the thing must do the second action as a means in order to do the first action. Or third, by the second action being a part of the first action, in other words, being one of the acts of which the total action consists. In the first case, doing the first act causes a thing to do the second act. In the second case, doing the second action is the means by which the first action is done, such that the first act requires doing

151

the second act. And in the third case, doing the second action is part of doing the first action. In all cases doing the first action requires a thing to do the second action, although in the first case the first action directly causes the second action, in the second case it is made necessary by a requirement, and in the third case to do the first act is to do the second act.

Take for an example of the first case the act of being lead and the act of sinking in water. If you take a thing which is lead and put it in water then being lead will cause it to sink in the water. The act of being lead causes lead particles to do the act of being heavy and that act causes lead particles to sink in water. For something to sink in water when it is placed in water is a consequence of the essence of being made of lead.

Take for an example of the second case the act of cutting glass and the act of being harder than glass. The act of cutting glass requires the thing doing it to do the act of being harder than glass, as you would see if you examined the act of cutting. The cutter's particles cutting the particles of glass need the act of being harder than glass as the means by which they cut through the particles of glass. Being harder than glass is a consequence of the essential act of cutting glass.

Take for an example of the third case the act of being a red square and the act of being red. If a thing does the act of being a red square, being a red square consists of being red and being square. The act of being red is part of the act of being a red square so when the thing is a red square it does the act of being red and the act of being square. Being red is a consequence of the essence of being a red square.

Here it is worth noting that the use of the word "is" for being has two similar but conceptually distinct meanings, "is-plus" and "is-equal". For example, a square is-plus a rectangle, because a square is a rectangle in that it does the rectangle act, i.e. the act of having four parallel sides connected at right angles, but has certain additional specifics, i.e. having four sides of equal length. But a rectangle is-equal a rectangle, because a rectangle is perfectly identical with a rectangle. When one says "X is Y" one can mean that Y is a part of the total act of being X, or that Y is exactly and precisely equal to the total act of being X, such that no difference exists between X and Y. In both cases the "is" in the statements is the same is the sense that "X is Y" means that X has the property of being Y, but the distinction between is-plus and is-equal is useful for avoiding needless philosophical confusion. For example, a classic argument of philosophical trickery asserts that "a white horse is a horse" but "a white horse is not a horse," which proves that contradictions are possible. The solution to this problem is to be clear about what one means using the is-equal vs. is-plus distinction. A white horse is not equal to a horse, but a white horse is a horse plus more, i.e. it is a horse plus being white.

In all cases, what the second action is arises from what the first action consists of, and the first action requires the thing doing it do to the second action. What the first action is and how it requires the second action can be understood in all three cases. The consequence of the first action is that the thing doing it also does the second action. The consequence of the essence is identical with this situation, with the addition that the first action is the essential action. This is how the consequence of the essence, the action consequential of doing the essential act, is required of the thing doing it by the essential action which it does.

Now let me explain how it is possible to use empirical essential reasoning to know that all things which are a certain something are also a certain something else, universally and necessarily, at all places and times. In other words, this line of reasoning achieves knowledge of truth which is necessary and universal. I shall describe this in terms of reasoning that all X are Y. One can substitute anything for X, including a chair, a tree, an atom, a planet, a human, or anything else, provided that one has come to know X's essential act through reasoning from perception (which I shall explain later), and that one has reasoned that Y is the consequence of X. The only premises of this line of reasoning are that X does the act of being X, Y does the act of being Y, and the Y act is a consequence of doing the X act.

1. To be a thing X is for that thing to do the action which is the essence of X, the X action.

2. Doing the action X causes the thing doing X to do the action Y, in one of the three ways I described. In other words, Y is the consequence of the essence of X.

3. Every X must do the X act because for a thing to be an X is for it to do the X act. Because every X does the X act and doing the

X act causes the thing doing it to do the Y act it follows that in all things which are X the act of being X will cause the thing to be Y. To do the Y act is to be a Y.

4. Conclusion: Being X causes the thing being X to be Y, and all X are X, therefore all X are Y. If Y is the consequence of the essence of X, then every X is Y, every real X is a Y, an X cannot not be Y, and all X are Y. As a further extension of this reasoning, we can deduce that if a thing is not Y then it cannot be an X because if a thing is not Y and if Y is the consequence of X then the act of being X must be absent in the thing for Y to have not been required by X so it would then not be an X at all.

X causing Y must be reasoned from the X action and nothing else. If a thing doing X is considered, then only the X action and no other actions which it does may be considered. If it is reasoned that the X action causes a thing doing X to do Y, then all things doing X must do Y, because if a thing does X then doing X will cause it to do Y. The other actions that any real thing doing X does besides the X action will not be able to prevent X from causing Y, because what X is causes it to do Y and so the only way other acts could do this would be to change what the X act is or to prevent the thing from doing the X act, and in both cases this would cause the thing to no longer be an X, since it would no longer be doing that precise act which is the X act, which makes it be an X in the first place. Therefore a thing doing the X act can do any other act and must still necessarily do the Y act.

Thus, so long as a thing is, in fact, an X, it must do the X act, and because the X act is such that it causes the thing doing it to do the Y act, so long as a thing is X it must be Y also. Because for a thing to be an X is for it to be a thing being X, which means, is for it to do that act which is the essence of X, doing that act is a requirement for a thing to be an X, and therefore necessary for every X to do. Since doing that act requires the thing doing it do to do the consequential act, it is impossible for any X not to be Y, and it is necessary for every X to be Y. That every X, throughout the entire length and breadth of the universe and at every moment in time, with any specific properties, is a Y, can be certainly reasoned by empirical essential reasoning, which gains knowledge of actions from perception and then uses essential reasoning to deduce the consequential actions of the essences.

Now let us consider the above line of reasoning in terms of essential things. The essential thing X is a thing doing the X act and only the X act. All other actions done by the essential X are variables. It is possible to conceive of the essential X by thinking of a thing doing X and not thinking about anything else that it is doing. Since doing the X action makes the thing which does it be an X, and no other action it does can prevent an X from doing the X act (because if it did then it would no longer be an X), one can see that a thing can both be X and be anything else, by doing the X act and doing other actions, so long as it is not a consequence of X that a thing cannot both do X and do the other act. A thing can be X and anything else that X can also be. For example, if X were red, a thing could be both red and square, or red and circular, or red and rectangular, or red and trapezoidal, because nothing in the essence of red affects shape. The act of being red and the act of being a shape are two non-mutually interactive actions. Being a color neither causes shape nor prevents shape, and vice versa.

In light of this, and considering that an essential X is a thing conceived to do the essential act and no other acts, if you reason something of the essential X, you must have reasoned that it is so because it does the X action, in which case it must be part of the consequence of the essence of X. Therefore, since it is so because it does the X action, it must be true of every real thing which does the X action (from the above "All X is Y" line of reasoning). If the essential X has the essence of X it must have the consequence of X also. If the essential act X causes the thing doing it to be Y, then the essential X is Y. Thus, what you reason of the essential X must be true of all real Xs.

It is in this way that thinking about essential things is properly used as a device to reason about the perceivable world. This is why what one reasons of the essential X is true of X with any specifics even though the essential X has no specifics. Thinking in terms of essential things is merely a device for doing essential reasoning. No real essential things exist in physical existence, but the essence, that action one is thinking of, is real (if any things of that kind exist) and the real things doing that action are real, and can be required to do actions by that act in reality. Essential things are not Platonic Forms which exist in a world of spirit. Instead they are a way of applying essential reasoning to the physical world as known by science and reason.

The matter of plugging specifics into variables in essential reasoning is not well understood. Whenever, while considering essential things, you conceptually add the act of being B to a thing being A, you get a thing doing all of A's acts, but also doing the B act. Thus this thing is both A and B, and it is the essential AB. AB has the consequence of doing A, the consequence of doing B, and also has the consequence of doing both A and B at the same time. If doing B is removed from AB, you are left with a thing being A, with A's consequence only.

A specific thing is a thing doing both the essential action and other actions. For example, a specific A is a thing both doing the A act and some B act. The B act is any act which A does which is not the A act. When considering a specific A, I call the A action the essential action and the B action the specific action. A, the essential act, is simply the action which we have chosen to consider first and reason the consequence of, and therefore, for any specific thing, the actions A and B could be reversed, but one would have to reason the consequence of B and then consider the thing doing both B and A as something doing B first and A specifically. Your choice of how to consider AB only affects whether you learn about a thing as an A or as a B, not whether it is fundamentally more A or B, for a thing which is both A and B is equally A and B. This explains why, although the choice of which act to consider as the essence is arbitrary, and it is the essence only in relation to all other acts which the thing does as specifics, essential reasoning's discoveries are necessary and neither arbitrary nor relative. For example, if C is the consequence of A, and D is the consequence of AB, then if you choose to reason about A(e) then you will deduce C, and if you choose to reason about AB(e) then you will discover D. The fact that all A are C and all AB are D is not arbitrary or relative but is absolute truth, despite the essential reasoning starting with you making the (perhaps arbitrary) choice of which act to think about as the essence, and also despite the fact that A is the essence of a thing which is AB only relative to B.

Variables become specifics as those aspects of the thing enter consideration and are specified (i.e. added) or deduced. To specify is to consider previously unconsidered details in the object of consideration. In reality only specifics exist, and real things have only specifics and no variables, because variables are a matter of thought, not reality. Variables are the parts and aspects of real things which are not being thought of. I have already explained how variables are used to reason about real things via essential reasoning. A thing with no variables, a fully detailed thing, would only exist if that thing really did exist: full detail is a consequence of existing, not the essence of it.

Since I just defined essential things and specific things, I may as well define real things also. What other people mean by "real" varies, but what I mean by a real thing is something whose essence is the act of existing, and the act of existing consists of objectively being at a specific place and time. Given this, the existence of all real things can be perceived, since they are at a specific place and time in the physical reality where you exist with eyes, ears, etc. capable of perceiving them. I believe that it is consequential of the essence of existing that a real thing is composed of physical matter and physical energy, since to be physical is to be in space and time.

In response to the objection that, for example, the pink elephant in front of me right now is at a specific place and time but does not exist? No, it is not at the place and time where it is thought to be, precisely because the matter at that place and time is not doing the act of being a pink elephant. To exist is to objectively be at a specific place and time, not merely to be thought of in one's mind as being at that place and time. If a person believed that Elvis were still alive, and they are talking to you about Elvis, this would not make the thing they were talking about live, for although they believe that the thing that they are talking about is alive, the essence of the thing that they show with that word is that it is Elvis, not that it is alive, and if Elvis is not alive then they are not talking about something alive. Something similar is true of talking about a pink elephant existing right now in the room with you: if it is a pink elephant then it is not here, because the objective reality here in this room right now does not do the act of being a pink elephant.

Essential reasoning has three steps. The first step is identification of an action (i.e. an act of being, also known as a property) from perception. The second step is considering this action as the essence and reasoning the consequence of the essence. The third step, which can be called essential identification, consists of perceiving a real thing and identifying it as doing the essential action and thereby identifying it as a thing of that kind because it does the essential action. Once the being of a perceived thing is identified one can use essential reasoning on it. This third step is completed by knowing that this perceived thing does all of the consequential actions as a result of knowing that it has the essence. For example, you could take one apple, slice it open, examine its seeds and flesh, and conclude that it is a piece of fruit. If you later see another apple and you identify that this thing is an apple then you can know that the apple is a fruit without figuring it out for this specific apple. Instead of reasoning that this specific apple is a fruit over and over again each time you find a new apple, your essential reasoning gives you knowledge of every apple as an apple. It is for this ultimate purpose that essential reasoning exists. The ultimate benefit of knowledge you derive is knowing that the consequence results from the essence so all things of that type must have the consequential properties. The consequence is what a thing is necessarily required to do by the essence and not merely what it is likely or generally probable to do if it has the essence. Therefore essential reasoning cannot be wrong.

All of visual identification, auditory identification, sensory identification, etc. of anything perceived is done by means of essential identification. It is only by identifying things by their perceived actions and using the knowledge one has gained from past essential reasoning about that essence that one is able to respond to the new things which enter one's experience.

Here it may be useful to explain the relationship between essences and kinds of things. To be a certain kind of thing is to have a certain essence. Specific things do not "participate in the essence," as Plato might have said, rather they do the essential action. For example, of triangles, there are three kinds, equilateral, isosceles, and right. A triangle(e) does the act of having three sides. Each different kind of triangle does that act, and also does the specific act of that kind. For example, equilateral triangles also do the act of having all three equal sides and angles. You can take the essential property, take one variable and make it specific, and create a new essence to reason from. A triangle has the properties of a triangle(e). An isosceles triangle has the properties of an isosceles triangle(e), which includes the properties of a triangle(e) and the property of being isosceles. An essential thing does not contain "all and none" of the properties of the specific things of that kind. The essential thing contains only its essential and consequential properties. Where do the specific properties go? They don't go anywhere. Instead they are simply not reasoned from.

Also, having explained kinds, we can explain genus and species. For example, a square is a kind of rectangle, and a parrot is a kind of bird, and a bird is a kind of animal. Referring to an essential thing AB, one can say that AB is a kind of A, it is an A with specifics B. A parrot is a specific bird, and a bird is a specific animal. In this way things are categorized according to their essence.

It is worth mentioning that a principle is an essential thing. For example, the principle that a logical contradiction is impossible, in other words the principle that a thing is itself, consists of a thing(e) being itself. The principle that "what goes up must come down" is a thing going up(e) later coming down, although I must note that the effects of gravity and escape velocity to leave the atmosphere would be details if one really reasoned from that as an essence. The scientific principle that lead sinks in water consists of lead(e) and water(e), which could also be stated as (lead in water)(e). Similarly, all ethical principles, such as "being nice to other people is good," are aspects or consequences of the essence of being human, and the principles of politics and economics are also things which have their metaphysical existence as consequences of the essence of being human.

Of course, I am sure that my intelligent reader is full of objections to this theory, but let me note that here I have only provided an overview of essential reasoning. I will explain critical details about how the human brain performs essential reasoning in my upcoming sections discussing perception, concepts, and certainty. First I would like to deal with some specific issues regarding the universality of essentially reasoned truths.

Chapter Thirty Two: Metaphysics: Essential Reasoning

It might be interesting to go off on a tangent here and consider mathematical reasoning. What I have said about essential reasoning above can shed light on how mathematical reasoning works. First of all, numbers are actions which real things do. For example, if I have five apples, being five is an act which this group of apples is doing, and it consists of this thing's act of being an apple, that thing's act, etc. In other words, the five acts of the things are the act of being five things, and each of the five apples' act of being one apple combines to form the act of being five apples, which with the essence of being apples removed leaves behind the act of being five things, i.e. the number five. Also, the act of being five things and being five fifths of the whole is the same act. The whole is the group, and the only act it does in common is that each individual is an apple, thus being apple is an act of the substance and being five is an act of the form, in light of what I described about substance and form.

The numbers that people consider when doing mathematical reasoning are essential numbers, which means, essential things doing the act of being a certain number. For example, 1 is one thing(e), 2 is two things(e), etc. The only act considered of the thing is the act of being that number. The variable in algebraic reasoning is the essential number: a number(e), whose only essence is that it is a number and that what the equation says is true of it. Finding Y for X in algebraic reasoning is finding the consequence of the essence. The equation with X specified as a specific number is the essence, and what number Y equals is the consequence. An equation or mathematic function is itself just an action that numbers can do, and these actions can be reasoned about essentially. To "let X be Y" is to specify X as doing the Y act, to reason from the essential XY. One can see this in an example such as "$Y=1/2X+2$." The essence of Y is that it is the number which results from a number halved plus two. With the act of being 6 added to X, it becomes $1/2(6)+2$, which is a kind of $1/2X+2$, and then the consequence is Y being 5.

Or take the infamous example of "$5+7=12$," which Kant made use of. There has been some debate about whether this truth comes from analytic knowledge of concepts, or from synthetic a priori intuition which equates two different concepts. Both ideas are absurd. This truth has nothing to do with concepts, because it is concerned with essential numbers and not concepts (this will become clear in my section on concepts). Also 5+7 and 12 is the same thing, merely named two different ways. This goes back to what I said about "a thing" vs. "what a thing is." 5+7 emphasizes this thing as the product of adding five ones to seven ones, and 12 emphasizes the aspect of this thing as consisting of twelve ones. One can reason that five plus seven is twelve by adding five perceived things to seven perceived things and perceiving the result. Once one perceives enough numbers, and one reasons concepts of the essences of them, one can then use essential reasoning to reason beyond that point, to reason about more abstract numbers.

Now let us turn to geometrical reasoning. Since the time of Plato (who contributed a great deal to this) it has been believed that the objects of geometrical reasoning are ideas, namely, the ideal lines and the ideal shapes. Since the ideal line is perfectly straight and has no width, and the ideal shapes are perfect in their proportions and the perfection of their lines, it was thought obvious that these things are ideas and have nothing to do with the flawed lines and messy flesh-and-blood shapes perceived in the physical empirical world. It was believed that geometrical proofs provided knowledge of these ideals that was 100% certain, in contrast with the uncertain beliefs and opinions of the perceived world.

This theory stems from ignorance of how empirical essential reasoning works. The line that you reason about has a width but the width is variable. Because you are reasoning from its length and not considering its width what you reason about a thing being that length(e), the real object of reasoning, must be true of all things being that length with any width. Geometers did not understand how reasoning works, and they could find no width-less lines in real things, so they thought that they were thinking about lines with no width, or shapes too impossibly perfect to be real. What they were really thinking about was essential lines, essential squares, essential circles, etc., all along. Geometrical essential things do the one act reasoned from with all other acts not considered. Every real line, every line in the real, physical, perceivable world, is a perfect line to the extent that it defines the area which it does, if one looks at it

161

from a human point of view and not under a microscope, and every real shape, if it fills an area of extended space with those exact proportions, is a perfect shape in the physical space which it occupies. In other words, if I look at a rectangular-shaped book then the book is a perfect rectangle to the extent that it really looks like a rectangle, and if I graphed it in a Cartesian plane with each point being a one inch by one inch square then it would perfectly define a rectangle, although seen under the lens of a microscope it may have minor deviations in the straightness of its lines.

Geometric proofs are all excellent examples of essential reasoning. If one considers a thing, and thinks only of it doing the act of having four equal sides, and reasons that its sides are joined at four right angles only from that, then one will have reasoned the consequence of the essence of a square, and this consequential action must be done by all real squares. Mathematicians have come very close to the discovery of essential reasoning. Their only mistake has been misunderstanding the objects of mathematical reason, which are essential things, and also an ignorance of the origin of the concepts of these things, which comes from reasoning from perception, as I shall explain. But the necessity and universality of the truths discovered by the mathematical line of reasoning or the geometric proof come not from a priori intuition, but from essential reasoning discovering the consequence of the essence. If you reason something to be a consequence of the essence, then it must be true of all things. That is why if a geometer demonstrates of a specific triangle with one side defined at a length of 5, one side of 7, and one side of 12, that its angles have a total of 180 degrees because it has three sides, then one has proved that all three-sided figures have summed angles of 180 degrees. This is known to be true for all triangles and not only for specific triangles with sides measuring 5, 7, and 12, because one reasoned the consequence from the essence of a triangle, which is having three sides, and the specific lengths of the sides were held variable in the essential reasoning because one did not reason from them, or if one reasoned using the numbers 5, 7, and 12, one reasoned the conclusion because the thing had three sides and not because of the specifics of those numbers.

Essential reasoning is absolutely certain, since if one essentially reasons that all X are Y then a thing cannot be X and not be Y, and it is therefore impossible for the reasoning to be wrong and it must always necessarily be right. Of course, a skeptical critic could raise objections to this claim of certainty. Next I will consider four arguments that, even having reasoned that all X are Y, it is still possible for there to exist a real X that is not a Y. These arguments are:

1. Claiming that universal ideas either do not exist or have no physical existence and therefore have no valid application to the physical world.

2. Finding something that is not Y and calling it an X when in fact it is not an X, to artificially create an example showing that not all X are Y.

3. Claiming that empirical reasoning consists of associative reasoning rather than essential reasoning.

4. Attacking experience and empirical sensory perception as the source of the ideas of essences.

I shall deal with the first three objections here, and deal with the fourth in the next section, which focuses on perception.

In response to the first objection, let me say this. The nature of universality is evident from my theory of essences. Only individual, specific, physical things at a specific place and time exist. These are all that one can sense-perceive. But universal things, things that are one and the same in all places in the universe, things that are as they are without regard to where they are, also can be known. Some things are eternal, which means that they are as they are without regard to when they are. These are essential things. Essential things must be the same at all places and times, for their place and time are variable. What is true of essential things must be true of all real things with that essence at all places and times. Essential things do not physically exist, but what is true of an essential thing is universally and eternally true of all real things which have that essence.

The only physical things that can be proven to exist universally and eternally are those that exist consequentially of their essence, and given what I believe the act of existing consists of, the only two things that necessarily exist at all places and times are space and time themselves. Furthermore, what is true of an essential thing can never change, for if it changed then the essential action would have to be different, and for the essential action to be different would be for it to not be the same action, in which case it would be not the same action changed but a totally different action, in which case the essential thing would not have changed, rather one would be considering a different essential thing. In other words, if X is Z, then X is not X, so if X is Z at a place and time then the thing at that place is time is Z, not X, and what is true of X would then not be true of it because it would no longer be X.

If one looks at things in terms of reasoning about essential things, as reasoning about the aspects of real things as the doer of the essential action, then all question of the validity of essential reasoning for the physical world can be answered. To understand the existence of the essential thing as the physical existence of the particular objects of that kind, one must remember the nature of identity. Identity is action, to be a thing is to do the action which that thing does. Doing what it does makes it the thing that it is, and its identity is what it is as the doer of the totality of its action. That all X are Y must be true across the universe at all places and times if all the things in the universe are X by being X, and being X causes the thing being X to be Y.

The essential square, which is also the universal square, does the square act, the action of filling a space with four equal sides, and no other action. The square act is the essence of the universal square. Now take an existing, specific square object, like, for example, a wooden box lid. If it does the square act, then it must do all the acts that doing the square act causes a thing to do. Therefore it does all the actions that the universal square does. It does the square's essential action, and all of its consequential actions. That wooden box lid doing those actions is the physical existence of the universal square, the only existence it has. It is true in a qualified way to say that the real square is the essential square, the qualification being that it is a square(e) but is not only a square(e), it is-plus a square. It is a square because it does the square act, and it is the specific thing it is because it does all the specific acts it does. The real square as the doer of the square act and only that act is no different from the essential square, but as I have said, one becomes aware of that aspect by not fully considering the real thing, but only thinking of one action which the real thing is doing. The real square is a square(e) in the same way that a square(e) is a rectangle(e): because the former does all the action that the latter does, but does more action as well. A specific square does the act of being the essential square plus its specific actions of existing at its place and time, and also being made of wood, etc.

The application of knowledge of the essential things to real things matters because this is the only way it is possible to empirically reason certain knowledge of anything beyond what one is immediately sense-perceiving at the current place and time. If one can reason the essences from the things that one sense-perceives, and then reason the consequence of the essence from the essence, one can then reason certain knowledge of all things which do that action, including all things that one is not currently perceiving, and one can also apply that knowledge to the things that one perceives and identifies as doing those essential actions. In fact, one can only identify any perceived thing as what it is by identifying it doing the essential action it does, through the use of the human brain's unconscious reasoning in processing the data received by the eyes and other sense organs, as I will explain in my section on perception and concepts. Obviously empirical essential reasoning not only applies to the situation that you perceive right now, but it is necessarily true of all perceivable things and is a vital source of knowledge for humans. In other words, essential things exist but only as real things with that essence doing that essential action.

My above explanation is a sufficient answer to the first objection I listed. In response to the second objection, in which one misuses words to call something an X and not Y, or vice versa, which is neither an X nor a Y, I must talk a little about the nature of words, as an elaboration upon my reply to this objection which I considered earlier. Philosophers have made so many attempts to establish uncertainty through the fluidity of definitions that I will not bother to give any further examples of this; the black swan suffices. The essence of X is not that it is that thing which is called X. It is not that humans use the word "X" to name it. Instead the essence of X is that it is that thing which does the X action, i.e. the act of being X. A word is an auditory sound or written mark that shows a thing besides itself to the mind of the person who hears or reads that word. Words are mentally associated with their meanings, so a reader or listener who exercises reason, unconsciously or consciously, to identify what thing the author of the word intended to show, can understand the communication by recalling that thing into awareness from memory, or by reasoning it by putting together the various things shown by the words into one thing if it was not previously known.

That thing which the word shows to the person who hears or reads it is the meaning of the word. I might say that words refer to things "outside themselves" or that words "correspond" to things, but it is better to say that if a thing is a word then it shows something other than the word itself to the minds of the people who use it. For example, the letters "c," "a," and "t", when combined, form a word that shows people "cat," but cat is not a word, cat is a four-legged tailed feline animal shown to people through the use of the word that names it. To speak precisely, the word used to name cat can only be referred to by saying "the word that shows cat" or separating the letters so that the mind does not unconsciously jump to the meaning upon recognizing the letters.

It may seem odd to say that cat is not a word, but if cat were a word it would be an auditory sound or written marks, when in fact a cat is clearly not an auditory sound or written marks, a cat is a feline four-legged tailed animal. Therefore, cat is not a word, cat is a thing named by a word, and the letters and sounds are not "cat" the word, but are the word that names a cat. Needless to say, a great deal of confusion in philosophy and other academic studies has arisen from confusing a word with that thing that the word shows or by equating the two. Some philosophers assert that "the Sun" is a word, not a star that shines during the day, which leaves humans trapped in a world of language without access to external reality. In fact, the act of showing something different from itself through the use of signs which do not resemble what they show and which are given meaning through mutual use by humans is the essence of a word. A word(e) is something that shows something other than itself to people by being used to show it such that it becomes associated with its meaning in the minds of the speakers and listeners.

Any word that does not name a real thing names an essential thing, to name is to show, and if a word does not name anything, if it doesn't have a meaning, then it is not in fact a word, since it does not do the essential act of being a word. Sentences are groups of words which function as a whole to show a thing composed of all the things that the individual words show, such as for example the sentence which shows "the girl kicked the ball" contains words that show a specific girl, kicking, and a ball, with the relative position of the

166

words showing that the girl does the act of kicking to the ball. The meaning of a word is the thing that it shows. It is appropriate here to mention that, of words and language, a sentence is true if it shows something that exists, and a sentence is false if it shows something that does not exist.

Given this theory of language, which could be called a correspondence theory of language, we can see that words show through a purely associative connection, and which word shows which thing is purely arbitrary, since the sounds or marks and the things they show have no actual resemblance and the words show only because they are used to show and the intention is understood by the reader or listener. The only non-arbitrary aspect of this act is that if a person intends a word to show some definite thing, and he or she uses the word to show that thing, then the word, if properly interpreted, would show that same thing to the person who uses it to receive a communication about things from the person who spoke or wrote. Which word shows which thing is arbitrary and is decided by humans, but the things themselves which are shown, i.e. things as things, are not arbitrary at all, and what a thing is consists only of what action it does, not what words are used to show it.

For example, there is a thing, the apple(e), which is an essential thing. There exists a set of letters and sounds, both of which are words for apple. But the apple is not a word. The apple is that thing which the word shows to the mind. What an apple is, its essence, is not arbitrary, but which letters and sounds show an apple to people's minds is arbitrary. No one decides or creates the essence, while the letters and sounds of words are decided and created by humans. I could create a new language with my friends and decide upon a new word for apple, like "elppa," and use it to name an apple to others, and think of an apple when I recognize that word in letters written to me, and I would have created a new word without changing the shown thing. The essence of a thing is that it does a certain action, not that it is named by a certain word. The only exception to this is (a thing named by a word)(e), but its essence as a thing named by a word is that it is a thing named by a word, its essence as a thing is not that it is named by a word. There are things for which there are no words, and things named by different words in different languages. The thing is separate from the word. The word is only the means by which the thing is shown to the person with whom one communicates.

An X is a thing which does the X action, and one can choose a word, say for example two crossed diagonal lines, and use it to show X to people. One can proceed to call an X by a different name such as Y, Z, A, B, C, or anything else, but one will have only named a number of different things, and one's verbal actions will not have changed the nature of X itself. For a thing to be X is for it to do the X action, and not for it to be named by a certain word, even if that word is used to name it. Being named by a word is a specific non-essential property. This is my answer to people who take something that is neither X nor Y, call it an X, and use it as proof that not all Xs are Ys. We should scold everyone who uses verbal confusion or switches their definitions in their philosophical arguments and reject those philosophers who claim that language is self-referential and that one cannot use words to talk about anything besides words.

Words consist of sounds or written symbols used to show something other than that word to the mind of the listener/reader. Words show because the mind is trained to recall the thing to which it refers when the word is seen or heard, as the result of the habit of association gained when one learns a new language. One reasons the meanings of new words by putting old words together, or seeing a thing and being told what word refers to it. The trick of equivocation, or changing the meaning of a word halfway through an argument, is what I am warning my reader against.

That is my response to the second objection. My answer to the third objection must be to describe how associative reasoning works and to explain the difference between associative reasoning and essential reasoning.

Since the beginning of human thought people have tried to categorize the objects of their experience, to divide things into groups. Such categorization, in order to not be arbitrary, must be done on the basis of some quality which all things in a group share. Thus everything with this quality is a thing of this kind. Because people were ignorant of the essences of things and did not realize that they were categorizing on the basis of grouping things with the same essence together as the same kind of thing, they thought that the basis upon which things were grouped was what they had in common. And of course, since everything in the same group had something in common, this was fine. But because they did not know about essences or look for an essential act of being to group things together through, they were reduced in the end to trying to group things together on the basis of their similarity and to then look for the essence of the group in this similarity, rather than first looking for the essence and then seeing what all things doing the same act had in common. This is the first kind of associative reasoning, the attempt to deduce what the essence of a kind of thing is from what all items in a group have in common.

These same people used essential reasoning unconsciously, but they were not conscious of using it and they sought a conscious explanation of how they knew what they knew. They knew that they had knowledge of what things would be in the future from the perception of what things had been in the past, but they did not know how this was reasoned. They supposed that they were able to reason that all X would be Y in the future on the basis of all the X they had ever seen in the past having also been Y. The second kind of associative reasoning is the attempt to reason that all X are Y from having only ever perceived X that were Y.

Lastly, people knew that they had knowledge of cause and effect, but they did not understand how they had this knowledge, and they knew that when they knew that something caused something else that in their past experience they had always perceived the cause followed by the effect, so they supposed that they reasoned cause-and-effect from having always perceived one thing followed by another in their past experience.

Then along came a philosopher named David Hume, who formalized all of these assumptions and popularized the notions that reasoning about essential things is impossible and that all empirical reasoning is empirical associative reasoning. Hume understandably concluded that all empirical reasoning is uncertain. (I refer in particular here to Hume's "Treatise of Human Nature," Book I, Part I section VII. and Part III sections II and VI).

It is true that empirical associative reasoning cannot possibly produce certain knowledge of anything. Consider an example of the first kind of associative reasoning, in which one tries to deduce the essence from what the members of a group have in common. For example, I could collect all of the things in my house which are pens and put them into a group. I could try to reason what the essence of being a pen is from this group, to get to the basis upon which these things were selected. If I believed that they were selected because of what they have in common, I would see what they all have in common, and I would conclude that this is the essence of a pen. All of my pens can write with ink and are black near the tip and white on the shaft and are made of plastic. All of the pens in my house have this in common. Associative reasoning would tell me that the essence of a pen includes being white and black plastic, but such reasoning would be mistaken. All cases of a type of thing can have in common a property which is not the essence as a result of a coincidence or accident. All things of a kind or type have in common that they do the action that is the essence and the consequence of the essence. But it is not necessarily true that the essential action is the only action that all things of a certain kind have in common because all real things of a certain kind could incidentally have other actions in common. As my example shows, if I equate essence with commonality then I would think that a green metal pen would not be a pen because it is not black and white or made of plastic. If I used empirical associative reasoning to study pens and I then discovered a green metal pen I would be surprised. It would be natural for me to doubt that reasoning is capable of achieving knowledge and to collapse into doubt and skepticism.

Things can be grouped according to a basis of being a certain type or kind of thing. If things are grouped on the basis of doing an essential action then one can see what they all have in common as an indication of what is likely to be the essence since the essence must surely be one of the actions that they all have in common. Humans do sometimes learn about things in this way. For example, young children can learn to identify a property by observing multiple items in a group which all share that property and they can learn to distinguish X from Y by observing the differences between a group of Xs and a group of Ys. But this only yields knowledge of probability or likelihood of what is essential, not certain knowledge of the essence. When things are grouped based on what they are, inclusion in the group is a result of having an essence. Having the essence is not the result of being included in the group. In other words, when things are grouped by essence having the essence causes inclusion in the group, not the other way around. Having an essence and being a member of a group are not identical, although having an essence causes a thing to be a member of the group of all things of that type.

Considering this same pile of pens that I collected for my last example, we can find an example of the second kind of associative reasoning, believing that what happens in the future will resemble what happened in the past. This associative reasoning would try to deduce, on the basis of having only seen pens that were black and white, that all pens are black and white, and all pens which one finds in the future will be black and white. Thus I would not think that a pen could be green. Someone who used empirical associative reasoning in this way would be surprised many times, when they first encounter a green pen, and he or she might conclude that empirical reasoning is incapable of certainty, and come to believe that knowledge of the physical world cannot come from experience and empirical observation. However, if one essentially reasoned that the essential act of something was such that if a thing did it then that thing could not do something else, and one did not expect to encounter things of that kind doing that something else, then one would never be surprised, because one could never run into anything of that kind in the future doing that impossible action. The products of associative reasoning are mere probability, since presumably it is likely for things in the future to be like all the things in past experience, but it gives no indication of certainty, whereas the use of empirical essential reasoning to deduce that all X is Y cannot be wrong, and if one reasons essentially then one can be certain that all X everywhere in time and space are Y.

Now let us consider an example of the third kind of associative empirical reasoning, which seeks to find cause and effect from having seen one thing followed by another thing. Say for example that you lived in a house, and you went to work every morning at eight o'clock sharp, and every morning as you walked out the front door and opened your car door to get into your car, you saw your neighbor across the street open his door to go to work. You might think that the conclusion that your neighbor goes to work because you open your car door is a ridiculous one. But empirical associative reasoning of cause-and-effect could easily produce that conclusion. Of course, just because you always perceive something followed by something else is no reason to believe that the second thing must always follow the first thing or is caused by it, and a person who used that kind of thinking would be easily surprised.

Interestingly, empirical associative reasoning is precisely what a certain kind of operant conditioning mimics, as for example the pigeon who believed it was fed because it jerked its head, because it was given a food pellet after jerking its head a few times, and thereafter jerked its head in an attempt to get more food (B.F. Skinner, "Superstition in the Pigeon," Journal of Experimental Psychology, 1948). If humans evolved from apes and those apes had this kind of conditioning then conditioning may have developed into empirical associative reasoning in early humans.

In any case, if something always follows something else, that indicates a certain probability that the first thing causes the second, but no certainty of cause-and-effect. On the other hand, if one reasoned that, as a consequence of the essence, doing a certain action caused the thing doing it to do a certain other action, or that as a consequence of the essence of the action of a thing acting upon something else causes the other thing to do some action, like striking a match causing it to catch fire, then one could empirically reason that the first thing was a cause and the second thing an effect, and that the first thing would always produce the second thing where that action was done, with absolute certainty, as a consequence of the essence of that motion.

This concludes the explanation of the flawed nature of empirical associative reasoning and the distinction between associative reasoning and essential reasoning. In response to the fourth objection I must discuss how the human sense organs, e.g. eyes, ears, etc., and the human brain operate.

Chapter Thirty Three: Perception: The Objectivity Proviso

Let me begin with an overview of the senses. There are five physical senses: sight, hearing, touch, taste, and smell. These are commonly called the five senses. Although there are nerve cells which provide the brain with feedback from the internal areas of the body, one is never aware of this and it need not be considered. I shall limit my study of sensory perception to the five senses just listed, which my reader is surely familiar with and conscious of using. Since this essay is concerned only with empirical reasoning, the only kind of perception I shall consider is empirical sense-perception, so when I say perception I always mean sense-perception.

The senses are the means by which one sense-perceives the things that one perceives. The organs of the senses in the body, namely the eyes for sight, the ears for sound, the skin for touch, the nose for smell, and the tongue for taste, are the organs that pick up the signals in the medium of that particular sense, and the nerves in those organs convert those signals to electrical pulses that are carried by nerve cells to the brain. The means of perception begin with the light bouncing off the object, the sound-wave being created by the object producing vibrations in the air, and the excitement by something of the nerves in the skin, olfactory nerves or taste nerves. The signals continue from medium to medium until the nerves deliver the signals to the brain. Those signals, when taken into the brain from the nerves adjoining it, are the brain's means of perception.

The sensory perception of a thing is the means by which it is perceived. The set of signals in the brain, which the brain received from the nerves coming from the sense organs, is a sensory perception. The act of perception, from beginning to end, is the act of the creation of the means of perception in the brain, and the act of perceiving is the means by which the sensory perception in the brain is created. The sensory perceptions in the brain are the means by which the brain performs the act of sensory perception, that is, the act of perceiving the things that it perceives. I call the things that one perceives, the objects of the act of perceiving, those things that you perceive right now and everything else you have ever perceived, by the name "perceived things". I call that thing that the sensory perception is of, that thing which creates the first signal in the act of sensory perception, for example, the thing that reflects the rays of light, or vibrates the air, or touches the skin, etc., and is thereby the ultimate cause of the sensory perception in the brain, by the name of "the object of that sensory perception".

Let us now answer the question of whether perceived things, that is, what a person sees, hears, smells, etc., are the objects of the sensory perceptions, are the sensory perceptions in the brain, or are a representation of them that the brain creates, or something else entirely. Note at the outset of this inquiry what is at stake: if perceived things are the objects of perception, then the perceived things exist in objective reality, and then one could conclude that perception produces knowledge of objective reality. On the other hand, if the perceived things are the perceptions in your brain, then you see and hear and smell things inside of your own mind, in which case perception is subjective, and the senses are showing you a subjective world within your own mind.

Here I will argue that perceived things are the objects of perception, and the perceptions in your brain are your means of perceiving perceived things, but the perceptions themselves are not the perceived things. From this one can conclude that the things that you perceive exist in physical objective reality, separately from and independently of your perceptions of them. Your perceptions exist in your brain, but the things that you perceive exist in objective reality. My main argument here is the Objectivity Proviso, which states that a thing exists objectively if and only if the condition is met that the thing's existence is separate from and independent of the perceptions of it and the act of perceiving it. In other words, a thing exists objectively provided that it is separate from and independent of perception, belief, and the mind. I would have called this the "Adlerian" Proviso, because it is based upon the work of philosopher Mortimer Adler, if not for the fact that the term "Adlerian" already has a different and unpleasant usage.

Adler argued that anti-science philosophers take advantage of linguistic confusion regarding the word "of", in the sense that when they say "the perception of a cat" they are referring to what should be called "a perceived cat" or "the cat as the object of perception".

174

My original contribution to this area is my extension of the logical conclusion of the Objectivity Proviso, which consists of my argument that you can directly perceive that perceived things are the objects of perception and not the means of perception because, for example, when you look at a black cat you see a black cat, you do not see your own eyeballs or the neurons in your brain.

What one perceives is that which is perceived. You perceive things, and each thing that you perceive is perceived by means of the means of perception. Perceptions are the means of perception. What you see is what you see, and not the means by which you see it. It must be so, because the essence of the means is that it brings the perceived thing to the perceiver, and it is not itself a perceived thing. The essential act of the means of perception is that it carries something which is not itself (its object) to the mind's eye of the brain for which it is the means of perception. Consequently the means are everything between the perceived things and the perceiver. The means of perception brings the perceived thing to the perceiver, and is the means by which the perceiver perceives the perceived thing itself. Perceptions cannot be perceived, because they are the means of perception and are therefore between the perceiver and the perceived thing always.

The essence of the perception in the mind is that it is the means of the act of perception, the agent in the brain that brings the perceived thing to the perceiver, that thing which shows something other than itself to the mind's eye in the act of perception. Specifically, the perception in the brain is the means by which the brain is able to perceive that thing which is the object of the perception. To understand how it is possible for the perception in the brain to show the brain the object that originated the means of perception, consider the dominoes example I offered in the section on metaphysics. Just as the hand was the ultimate cause of the ball being hit and the dominoes carried the hand's action to the ball, so too the objects of perception, the things that you perceive, are the ultimate causes of the chain of signals that transfer the action of those things to the brain. The perception in the brain, the final signal received from the nerves from the sense-organs, which is stored in the brain, is the immediately caused thing that can be used by the brain to perceive the ultimate cause, when it shows its object to the mind's eye. (How the ultimate cause is reasoned from the immediate cause to enable the conscious perception of the ultimate cause is a matter involving unconscious reasoning, which I will deal with later in the essay.)

Therefore, a perception itself can never be perceived, and its entire existence consists of it being used to show to the mind a thing other than itself. The perception of a thing, the perception in the brain, cannot be perceived because it is the means of perception, while the object of the perception must be that thing which is perceived, since it is that thing which the means of perception brings to the mind. The object of the sensory perception must be the perceived thing itself. The objects of perceptions must be the perceived things precisely because they are what you perceive. For example, you see a physical world of houses and streets and cars and people, etc., and this is a physical world of science which exists in physical space in the area outside of your eyes. When you look around you are not seeing your own eyeballs or the gray and white matter in your brain. Therefore the objectivity of sensory experience is self-evident from the perceived things themselves.

One can also reason this linguistically from the objects being of the perceptions, for the implication when one says "the perception of the square" is that the perception and the perceived square are two different things, and that the former is the means of perceiving the latter and is the thing in the mind that enables the mind to be in contact with things outside of itself by means of the perception which shows it that thing. That was Adler's original argument. The conception of perceptions as the means by which the perceived things, the objects of perception, are perceived, was first introduced to me in a paper by Mortimer Adler called "Consciousness and Its Objects." It will be useful to quote the key passage in that text here, to make my argument clearer.

Adler begins by asking what the objects of consciousness are. He opposes the view, proposed by Locke, that the objects of consciousness are ideas in the mind, and he argues that Locke's position leads ultimately to skepticism regarding the objective world or subjectivism and solipsism, since one would be in contact only with ideas in one's own mind and one would not have any direct access to the things which caused the ideas. Taken to its logical conclusion, the Lockean error produces Hume's philosophy, which naturally collapses into the Kantian epistemology that the mind perceives the phenomena, the perceivable world, but the mind lacks access to the noumena, the things in themselves which produce the phenomena by acting upon the mind. If perceived things are perceptions, not the objects of perception, then the mind perceives perceptions and never experiences the objects of perception, so Kant follows from Locke. To oppose Locke (and Kant), Adler puts forth the theory that the ideas in the mind are the means by which the objects in objective reality are perceived. The quote from Adler follows:

"Let me spell this answer out in all of its significant details. It means that we experience perceived things, but never the percepts whereby we perceive them. We remember past events or happenings, but we are never aware of the memories by which we remember them. We can be aware of imagined or imaginary objects, but never the images by which we imagine them. We apprehend objects of thought, but never the concepts by which we think of them.

"Do you mean to say (readers may ask) that I am never conscious or aware of the memories or images I am able to call to mind, and that I cannot directly examine the concepts or conceptions my mind has been able to form?

"The answer to that question, however contrary it may be to our loose habits of speech, is emphatically affirmative. A cognitive idea (including here percepts, memories, images, and concepts) cannot, at one and the same time, be both that which we directly apprehend and that by which we apprehend something else—some object that is not an idea in our own minds, but unlike our subjective ideas is rather something that can be an object of consideration or of conversation for two or more individuals.

"Let us go back for a moment to the table at which you and I are sitting with its bottle of wine and its glasses. We noted earlier that our awareness of these objects was a public or communal experience, one that we both shared. It could not have been that if each of us was aware of nothing but his own perceptual ideas—his own sense perceptions. Its being a communal experience for both of us, one that we shared, depended on our both apprehending the same perceptual objects—the really existing table, bottle, and glasses—not our own quite private perceptions of them…. What is true of one type of cognitive idea, our perceptions, is true of all the other types of cognitive ideas—all of them the means, not the objects, of apprehension; that by which, not that which, we apprehend." (Adler, 14-16, "Consciousness and Its Objects," Ten Philosophical Mistakes.)

The act of sensory perception shows the sense-perceived thing to the mind's eye. The sensory perception itself shows that thing which it is the sensory perception of, the perceived thing itself, to the mind's eye. For a thing to show something is for it to bring some other thing that is not itself, the thing it shows, to the mind's eye.

One does not perceive sensory perceptions. The things that one perceives, like the words on this page, are the objects of the perceptions brought to your brain at this moment. It is impossible as a consequence of their essence for perceptions to be directly perceived. The things that you see, hear, touch, smell, and taste are the objects of your perceptions, shown to you by the signals in your brain. Furthermore, if one says "the sensation of this book" or "the impression this book makes upon the senses" the implication in both cases is that the book and the sensation or impression are two different things. If one wants to speak in terms of these things, the book is the perceived thing and the sensation or impression is the means of perception. Therefore, one does not perceive sensations or impressions, rather one perceives perceived things.

This also makes sense because the way that most people speak implies that the sensation is the sensory data from a thing received by the sense-organs. The thing itself that one sees, hears, smells, etc. is the thing that the sensations are of, the source of the sensory data. One says that one smells a rose, not that one smells one's own nose. That the things that one sees, hears, smells, etc. are perceived is the proof that those are the things that one perceives. Perceived things are self-evident in the sense that if you see a red apple then the perceived redness is the evidence that the thing is red. If one perceives a tissue box made of cardboard one can reason from this that the perceived thing is a tissue box. The tissue box is on my desk, and the perception of it is in my brain, therefore what I perceive is a cardboard tissue box on my desk and not neurons in my brain. I can clearly see this because I see the box on the desk and I do not see brain matter, which looks like bloody gray or white flesh. I can reason from this perceived thing that the perceived thing is on the desk and not in my brain.

I can point at the tissue box on the desk, and then point at my brain in my skull, and see with my eyes that the tissue box is not in my head, and so empirically reason that what I am perceiving is a tissue box outside of my brain, not in my brain, or in other words, it is in objective reality, not in my mind. This is especially true since I perceive this tissue box doing all of its essential actions as a tissue box on my desk, and these are all physical objective properties, such as dispensing tissues, occupying space, and being made of cardboard. I can reason that it is a thing on my desk, not a thing inside my head.

For a thing to exist objectively is for it to exist separately from and independently of the act of perceiving it and the means of perceiving it. For a thing to exist subjectively is for it to be created by or altered by the act of perceiving it or the means of perceiving it. Based upon the Objectivity Proviso and my arguments presented below which prove that perception is caused by the effect of reality acting upon the sense organs, I deduce that perception does not create or alter the perceived things. From this I conclude that the things which we perceive exist objectively, and perceived things are objective, not subjective. Therefore the world of our experience and the external world are actually one and the same thing.

One can reason that the things that one perceives exist objectively, that is, are neither created by nor altered by their means of perception, from many things. Four arguments for the objective being of the objects of perception follow.

1. The means argument: sensory perceptions are the means of perception. They show their object to the mind's eye. To show is to bring a thing different from that which shows to the mind's eye. If the act of perception created or altered that which is perceived, then the means of perceptions either would be perceived (if they created perceived things) or would be part of perceived things (if they altered perceived things, for what they changed would be a perceived addition). If perceived things were subjective then the means of perception would be perceived, but the means of perception's essential act is bringing things other than themselves to the perceiver to be seen, so they cannot be perceived either in whole or in part as a consequence of their essence. That which shows cannot be that which is perceived consequentially of its essence. Thus, if perception (of the physical world) exists, as distinct from

intuition or feelings or magic, then perception must show an objective world to the mind.

2. The closed eye argument: If the act of perception altered the object of perception, then it would add something to perceived things in addition to the objects of perception. This addition would be seen regardless of them, since it comes not from the objects but from the means, and the means of perception are always there, so it would always be seen. Thus, when one's eyes are closed, or when no object was producing sound waves that reach one's ears, one would see or hear what the means of perception adds to perceived things. But when one's eyes are closed one sees nothing, and when no noise is made one hears nothing, so nothing is being added by the eyes. And so on for when the other sense-organs show nothing.

Also, if the act of seeing created that which is seen, then closing one's eyes would destroy the world one perceived. Thus if one sees the world, closes and opens one's eyes, and sees the same world, one can reason that closing one's eyes did not destroy the world. If not seeing it did not destroy it then it continued to exist while unperceived. Because what is perceived existed both while perceived and while unperceived it exists separately from the act of sight and is not created by sight. This cannot be explained by others perceiving the world while you do not, for if perception created existence then you would need to perceive those other people for them to exist in order for them to perceive the world.

3. The argument from the act of perception: This consists simply of an examination of the act of perception, specifically the signals between the perceived thing and the sense-organs, such as the light waves and sound waves, and of the sense organs, and of the neurons leading from the sense organs to the brain, to reason that the being of the sensory perceptions in the brain are directly caused by the being of the perceived thing. For example, the signal that enters the brain corresponds to the signal sent from the eye to the brain, and this signal corresponds to the energy that affected the nerve cells in the eye, i.e. the rod and cone cells, which convert energy to nerve pulses, and that energy was the energy that the original object affected through its act of being, when it reflected the light which traveled through space and entered the eye.

In each case the agent "corresponding" to the thing before it consists of its actions having been caused by the actions of the thing before it. The signal moves from the perceived thing, to light, to the eye, to the brain, like a motion traveling down a series of dominoes. The perceived thing is the ultimate cause of perception, and the perception in the brain is the immediate cause of perception. From studying the physical means of perception one can see that the act of perception brings something capable of showing the perceived thing to the mind's eye. Something, namely, the sensory perception, that enables the brain to know the perceived thing is carried into the brain by the act of perception.

4. The tracing a line of sight argument: One can reason what the act of perception consists of from an empirical examination of its perceivable components, and then reason that the essence of this action is that it shows the perceived thing to the brain. One can do this simply by tracing a line with one's finger between any perceived thing to one's eye, to perceive the space between the two and reason from this that the perceived thing exists outside the eye. One can turn on a source of light, such as a TV, and walk over to it and see the line that goes from the TV to the viewer's eyes, and conclude from tracing the line of light that the perception comes to the mind from the TV which exists in physical space outside the brain. Or, if one plays music on a device, one can trace the line from the speaker to the ear to evaluate whether the perception is coming to the ear from the sound.

In the example of sight, the being of the object affects the light in a specific way, which corresponds to how the light affects the eye, which corresponds to the signal sent to the brain from the eye. In this way the perception in the brain has a being that corresponds to the being of the original object, enabling the direct perception of the original object outside the brain by means of contact between the brain and the perception of the object in the brain. Since the perceived thing causes the perception, the perception cannot be the cause of the perceived thing.

Any of these four arguments can be used to reason that the act and means of perception do not cause the existence of, alter, or create the things that one perceives, and that therefore the things that one perceives exist objectively.

One can reason from this that perceived things are not sensations and ideas in the mind, nor are they "representations" of external things that the brain produces from the analysis of sensory data, rather, perceived things are the external things in reality themselves. In other words, much to the horror of a philosophical Kantian, I am saying that perceived things, e.g. a cat or a book or a desk or a box of tissues, are things in themselves. The idea of representation, that perceived things are representations of external reality created by the brain which exist in the mind, is absurd. If one did not have direct knowledge of things outside the mind, then one could never reason that the things outside the mind correspond to one's sensations or ideas of them or one's representations, or even what they were at all, since something different from them would be one's only source of knowledge about them. An example of this is Kant's noumena, which are unknowable on the basis of the phenomena.

It has long been assumed that the paradox of the inability to know that one's ideas match objective reality arose from the assumption of the existence of things outside the mind, but nothing could be farther from the truth. The philosophical problem is, and always has been, caused by the assumption that the things one sees are sensations and ideas existing in the mind. The things that one perceives are the objects of sensations and of ideas, whereas the sensations and ideas exist in the mind, while their objects, which one senses, exist outside the mind. If one does not assume that sensible being is created by the mind or brain or mental interpretation, in other words by the means of perception, the representation theory cannot be seen in perceivable evidence and we have no reason to believe it. The brain's interpretation does not create what one sees. Instead the brain's reasoning enables the mind's eye to see the source of the interpreted sensory data, which are the objects in physical reality.

It is assumed that because one perceives by means of one's own personal senses that what one perceives must be subjective, that to see an objective thing would be to see an unperceived thing, and that to see an objective thing one would need to see it by means other than one's means of perception. The idea that one's point of view defines one's vision, and that color is a subjective creation of the eyes, support this theory. But this all assumes that the senses are subjective. If one does not assume that the means of perception create or alter what they are used to perceive, then it is possible to reason that one can perceive a thing exactly as it would be if unperceived, by means of one's personal means of perception, one's very own sense organs.

It has been said that to prove that there exist things unperceived, one would have to learn of them by a means other than perception, and thus it is impossible to be certain that unperceived things exist, because everything that you perceive is perceived by means of your means of perception. One can never perceive an unperceived thing, or perceive an act of existing as separate from the act of being perceived. An example of this is the belief that the question "if a tree falls in the forest, and no one hears it, does it really make a sound?" is unanswerable.

This theory rests on the equation of the unperceived with the objective, which itself rests upon the assumption that the means of perception creates or alters the objects of the act of perception. To be objective and to be unperceived are different. To be unperceived is to not be an object of the act of perception, and to be objective is to exist separately from the act of perception. As I have explained, the act of perception and its means cannot create or alter the objects of perception, which are the perceived things. Thus I can say in answer to the above objection that it is true that one can never perceive an unperceived thing, but one can perceive an objective thing. An objective thing is the same whether perceived or unperceived, and one can reason the existence of the unperceived from that which is perceived. It is not the ability to perceive the unperceived, but the ability to perceive the objective, which is needed to be completely certain of the existence of the unperceived. One can also reason that the unperceived exists directly from the closed eye argument or the tracing lines argument, as I explained. Thus, if a tree exists objectively in a forest, and it falls, and nobody hears it, then it makes a sound, because the sound is the tree itself causing air vibrations when it falls. In other words, the sound is the falling tree, so if the tree falls then the sound exists.

A subjectivist might object that, when I say that one can reason the objectivity of perceived things from experience, I am assuming that the perceived things exist objectively. But my reasoning does not make assumptions. My epistemological approach is to take sensory experience, to take the perceived things themselves, and to then apply reason to analyze them, and to see what the perceived things indicate. Such reason makes use of no assumptions, because its only premises are the perceived things themselves. As such, the objective existence of the perceivable world is self-evident from sensory experience, and does not require making the assumption that perception is objective or that empirical reason is valid.

I will conclude this section by first responding to a more sophisticated objection to my theory of the objectivity of the senses, and then dealing with the question of whether or not the senses are capable of deception.

If being consists of action, and to have a perceivable property, i.e. to be something perceivable, is to act upon the sense organs in a certain way, then how can things have perceivable properties when unperceived? For example, if to be rough is to affect the skin in response to touch in a certain way, and to be blue is to absorb and reflect light in a certain way, is a thing rough when it is not being touched, and is a thing blue if no light hits it? Also, in a related question, if all being consists of active actions, then how can anything be able to be something, since potentiality consists not of action but of the ability to do action? For example, if the essence of a pen is to be able to write, then what does being a pen consist of, and isn't it really true that the essential act of a pen should be writing itself, rather than the ability to write?

Both of these questions happen to have the same answer. In response to the questions, how is the potential of a thing, i.e. what it can do, an active action done by it, and also, if to be is to do, how do you know what things are when they are not acting upon the senses, I reply that a thing responds how it does because of what it is. The very same act that a thing does continually, which is its act of being, is the very same act that acts upon everything which comes into contact with it, at which point that act is its response.

Let me provide two demonstrative examples of this. First, consider a ping pong paddle and ball. Take a ping pong paddle, and swing it. As you swing it you can watch it move through the air. Swing it enough times to see that each swing is the same action repeated. Now, get someone to throw a ping pong ball at the paddle, and swing the paddle. The very same action that the paddle was doing when it was just swinging through the air will hit the ball and knock it away. It is the same action which the paddle was doing before that it does when it hits the ball. The only difference is that if a ball is in contact with it while it is swung then its motion knocks the ball away.

Along the same lines one could consider a spinning windmill. If a tennis ball is thrown at it, and comes into contact with its arms, the arms will knock the ball away, in the direction in which the spinning motion was going. The spinning, the act it was always doing, knocks the ball away. Its response is the act it was always doing, but that act becomes a reaction when it acts upon other things that come in contact with it. It isn't always responding only because there isn't always something in contact with it. But when it responds, its response consists of the act it was perpetually doing, when that act affects the things that come into contact with it.

For another example, consider that toy that is used to teach children about shapes, which consists of a board with differently shaped holes and differently shaped blocks, each of which is the same shape as one of the holes. The child is supposed to learn that only the right shape can fit into the right hole. Why is only the same shape able to fit into that hole? What does the ability and lack of ability consist of? As can be seen, it consists of nothing besides the block being the shape it is (occupying an area of space of that shape) and the board occupying space around a hole of that shape. Because both the board and the block would have to occupy the same space for a piece to fit into the wrong-shaped hole, and two objects cannot occupy the same space, it cannot fit into the wrong holes, and so it can only fit into the hole of the same shape. (Why they both can't occupy the same space is consequential of the essence of occupying space, but that is specific to this example). The ability and inability of the blocks to relate to, act upon, or interact with the board consists of the action the blocks do and the action the board does, which they are both always doing as long as they remain the same shape.

Going back to the windmill example, an analogy applies to perceptual properties. If the windmill were a surface, spinning was the act of being which the surface does, and if the tennis ball were a ray of light, it is the very same act that the surface is always doing that affects the ray of light when the ray of light is in the right place to be acted upon by its action. This is why one is able to perceive things as they are when unperceived by means of the senses. The act of being rough consists of the surface having an uneven or jagged texture, and the microscopic sharp edges of the surface, which are always there, which is that surface's textural being, are what is felt when it responds to contact by scraping when a finger or anything else rubs across it. But the act of the response is simply the act of being a jagged surface when something is present for it to act upon.

To apply this to the other question, namely, how can potential be an essence, e.g. how can the ability to write be the essence of a pen if all being actions are active actions? A pen always does a certain act of being, which consists of the being of its parts and their proportion and interaction. Paper rolling across the point triggers this act to leave ink on the paper. Thus the essential act of being a pen is being able to leave ink on objects, and that potentiality is an active act that the pen always does. The pen's potential to act consists of the acts that it does. The act of writing, when it occurs, is the act of being able to write, acting upon an object in contact with the pen's tip. Also, a pen that is out of ink is still a pen only because it would be able to write were ink to be added to it. Its act of being a pen consists of the potentiality to write with the presence of ink and an object on contact with its point, which consists of the being and interaction of its parts.

What you perceive is the doer of actions. It is a mass of particles of matter, e.g. atoms, in a certain arrangement, interacting with each other as they do. This same action is the act by which the particles form a whole, and it is also the same action that affects light, produces vibrations, exerts friction, etc. Each action that it does is a part of that action it does as a whole, which is its total being action. You see different aspects of the same doer, the same doer as the doer of different actions, when you perceive it by means of its different effects, by means of the different senses, as for example when you see and hear the same thing.

It is actually not correct to speak of the differently perceived aspects of the same things as if they were different things. For example, it is incorrect to say that you hear "the sound of hands clapping" and that you see "the sight of hands clapping" or "the image of hands clapping," or that you tasted "the taste of an orange." To speak precisely and accurately one would have to say that one heard hands clapping and saw hands clapping, or that one tasted an orange. One should speak about things in their aspect as the doer of certain effects, such as for example to speak about hands clapping as the affecter of light, one should say "clapping hands seen," and to speak about hands clapping as the producer of vibration, one should say "clapping hands heard." If what I have said is true then one is really talking about things as seen or things as heard, etc., the aspects of things, but the current terminology arose from the idea that perceptions are themselves the objects of perception. (By the way, it is possible to reason that the same things are both seen, heard, touched, etc., from a thing perceived by means of one sense producing an effect perceived by means of another sense.) Thus, an ideal language would not have the two words "thunder" and "lightning," but would speak of one thing in its two aspects as lightning heard or lightning seen.

Now let us turn out attention toward the question of the ability of the senses to deceive. My above arguments that it is possible to reason that perceived things exist objectively can be used equally well to reason that the senses do not deceive, since if they did then perceived things would not objectively exist, and one is only capable of perceiving things by means of accurate perception. One can reason from this that, since the act of perception is accurate consequentially of what it is, that human sense perception can never be wrong. For the last two thousand years no position in the history of philosophy has been less popular than the infallibility of the senses, and indeed for some time the vast majority of philosophers have held the evidence of the senses to be fallible. However, for everyone to believe something is no evidence that it is true, and perhaps the very magnitude of opinion on the side of the fallibility of the senses has led to the arguments in favor of the fallibility of the senses being taken for granted and not being properly considered.

I shall conclude this section by examining and attempting to refute the two kinds of arguments for the fallibility of the senses that I know of, these two kinds being (1) the "trompe l'oeil" and (2) the argument which I have named "the infinite doubt loop".

Before I proceed to my refutations, I must make something clear, namely, that I believe that the senses of a human being whose body and brain are functioning properly are infallible, but if the sense organs were to malfunction or be sick, or if the brain were to function improperly or abnormally, then the senses and reason would be fallible. For example, a color blind person will not see color, and a crazy person who is hallucinating will not use his eyes to perceive reality. One might call the proposition that the senses and reason are only infallible if the sense organs and brain are healthy by the name of the "Sanity Proviso," which states that knowledge is only possible provided that the brain is sane enough to perceive and reason.

The Sanity Proviso is my answer to the strongest argument against the accuracy of sense perception, namely the "phantom limbs" case that a person with an amputated limb may feel pain in the limb which is not there. I say that this is a brain malfunction, not perception where a means of perception comes from an object and acts upon the sensory organ to produce a perception in the brain. Because it is not perception it does not prove that perception lies. However, the Sanity Proviso does not render it impossible to achieve certain knowledge of objective reality, because any defect or sickness of the brain or sense organs could be identified in order to distinguish accurate perception and reason from error or hallucination. A color blind person can figure out that they are not seeing specific colors. A person with phantom pain knows that it is not real. Although a crazy person will probably not know that he hallucinates, nonetheless a noticeable difference exists between physical reality as known by reason and science, and a world of figments of the imagination such as a crazy brain will perceive.

Thus, any sane person can know that he is sane, although crazy people will not always know that they are crazy, because if one were to reason what reality is like at some earlier point in time, then at a later point if one's brain or sense organs were to malfunction one could distinguish truth from fiction by using one's previously reasoned concept of reality, although obviously a complete and total incapacity of the brain would render this, and all thought as such, to be undoable. The Sanity Proviso does not render the senses or reason to be generally untrustworthy, nor does it mean that perfectly trustworthy knowledge of physical reality is impossible. Note that "reason" and "perception" are not what a crazy hallucinating person's brain does because hallucinations come from the imagination and not from perception. The possibility of hallucination is not an argument against perception, and insanity is not an argument against reason as a method of achieving knowledge. In other words, if you, my reader, if you are sane and if your brain and sensory organs work properly then you can achieve knowledge of reality.

Chapter Thirty Four: Perception: Tricks of the Eye

The first kind of argument for the fallibility of the senses consists of finding a perceived thing that seems paradoxical and arguing that it could not possibly really exist and is therefore an illusion, and using this supposed illusion to cast doubt upon the reliability of all sense perception. I call these "trompe l'oeil" arguments (pronounced "tromp loy"), after the French term meaning "trick of the eye." I shall now proceed to examine and refute six "trompe l'oeil" arguments presented by Thomas Hobbes in his book "The Elements of Law Natural and Politic" (Part I Chapter II), eighteen "trompe l'oeil" arguments I have encountered in reading and conversations with people, two "trompe l'oeil" arguments used by Plato, and one shared by Plato and Descartes. I will then refute the subjectivism loop argument in the next section. In total, this section will conclude by systematically refuting twenty seven different arguments regarding the ability to deceive of the senses.

Subpart 1. The "mirror" argument: Take any object, and place it next to a mirror. You can see the object in front of you, and you can also see it in the mirror. But if what you see is really only one object at one place, then it cannot be both where it is and where the mirror is. What you see in the mirror is an illusion, especially since the mirror has none of the qualities you see in it but is merely silvered glass, and so what you see in a mirror is an illusion.

A person could only believe this if he was ignorant of how the human act of perception works, and in particular, of the existence of the means of perception, of those things that show perceived things to the brain. The mirror, because it reflects light without altering that light, has no perceivable qualities. When light from a mirror enters the eye the mirror shows those things that last affected the light that bounces off of it to the eye. The mirror is no different from the rays of light themselves in carrying the signal from the perceived things to the eye. Light hits an object, that object affects the light (at which point the light becomes a signal that can show that object to any eye which absorbs the light), the light bounces off the object into a mirror, bounces off the mirror without being affected, and enters your eye. Thus it is, in fact, the same object that you see in front of you and in the mirror, the only difference is that you see the object purely by means of light in the first case, and you see it by means of light reflected off a mirror in the second case.

With respect to the object's position, it is easy to reason from this that you see the object where it is, and that when you look in a mirror you are seeing the object where it is again (for the same object is always at the same location, where it is) but because the light rays showing it to you are coming from the mirror, you see the place where the object is surrounded by all the places that surround where the mirror is. Note that generally, location is measured relative to another location, like points on a Cartesian plane measured relative to the origin at X=0;Y=0. The thing that you see in the mirror is not an illusion, it is a thing seen by means of a mirror, and since you can reason how mirrors operate from perceived things the case of the mirror is no indication that the senses are capable of deception.

Regarding the fact that what is seen in a mirror has left and right reversed, it is easy to see that left and right are relative, so that what is in the mirror is proportionally the same as things seen outside of the mirror, and simply shows things from a backwards perspective. When a person turns backwards and looks behind them then what was on their left is now on their right, so it can be concluded that left and right are properties belonging to the point of view of the perceiver, whereas the proportions of the physical object, e.g. if one side is bigger than the other, are objective properties which do not change based on whether they are on the viewer's left or right. In other words, "left" means the left side of the viewer, and so with "right". On a baseball field, right field from the point of view of the center fielder is actually on his left, although it is on the batter's right. The right field itself does not change in its objective qualities, e.g. its grass is no taller or shorter, depending on whether it is on someone's left or right side. When you look at a mirror you see left and right from the point of view of the person looking into the mirror and not from the mirror's point of view, e.g. if you wear a shirt with words on it and you look down at your shirt the left side of the words on the shirt will be on your right.

Subpart 2. The "concussion" argument: This argument says that the lights you see when you close your eyes and press on them, or rub them, or when your head is violently hit (as for example when one "sees stars"), do not exist, and that this is a case of sight deceiving you. While it is true that this light does not exist, it is not precisely true that it is perceived, either. This light is "seen" as a result of the optic nerves or neurons in the brain misfiring as a result of being hit. It is not perceived as a result of the act of perception, whose nature was described earlier. This light, speaking technically, is not perceived, it is imagined (hallucinated, in fact) as a result of the malfunctioning of the neurons. It is not seen by means of the act of perception so it does not prove that perception is fallible.

Subpart 3. The "color" argument: Hobbes says that color is perceived light, because "where there is no light, there is not sight, and therefore color also must be the same thing with light, as being the effect of lucid bodies," and extends this by saying that what is seen is light when it comes from a light source or is reflected off something that does not alter light (e.g., when one looks at a lit light bulb, one sees light) and what is seen is color when it is light affected by what reflects it (e.g. when one looks at an apple, the red color one sees is the red light the skin is reflecting. He concludes that color is not inherent in the objects of perception, but is an illusion imposed upon us, caused by the object's motion. What this means is that light rays are all that one sees, and to see a color is to see a certain frequency of light.

This is exactly what someone who was ignorant of the intermediary signals by means of which objects are perceived would think about color. This position is paradoxically contradictory, since if it were the case one would perceive both the object of perception and the means of perception, which is impossible, and indeed this theory equates the things seen with the light between the things seen and the eyes.

Light is the means by which objects are perceived, but the perceived objects themselves are the things of which you and I have visual experience, and these things are, quite visibly, not light, but pens, dominoes, tissue boxes, cars, people, and so on. The color that is perceived, i.e. the quality of the perceived things that is seen and is called color, is the quality of what frequency of light the perceived thing emits and absorbs, and consists of the act of reflecting and absorbing light. It is the doer of this act, not the light acted upon, that is seen, and the thing as the doer of the action it does upon light is the seen thing. Precisely speaking, color is the property of emitting only a certain frequency of light, and so both light sources and reflecting objects have color, but the reflecting object absorbs all other frequencies of light consequentially. When you look at a light bulb, you see the source of light, the glowing metal in the light-bulb, not the light it emits, and when you look at an apple, you see the skin of the apple by means of the light it reflects, not the light itself. One sees the quality of reflecting light, but it is not done by seeing the light itself. Along these lines, brightness is the quality of reflecting more light. Light is the means of sight, not that which is seen.

When a scientist says that the color you see is your brain's interpretation of the arrangement of light-affecting particles, he is wrong, because color is in the things themselves in reality, not in the brain, and this can be clearly seen. For example, green is what the arrangement of particles that reflects that frequency of light looks like. Green is what it looks like because green is what it is, and the green things you see are the things which produce that frequency of light which enters your eyes. This error of scientists is easy to explain. The widespread acceptance of this view of color pre-dates the development of modern human biology. Hobbes and Locke both wrote in the 1600's, and they made this view popular. Scientists assumed this as true about color and used this theory to interpret their findings on human optics, such that they saw the brain reasoning the perceived thing from the perception and interpreted this as constructing something in the imagination, "color," to represent the effect of light upon the eyes.

Most people believe that color is subjective, and that it is not in the objects as they exist in objective reality, but this is not the case, as can be seen by looking at a red apple and seeing that the apple is red. Someone might say that a perceived red ball is not red, because it reflects red light and absorbs all the other frequencies, so red is not in the ball, it is in the light, or in the brain's representation of the light, or in the subject. But the red quality is the quality of emitting that frequency of light and absorbing all the others, such that the frequency of light reflected shows the red object's color quality to the eye, which the brain then uses to show the red quality to the mind's eye. Thus the red is in the object, is a quality in objective reality, and is not subjective.

Subpart 4. The "echoes" argument: This argument says that when one hears the echoes of the same thing in different places one perceives in many places a thing that is really in one place. From this the argument concludes that the auditory sense deceives. This argument is the equivalent of the "mirror" argument for sound. The sound waves that the thing creates when it vibrates the air can bounce off any number of things, walls or mountain faces, before they enter the ears. If the sound wave is that thing which shows the heard thing to the brain, then it is the origin of the waves that is the heard thing, and where the means of perception comes from when it enters the ears doesn't matter, because it is the original object where that object is that is heard.

When one listens to a song on a radio, one is hearing the performance of musicians who might have played and recorded that song many years ago in a studio in a different country. Their sound waves were recorded onto data storage devices that can show those sound waves to a speaker for the speaker to re-create the sound waves that show the music. The radio station broadcasts or sends through the internet signals that show the song to your radio or computer for it to re-create those sound waves. But it is that original performance that one is listening to, by means of the sound waves coming from the band to the studio microphones to the data storage device to the radio station to the radio waves or internet data to your radio or computer to your speakers to the air waves where you are to your ears to your brain. This whole chain of showing signals brings the original sound, which is the performing band itself playing the song, to your ears by means of the means of perception that carry the motion of the musicians playing their instruments to your ears. You hear the musicians at the time and place where they recorded the song, not the computer speakers or the data signals or the sound waves.

Along these same lines, if something in front of you produces sound waves that bounce off a wall behind you and into your ears, what you hear is the thing in front of you. You are hearing it behind you because the means of perception is coming to you from behind, but it is the thing in front of you, in the place in front of you where it is, that you are hearing. For example, glass shatters in front of you and you hear the echo from behind you. What you are hearing is the glass breaking, not the wall behind you. To the extent that you identify locations from hearing sounds, if you hear echoes of glass in front of you from a wall behind you then you are actually hearing two things, the thing that originated the sound, e.g. the thing in front of you, and the location that bounced the sound waves to your ears, e.g. the wall behind you. Being behind you is an objective property of the wall behind you which you can infer from hearing the breaking glass. However, the sound and the location are two distinct properties, and in both cases you hear the object of perception, not the means of perception. If the object of perception is not confused with the means of perception, then this becomes clear.

Subpart 5. The "differing perception of the same things" argument: Hobbes says that "the smell and taste of the same thing are not the same to every man, and therefore are not in the thing smelt or tasted, but in the men." This argument seeks to prove the subjectivity of perceived smell and taste. Why does the same thing taste good to one person, and bad to another? Aren't the means of perception adding something to that which is perceived, a good or bad tasting quality? Isn't the perception of good or bad taste, e.g. an enjoyment of chocolate or coffee, in the neurons in the brain which process the sensory data, such that taste and smell are in the mind?

The precise answer to this involves the difference between perception and awareness, and how unconscious and conscious reasoning interact, which I will explain later in my essay. I think I can satisfactorily explain the flaw in the argument for now. Given the example of two people who taste exactly the same food, and one person says that it tastes bad while one person says that it tastes good (or one could say it tastes like apples, the other like oranges, or any difference), they are both tasting the exact same thing, they both experience the same taste, and the difference lies not in the thing they taste, which is perception, but in their evaluation of that thing, which is reasoning and conceptual identification and judgment.

Each person tastes the same thing, and then identifies it (either by means of conscious or unconscious reasoning) as a good-tasting or a bad-tasting thing, using his concept of the good-tasting(e) and his concept of the bad-tasting(e). If the perceived taste's properties match the properties that his concept of the good-tasting(e) is of, then he judges it as good-tasting and is aware of a good taste in his mouth. And the same if bad, but the difference is not the taste he perceives, but the aspect of the taste that he is aware of. The difference between the men is the concepts in their brain, not the perceptions in their brain or the thing they taste.

This is not an essay on aesthetics, so I will not explore the question of whether taste and judgment and preference, e.g. for food or art, is objective or subjective. I think that where a range of preferences match objective reality, e.g. liking romantic comedy movies vs. liking fantasy/science fiction movies, then preference becomes an expression of personality and individual identity, although I would need an entirely different essay in order to analyze whether this constitutes "subjectivity." What matters here is the argument that if two people see the same work of art, e.g. a painting, and one person feels it to be beautiful and another person finds it ugly, what differs is their judgment, not the painting. They both perceived the same image but one person judged it to be beautiful and another judged it to be ugly because their tastes and preferences for works of art were different, thus a difference existed between the concepts of beauty in the two people's minds, but one objective work of art existed whose visual properties were perceived by both people. For a more detailed example, two people both see a painting of a picnic on a field of purple grass below a pink sky and the humans have blue skin, and one person thinks this image is creative while another thinks it is weird, but they both saw the same colors. The same holds true for a chicken sandwich which one person thinks is salty and another thinks is bland.

A related objection is that five different people who see the same thing will tell five different stories about what they saw. This happens because of their differing judgment, not their different perception. No matter how far their accounts differ, if they are speaking truthfully their stories differ because of different reasoning, not different perception. When two different people look at the same thing they perceive the same thing, as one can reason that from the objective existence of perceived things, but anyone with different concepts will be aware of looking at different things when they see the same thing. Let me say tangentially that the classic definition of the objective as that which multiple people agree on is flawed, since one cannot reason the existence of something independent of perception from multiple people seeing the same thing, because one must rely upon one's own perception to learn of what other people have seen, nor can one reason that multiple people disagreeing is evidence of subjectivity.

For an example, I am looking at a piece of furniture right now. It is a little wooden piece with four legs and a drawer beneath a flat top surface. You and I could both look at it, and I might say that it is a desk, while you might say that it is a table. We both perceive the exact same thing. The difference in what we say arises because you and I have different concepts of what constitutes a desk and a table, not because we perceive two different pieces of furniture. Or, if I turn my head, I can look at my bed, which you could also see. I might say that it looks like a king-size bed, and you might say that it looks like a queen-size bed. We both perceive the same thing and we both perceive the same size, but we judge it differently based on our different memories of how big king- and queen-sized beds are. The difference lies in the judgment of what is perceived, not in perception. Note that, to elaborate on this example, in objective reality the bed is either a king- or queen-sized bed, so one of us is right and one of us is wrong, but the difference is not fallibility in one of our perceptions, rather it is a mistake or error in our judgment and reasoning and concepts.

Subpart 6. The "differing perception of pleasure and pain" argument: Hobbes says that heat feels pleasurable or painful, but in the fire there is no such thing, for our good or bad is not in the fire itself, but in us. First, let me say that the cause of pleasure and pain is the sensed thing, the thing itself that one feels. If one puts one's hand into a fire, there is no pain as separate from the fire that one is feeling. The thing that excites the nerve cells is the object of the perception, and one feels that object.

Hobbes is wrong because the fire itself is bad for the person. If the fire is harmful to your hand and it burns your hand then you feel pain, which is the sensation of something bad for you which damages your body, so your perception was of something that objectively existed, i.e. it was an objective fact that the fire was bad for your hand and damaged your hand, and this badness for you is the pain that you feel. The fire exists objectively, and the wound of burnt flesh on your hand exists objectively, and the thing that you feel is nothing other than the fire itself or your wound itself. So perceived pain is an objectively existing thing (and pleasure would likewise be an objectively existing thing that is good for you or beneficial to your body). Fire burning flesh is a physical thing in physical reality which is not subjective or relative or in your mind, so your feeling of pain is an objectively existing thing, it is the fire.

However, we must draw upon the distinction between perception and awareness, which is, simply put, that one perceives a thing, one judges it by using unconscious reasoning to match a concept to the sensory perception in order to identify the sensory perception, and then one is aware of the perceived thing's aspects as what one judged it to be from one's concept. (This will be explained in detail later). I can say that whether the thing that one perceives feels good or bad depends on whether one unconsciously reasons that it is good or bad, for the pleasure or pain that one feels is the aspect of the perceived thing as good or bad that one is aware of.

Goodness or badness is not a directly perceivable quality, but it is a quality that one is aware of, and, after the brain identifies the burn of the fire as good or bad, one feels the fire to be pleasurable or painful. This process of judgment consists of unconscious reasoning, so one is not aware of it and it happens very quickly. I will get to the unconscious conceptual identification of perceptions, which is a critical part of the act of conscious perception, later in the essay.

That concludes my refutation of the arguments of Hobbes. The arguments I have heard about from multiple sources include:

Subpart 7. The "perspective" argument: This is the argument that the angle of sight, i.e. one's perspective, influences what is perceived and makes visual perception subjective. This could also be called a "point of view" argument. Consider, for example, a box, with one side painted red, one side painted blue, one side painted green, and one side painted yellow. If four people were sitting around the box, each facing one side, then each person would be looking at the same thing and yet see something totally different. It is assumed that their position is influencing what is seen, and that therefore the subject itself is part of the seen thing, by influencing what is seen. The perceived thing visibly alters as the angle between the viewer's eye and the perceived object changes, and perspective is an aspect of the means of perception, so it is concluded that the means of perception alters what is seen.

But, examining perception without any assumptions, it can be seen that, in fact, each person is only seeing one part of the thing, one side, one surface of it, and that, while their location determines which surface they see, the point of view in no way influences, adds to or alters what that surface is, what it consists of, and hence what they see. Each person does not see the whole thing, but the part of the thing that they see objectively exists, and the objective thing is what they see. What changes when their angle of sight alters is which part of the thing that they see. It is not the thing itself that is being altered by the change in angle. Because which side is seen changes gradually and continually as the angle of perception changes, it can seem that the thing itself is changing, but this is a mistake of judgment, not perception.

Considering my example, if you walk in a circle around that box you will see first red, then blue, then green, then yellow. But your changing angle of sight is not altering the color of the object. It is only altering which side of the object you see. Since each side is a different color, you see a progression of colors. The entire box is red, blue, green, and yellow, but if the side that you are looking at is objectively blue, then you are seeing what exists objectively, although you only see one side and not all four sides at once. If you consider a flat plane in the same way, and rotate that plane as you look at it, the plane you see when your angle of sight forms a right angle with the plane is the whole plane, and as it rotates towards you less and less of it is seen, until when your angle of sight is the same angle as the plane none of it can be seen. Less of it is seen, but it is not a different thing which is seen.

The fundamental mistake of deducing the subjectivity of sensory perception from perspective comes from seeing part of a thing and not seeing other parts of the thing and confusing this with the perceived thing being influenced by the subject or means of perception. All those differences between what two people perceive when they perceive the same thing from different angles but seem to see two different things which cannot be accounted for by a difference in their judgment can be accounted for by the phenomena of a whole thing looking different than a part of that thing. The same thing looking different to different points of view should not be taken as proof of subjectivity. The experience is adequately explained by the analysis that one part of a thing looks different from another part of the thing, and the whole thing and each part exist objectively and are what they look like. Each side of a thing is a part of that thing. One can only see one side at one time, and one's angle of sight determines which side one sees, but perspective does not create or alter the side itself or the thing itself.

Subpart 8. The "glare" argument: This argument says that the glare that one sees is surely light that one perceives, which is an instance of seeing light. This is asserted to refute the argument that light, as the means of perception, cannot be perceived. This argument is flimsy unless one has already granted the assumptions that were refuted my section on Hobbes's "color" argument. The brightness that one sees is the quality of producing a lot of light. It is not the light itself that is seen, it is the producer that is seen by means of the light that it produces. If one looks into the Sun one is blinded by the Sun, by the producer of a large amount of intense light, by means of the light it produces that enters your eye. You are not blinded by the light itself, but by the Sun itself. The glare you see is not the light itself, but the Sun, the burning star. Seeing bright and dark is seeing the doer of the act of emitting more or less light. This also explains shadows, which are not illusions but which are merely areas which reflect less light than their surrounding areas.

Subpart 9. The "colored glass" argument: When one looks at something through colored glass, one sees the thing, but its color is not the color that it really is. For example, if one looks at a red apple through blue paper, one will see a purple apple. This argument says that the apple one sees is an illusion, since it is not the red apple.

Two things must be understood to grasp the error here. First, being a color consists of the acts of emitting all the frequencies of light which that act is composed of. For example, being white is the act of emitting all colors, and is an act that consists of parts, each part being emitting red light, blue light, green light, etc. Being purple consists of the act of emitting red light and blue light, and the act of being purple consists of being red and blue. Thus when one sees a purple thing, one is seeing something being red and blue at the same time.

Second, the means of perception show their object to the brain because their being corresponds to the being of their object in some way. With respect to light and vision, the being of the means is caused by the being of the object. The perception in the brain is caused by the pulses from the neurons, which are caused by the pulses the sense organ produces, and the being of the pulse is caused by the sense organ's response to being acted upon by the object, in the case of touch, taste and smell, or the intermediary means of perception, air vibrations for hearing or light for vision. The being of the ray of light is caused by the being of the thing that acts upon it. The being of the light then acts upon the sense organ, the eyes, to produce the nerve impulses. The brain then uses the nerve signals to think about the perceived thing, and when the brain decodes the nerve signals then the mind's eye sees the thing that sent light to the eyes. The properties of the light which the eyes register and convert to a neural signal are the properties which the object that last emitted or reflected the light caused those light rays to have. This is how things are seen by means of light and the eyes (and substituting air for light, how things are heard by means of air vibrations and the ears).

Regarding the purple apple example, the properties of light which communicate shape were last affected by the apple, and the glass it passes through does not act upon those properties of the light, and so one sees the shape of the apple by means of the light that reflected off the apple and passes through the glass. But the colored glass does affect the property of the light that shows color. Thus when one looks at the apple one is seeing, simultaneously, that apple's shape, that apple's color, and the glass's color. The purple one sees is the red of the apple and the blue of the glass perceived at the same time because one actually perceives the apple's shape and color and the glass's color at the same time. Blue plus red is purple, as any artist knows from mixing paints.

Subpart 10. The "prisms" argument: This argument says that the case of white light entering a prism and being split into the different colors of the rainbow clearly proves that color is perceived light, because one sees the colored lights. However, referring back to the understanding of the act of being a certain color being a composite act consisting of being the reflector of a certain set of frequencies of light, and to the issue of seeing less of a thing vs. not seeing a thing, we can see that when the prism divides the white light into light of different frequencies, and projects each frequency of light at a certain place, the thing at the place is only able to reflect that frequency of light because that is the only light hitting there. So each thing can only be seen to be the reflector of that color of light, even if reflecting that light is only a part of its total act of color, i.e. if it could reflect that frequency and others.

This is why prism-split light cast upon a white wall causes each section of the wall to be a certain color. The whole wall is still white, but reflecting each of those colors is the only part of the act of being white that the wall can do at that place, e.g. it looks red where red light is shining on it. In terms of seeing the colored light itself, the light rays can never be seen, only the things that it shines upon can be seen. When a ray of light itself seems visible this is only because one is perceiving the dust in that area of air that the light is reflecting off of. When one is in a dark room and one shines a flashlight at a white wall, one sees the circle of whiteness of the wall (which is the wall being white), one does not see a white beam between the flashlight and the wall. Try this and see.

It is worth noting on this topic that some scientists will surely assert against me that a rainbow is an optical illusion. They will argue that there is no physical object in the sky that the rainbow consists of. If one tries to approach a rainbow it moves away, and if one is located at the place in the sky where someone on the ground sees a rainbow one sees nothing and nothing is there. In reply to this I say that a rainbow is actually the water droplets or mist in the sky. Sunlight is refracted by the mist which separates the light by wavelength, and because a color is the property of emitting a certain wavelength of light one sees a rainbow as a band of colors. To be precise, when one looks at a rainbow one is seeing the sun seen through mist. Similarly, when one sees the light cast by a prism, one is seeing the source of the light as seen through the prism. Thus, the cases of rainbows and prisms do not prove that perception lies. Instead they support my theory that color is the property of reflecting or emitting a specific wavelength of light. This case also illustrates my answer to those scientists who say that color does not align precisely with the wavelengths of light that actually exist. For example, in the band of a rainbow that looks like a red stripe there is actually a continuum of different wavelengths, and the red light is not all precisely the same wavelength. Nonetheless, all the different wavelengths seen as red are roughly in the same place on the spectrum, and this general identification of color by the eyes as red is accurate although it is not precise or detailed.

Subpart 11. The "color-blind" argument: There are some people who cannot see color at all but who only see black and white, and there are some people who only see some colors and not others, and some people who see the same color when they look at blue and purple things. Is this not a clear case of the eyes of people deceiving them? Referring back to what I said earlier, for the means of perception to show some parts of things and not others without creating or altering what it does show is not an argument against objectivity, and is frequently misinterpreted to imply subjectivity. Along similar lines, colorblindness is not a case of the eyes of these people deceiving them. What they see does exist, the shapes they see are the shapes of real things and the colors they see are the colors of real things, but because of some flaw of their eyes they are not able to see all, or certain colors. This does not mean that what they do see is subjective, and does not mean that they are being deceived by what they do see, for what they see are objective things doing some of the actions that they do, but not their entire being actions.

Color-blind people see parts of the colors that are out there in the world, not a different set of colors. Regarding the person who looks at blue and purple things and sees the same color, let me refer back to the above, where I explained that certain colors are composites formed of the acts of reflecting a number of different frequencies. This person's eyes cannot see one of the frequencies that purples do reflect but blues do not, and so this person sees the same color, but the person is actually seeing the purple be blue, which it is, but is not seeing it be red, which the purple thing also is. A person who had studied the act of perception would not say that he was being deceived about what he did see, but is merely unable to see certain things. He is no more deceived by his eyes than a blind person is deceived by their eyes.

Subpart 12. The "hallucination or dream" argument: This argument says that hallucinations and dreams are two clearly existing cases of the senses deceiving the perceiver about the reality of the perceived objects. Most people believe that when they see things in a dream, or have a hallucination (if they are given to having hallucinations) that they are seeing a thing that looks exactly as it would if it were real, and that the only difference between the things seen in dreams and the things seen while awake is that the former does not exist while the latter exists. On the basis of this, dreams and hallucinations would indeed be an example of deceitful perception.

Let me begin by distinguishing two different arguments. One is the argument that dreams and hallucinations are acts of perception, which are deceitful, and hence perception can be deceiving. The second is that the possibility that everything one sees might be a dream or a lie renders it impossible to achieve something that can properly be called knowledge if derived from sensory experience. Here I will refute the first argument. In the next section I will refute the second.

If the eyes and ears, etc., function properly then the perception was not deceitful, because the data which went from the objects to the brain was accurate and truthful, but if the cortical centers in the brain which process sensory data are under the influence of drugs or mental illness then the brain will lack be unable to process the input and will lack knowledge. But this was truly due to what should be classified as an error of conceptual identification, not an error in perception. A person who was asleep and dreaming, or who was hallucinating due to drugs or mental illness, might not be able to tell the difference between imagination and perception, and imagination might look like perception. But imagination is not perception, and the difference is that imagination is constructed from within the brain, while perception comes into the brain from the external physical world by means of the senses.

My philosophy uses the Sanity Proviso, namely, that a good philosophy is designed for people with functional, sane, rational brains, and such people are the only concern of my philosophy. Specifically, the Sanity Proviso states that a rational philosophy can be understood and used only provided that the person who uses it and thinks about it is sane and rational and has a brain that is able to perceive and to reason. A person who is sane can look at the upcoming section which distinguishes perception from imagination, and can then know that they are sane. However, a person who was actually insane or asleep and dreaming or hallucinating due to drugs might not be able to distinguish delusion from reality, not because imagination looks like perception, but because their conceptual judgment would be unable to distinguish the input of perception from the creations of the imagination within their own brains.

If our approach is Analytic Realism, and not an abstract theory devoid of any basis in practical reality, then we can see that an infallible means of knowledge exists for sane people to know that they are sane and that their perceptions are accurate. This infallible means of knowledge is simply reason applied to perception, which consists of perceiving a physical, objective, real world of science, and observing that one is not looking at a strange abnormal world of magic and spirits where reality spins and melts and things don't make sense. Reason can tell the difference between perception and imagination, by examining what is experienced. The difference between perception and imagination may or may not be directly visible, but it can either be seen directly, or else it can be inferred from one's knowledge of reality in order to conceptually identify the real and unreal.

I would not say that this is "common sense," because that term lacks any clear content, but I will say that sanity vs. delusion, whether one is sane and perceiving or in a dream or hallucination and imagining, can be reasoned by a sane person, hence the possibility of delusions is not a proof that perceptions can lie.

Subpart 13. The "residual light" argument: If one stares at a bright light source too long, then after one looks away there are residual spots of whiteness that one sees, or one may continue to see the outline of the light source. This effect lasts for a period of time and then dissipates. This residual light is an optic illusion, and the spots are not real.

First, let us consider when what we see exists. What one sees exists at a certain time. For example, if one looks at a clock that says "6:26pm," and one then looks around, everything one sees exists at a moment that is named by that set of numbers. It is a common bit of "pop physics" to say that everything that one sees exists in the past, not in the present, because of the period of time it takes for light to go from the perceived thing to the eyes, and from the eyes to the neurons, etc. This is true, but only in a qualified way. Time can only be measured relatively (I'll deal with the issue of relative measurement later in this section), and so to be earlier or later is to be earlier or later than a certain moment, to be in the past or the future is to be earlier or later than a certain moment when something happens, which one calls the present.

But when is the present? What is its essence? One could consider two different moments to be the present. One can consider the moment when one's brain performs the act of deducing their objects from the sensory perceptions in the brain, which is the act that enables one to consciously see perceived things, and if that is the moment one calls the present then one is seeing the past. One could also consider the moment when the things that one perceives exist to be the present. Since one is only ever able to respond to the moment that one is perceiving, it is better and more efficient to consider that moment when perceived things exist to be the present that one measures time comparatively with. Then everything that one sees exists in the present, whose essence is that it is the moment when what one perceives exists, and the time when one's brain perceives perceived things is slightly in the future relative to the present. However, since the brain's perception happens so quickly, for most practical purposes the time when the perceived thing exists and the time when the brain perceives it are so close together that you will notice no difference between them.

What I have just explained is pertinent, because, if one keeps in mind that the thing that one perceives exists at the moment when it exists, and that moment is the moment that one perceives when one perceives a perceived thing (perceiving a thing in space and time is no different from perceiving an existing thing, which is no different from perceiving a perceived thing, hence being in time is perceivable), then the residual spots phenomenon can be understood. For as long as one's sense organ is sending the signals that show a certain object to the brain, the brain will perceive that object. If the organ continues to send the signals after the object has ceased to act upon the organ, it is the same means of perception that will be sent to the brain and it is the same object, at the same time that it acted to create the means of perception (i.e. the moment when it emitted light) that one will continue to see. Thus, the residual spots that one sees are actually the light source one was seeing, which are in the past relative to the perceiver at the time of perception, but in the present relative to when the perceived thing itself exists.

It is blurry and indistinct because as time progresses the organ sends less of the original signal to the brain, and less of it can be perceived. When one continues to perceive the outline for a little while until it fades away, this is again the case of perceiving less and less of something until one can no longer perceive it. On a similar note, of things remembered, the thing that you remember is not in your memory, although the memory of it is. The recalled thing exists at the time when it existed, and its act of being at a time consists of being at that time when it existed. The memory of it exists in the present but the thing itself existed in the past.

Subpart 14. The "hurt finger" argument: If one burns or cuts one's finger, and one then touches something, one will feel pain. But the "cause" of the pain, that perceived thing which, to be precise, is the pain, is the finger, not the touched thing. Is this not a case of the means of perception altering perception?

This objection is very easy to answer. The nerve cells in one's skin which are the means of the sense of touch are used not only to feel things that one's skin touches, but to feel the skin itself. This is how one feels when one is cold or hot, or when one's skin itches or is in pain from a cut. One is simultaneously feeling the thing one touches, and the skin around the nerve cells, by means of the nerve cells in the skin. One feels the wound in one's finger, which is the pain, and one also feels the surface that one touched, at the same time, but the two things are distinct.

As a somewhat similar matter, let me note that when one feels a pain in one's stomach this is not subjective, despite the fact that the perceived thing is in the body, because nerve cells exist which send data from within the body to the brain, and some trouble exists objectively in your stomach which the perception is reporting to you. Perceptions of things inside your body are what I would call a sort of "quasi-perception," in the sense that it consists of things acting upon nerve cells to send signals to the brain, but it is not directed toward objects outside the body in the external world. The sense in which introspection consists of the mind "hearing" its own thoughts and seeing the visualized things in its imagination can also be thought of as quasi-perception, assuming that if the mind is the brain then in a sense the thoughts and images exist objectively because something physical in the brain exists which the brain experiences.

Subpart 15. The "blurry vision" argument: When one is crying, or tired, or under the influence of certain drugs, one's vision is blurry, and the things one perceives are indistinct and blurry. But the things themselves, as they exist in objective reality, remain as clear as they were before. Surely one's means of perception is altering what is perceived?

One's means of perception is preventing one from fully perceiving what one would otherwise be fully perceiving, i.e. it is causing one to only see a small part of what is perceived, but it is not altering that which is actually perceived. This is the same misunderstanding as the perspective argument and the colorblindness argument. The only thing specific to this example is how a blur is actually a "part" of a clear thing. To be precise, it is not, rather, if one considers a perceived thing on a 100 by 100 grid, e.g. a Cartesian Plane, and one colors each square completely with the color that predominates in that square, and one then considers the perceived thing, then it will have clear visible detail. Now reduce the scale of the grid to 10 by 10, and color each square with the predominate color. One has gone from a clear thing to a blurry thing, and yet what has changed is that the scale of detail one is capable of perceiving is lessened. It is not a different or altered thing that one is now perceiving. Tears and drugs can make it difficult or impossible to clearly see what one is looking at. But the seen thing is not an illusion. It is merely a very small aspect of what objectively exists that is reflecting light at your eyes.

Subpart 16. The "buzz" argument: If one is in a large room with lots of people speaking, one hears a buzz. If one is listening to anything and one covers one's ears, one will also hear a buzz. This buzz is no thing in reality, and yet one perceives it. Similarly, if one looks at a bright light source and closes one's eyes, one will still see something, and yet one will not be seeing the perceived thing anymore. In both cases what one perceives is asserted to be an illusion.

The answer to this objection relies upon the idea of clarity and vagueness as the scale of detail that is perceived. In the room full of people one hears less detail, and hence hears the buzz, but it is still all those real voices that one is hearing, although one cannot hear them all fully. The reduced-clarity collection of voices one hears is the buzz. One hears less of the real thing, not something that is not real. One also hears less of the real thing with one's hands over one's ears, and one sees less of the light source with one's eyes closed, but it is still that light source that one sees, by means of the light from it that penetrates one's eyelids.

Subpart 17. The "two fingers" argument: If one looks at one's finger, and then moves one's finger until it touches one's nose, then one will see, not one, but two fingers, one on each side. But one has only one finger on one's nose, and one is seeing two fingers, i.e. each eye sees one finger so the mind's eye sees two fingers at once, hence one is being deceived.

As one can see by closing one eye at a time and looking while in that situation, each finger one sees is seen by means of a different eye. The finger on the right is the side of the finger that the left eye is at the proper angle to see, and vice versa. One is seeing the same finger, at the same place, by means of two different sense organs, and so one is actually seeing one thing, at one place, twice. One is seeing two sides of the same finger, not two different fingers.

For one to be deceived would be for that thing which one sees to not exist. The thing one sees in this case, i.e. the finger, does exist. One finger, at one place, is seen twice from two different sides. Thus the problem here is one of conceptual identification of the perceptions, e.g. the mind fails to identify that both fingers are the same thing, and the so-called deception is merely an error in judgment. The thing that one sees exists objectively, and it exists as it is seen, in other words each side of the finger that is seen exists objectively and looks like what is seen.

One could further argue from the example that the fingers one sees seem to be slightly transparent, in that one sees the background through that finger, but the finger in reality is opaque, and this is an illusion. What one is seeing, if one looks to one's left, is the finger by means of the right eye and the background by means of the left eye. If one actually looks at the finger closely, one will find that one sees not a transparent finger, but that one sees the finger and the background simultaneously. Thus what one sees is not an illusion, one is seeing two different things via two different eyes simultaneously, and one is only deceived if one misjudges what the perceived thing is.

Subpart 18. The "distance" argument: Things seen in the distance appear to be vague and indistinct, but in reality they are as clear as things that are close. Also, things that are large look small in the distance. These are asserted to be two illusions.

Refer back to the argument I made about the scale of detail that one can perceive in my refutation of the "blurry vision" argument, e.g. comparing a 100 by 100 grid with a 10 by 10 grid, showing that the same thing seen at two different scales will look different but it is the same thing. The things that one sees in the distance cannot be seen as clearly as things close up because of the distance between the perceiver and the perceived and the limit of the means of perception's ability to transmit detail over distance, because of the diffusion of light and air vibration and the angle of the rays of light in relation to the eye. Things in the distance cannot be seen in detail, and so are blurry, but that does not make them illusions. Rather they are real things that one is perceiving, but one is not perceiving very much of their detail. This too ultimately goes back to the distinction between seeing less of a thing and seeing an altered thing that I drew when refuting the "perspective" argument. One scale of perceivable detail applies for close objects, another for objects in the distance. One can see more of objects that are close, and less of objects that are far away, but the objects themselves are not perceivably different as they move closer or farther away.

To say that things that are big look smaller in the distance is to be ignorant of the fact that size is measured relatively. Nothing is just "big" or "small." Rather "bigger" and "smaller" are always bigger that something or smaller than something, that something being the standard that is used to judge size comparatively. Things look "bigger" up close because the scale of perceivable detail has increased and more of the thing can be seen, not because the thing has grown. If one was ignorant of the scale of perceivable detail changing with distance, then one might compare something seen in the distance with something seen close up, disregard the difference in the scale of perceivable detail, and then conclude that the thing in the distance "looked smaller" but was really bigger. If one had studied perception and understood the difference of scale, one would only visually compare the sizes of things at the same distance from the eyes of the perceiver, and one would not have this problem of conceptual identification.

Subpart 19. The "microscopic" argument: This argument says simply that the microscopic exists but is unperceivable, and therefore the senses are limited. This argument is actually very cunning, since it equates the senses being limited in the sense of having a limit to the amount of detail that is perceivable with being limited in the sense of being flawed. Those two things are not equivalent. The assertion is wrong because one does perceive the microscopic, but not individually, rather one only perceives it as the sum whole that it forms which the microscopic particles joined together are big enough to be perceived. For example, a tissue box is made of atoms, and when one looks at a tissue box one does not see individual atoms, but one does see all the atoms on the surface, acting as a whole, in the form of seeing a tissue box that is made of atoms. When one sees a tissue box one is seeing atoms acting together as a whole thing. Also, by means of microscopes and other devices that increase the amount of detail that the means of perception can transmit, the microscopic can be seen individually. This argument ultimately rests upon ignorance of the scale of perceivable detail.

Subpart 20. The "circus mirror" argument: One may accept that mirrors show things that are not themselves to the eye, but this does not explain circus mirrors, for what one sees in a circus mirror does not exist, and is a distortion of what does exist. Does the mirror distort light rays to create something that does not exist in reality, such that the things that one sees in the circus mirror are an illusion?

While it is true that things one sees in a circus mirror does not exist, those things are composed of things that do exist. To be precise, what one sees in a circus mirror is composed of bits of the real thing seen from many different perspectives, all put together. The curvature of the mirror bends the light so that all the different parts of the thing are seen from all sorts of different angles which depend upon the angle of the mirror's curve in relation to the angle between you and the mirror. So the thing seen in the mirror seems to form an illusion, but it is, precisely, a thing that does not involve the means of perception creating or altering what is seen.

Subpart 21. The "trompe l'oeil" argument: Trompe l'oeil art is a kind of art that consists of a two-dimensional artwork that has the illusion of three-dimensionality, so that a person who looks at it believes that he or she is looking at something with depth, which does not in fact possess depth. Also, all visual artwork is illusory, since, for example, one looks at a canvas smeared with oil paint and one sees, not a canvas and paint, but people having a picnic on a grass lawn beneath some trees. Are not all works of visual art in general, and trompe l'oeil in specific, cases of the senses deceiving the perceiver as a result of the artist's skill?

Let me begin with trompe l'oeil art and then proceed to art in general. In the case of trompe l'oeil art, the work of art looks like something with depth, e.g. the painting is a painting of spikes which seem to be coming towards the viewer, although the painting itself is really flat. The eyes see something that looks deep and the brain then processes the perceptions and evaluates that the seen thing possesses depth, so the mind is aware of the seen thing as something possessing depth. The distinction between perceived depth and depth which one is aware of matters because, while I am going to argue that sensory perception with a functional brain and healthy sense organs is infallible, and that sensory perceptions cannot be wrong, I am not going to argue that the conceptual identification computations of the brain, which I also call judgment, is infallible. (I will proceed to explain what can cause judgment to go wrong later.) The brain identifies depth in the painting from certain general qualities that the perceived things have, and something that is visually similar enough to some other thing to have the same general features can be conceptually misidentified. Things are visually identified by their general features, because they are identified by means of concepts of essences, not by matching memorized specifics. In this case, one perceives something real, the painting, but one is aware of it being something that it is not, an object with depth. The trompe l'oeil art resembles the visual features of something with depth enough so that the brain misjudges it, and the thing one perceives has a depth that one is aware of, even though the thing that one perceives does not really have depth.

Similarly, because the visual features of a painting so closely resemble the scene that it shows, the brain misjudges it and is aware of the scene, not the paint, even though the paint itself is seen. The painting is seen as the scene, e.g. one feels that one is looking at a picnic and not at splashes of paint, because the scene is what one is aware of. Let me very briefly describe how artwork works, to deepen this analysis. A work of art shows something other than itself to the mind's eye. In this sense artwork is like words, which show their meaning to the mind, and the means of perception, which carry their objects to the mind's eye. Art's mechanics of its act of showing, however, is different from words, which show by association, and perceptions, which show by corresponding to their objects in a way that can be processed by the sensory regions of the brain. Art shows by having some properties which are identical to the shown thing while having others which are not, such that the audience's mind notices only the properties which match the shown thing, and do not notice the other different properties, so that the audience thinks they are seeing the shown thing which the artist sought to convey, not the work of art which is the artist's means of showing. Art functions because the identical properties between the art and the shown thing tend to be the aspects of the thing which the mind uses to identify the thing by its essence.

For example, if one sees a performance of the play Macbeth by Shakespeare, one sees the actor who plays Macbeth say and do the things that the character Macbeth says and does. The real Macbeth of the story would say and do those things and look that way, so in his aspect as the actor of that essential being of his role the actor is the same as Macbeth. The audience sees the properties of the thing that is the character Macbeth, and the mind does not pay attention to the properties of the actor as actor, e.g. one notices the actor portray Macbeth but one does not see the actor out of character, so the judgment is aware of Macbeth and not the actor. For another example, if one sees a painting of a red apple on a blue tablecloth, comprised of a red apple-shaped sphere of paint on a background of blue paint, the thing that the artist sought to show to the audience is (a red apple on a blue tablecloth)(e). One sees the thing shown to the audience by the artist by means of paint that has seen properties similar enough to a red apple on a blue tablecloth that the brain's conceptual identification identifies the paint on canvas as a red apple on a blue tablecloth, so that when one looks at the painting one is aware of the apple and not of the red apple-shaped sphere of paint.

Art can only be understood if one already knows two things, the nature of essences, and the human brain's system of conceptual identification of perceptions. The work of art has some of the essential properties of the shown thing, although it has specific properties which are different, and the brain conceptually identifies the perception by the essential properties which match the shown thing, so that the brain is aware of the shown thing, e.g. the apple, and the brain is not aware of the specific properties of the art itself, e.g. the red paint. Generally, art functions both to communicate ideas and values, and it also serves an important role in a brain teaching itself to conceptually identify perceived things by their essences, by strengthening the concepts of the essences in the things shown by the art. This will become clearer later when I discuss concepts.

The illusion of trompe l'oeil art in particular, and of all art in general, is a mistake of judgment, not perception (assuming that art can be called a "mistake" in the sense of how it works, although the perception of art is correct in its role for human mental functioning). The sensory data in the brain accurately corresponds to the object

212

that originated it. Perception is infallible, but the brain can fail to correctly identify what the perceived object is from the sensory data, by making an error in analyzing the sensory data using the brain's concepts, so awareness is not infallible. The perception of directly perceivable qualities is infallible, but the awareness of the reasoned identity of things is fallible, and when one "sees" something the mind's eye sees the thing that the mind is aware of superimposed upon the thing that one perceived, i.e. one perceives the perceivable properties and is aware in the same thing as that thing's conceptually identified properties, so a mistake in judgment can look like a deception in perception, even though it is not. Later I will argue that awareness becomes infallible if it is the product of pure empirical essential reasoning, but pure empirical essential reason is a specific subset of essential awareness.

Subpart 22. The range of senses argument: According to this argument, one can consider the case of animals to illustrate the relative subjectivity of the human senses. There are animals able to hear a different range of frequencies of sound or see different frequencies of light than humans. The world surely looks very different to them than it does to us.

Let me explain the mistake in the argument. The senses of animals are a means of perception, just as they are for humans. Therefore, they must show an objective world to the mind of the animals, for the reasons explained above. The only difference is that the animals capable of a different range of frequencies see aspects of the world not perceived by humans. The difference is not that they perceive a different world, but that they see different sides of the same world. Consider the example of a human and a bat. A human hears less than a bat. But this means only that less of auditory reality is perceived by the human, not that a different or subjective reality is perceived. The world sounds different to a bat because a bat can hear more, not because the bat hears a different world. Thus, the different ranges of frequencies do not give animals access to a different, subjective world. The world that they perceive is the same world that humans perceive.

Subpart 23. The blind girl argument: This argument claims that, of a person who is blind, if the exact scientific details of the frequencies of light were explained to her verbally by a scientist, she still would not be able to imagine what light looks like, and the difference between seeing light and knowing light is a subjective element inherent in the experience of color. Regarding the blind girl, I reply that the direct experience of something and the indirect experience of something yield differing amounts and types of information. The experience of color directly is the color perceived, while the indirect experience of color is the description of light frequencies. No subjective element to the experience exists, but more information exists in the direct experience than in the indirect experience. As discussed above, for someone to perceive less of something than what someone else perceives is not the same thing as two people having two different subjective experiences. (This argument is presented in "What Mary Didn't Know," Frank Jackson, Journal of Philosophy 83:291-295).

A similar thought experiment which has produced much confusion is the scenario that when a biologist dissects a human brain, what he finds is a lot of gray and white matter and neurons and blood and veins, and he does not find perceptions of cars and chairs and houses, or ideas like freedom or philosophy or reason, or values like virtue and courage. This leads philosophers to mistakenly believe that the mind cannot be precisely the same thing as the brain, because the brain has a mere physical existence while the mind includes a "subjective" experience of reality. The solution which clears up the confusion and solves the problem is the Objectivity Proviso. The bits of matter in the brain are the means of perception, which are the means by which the brain experiences reality, but the "experience of reality" consists of the reality, not of the brain. To see the house that a mind experienced one would need to look at the house, not at the brain of the person who experienced it. Thus, when a biologist dissects a brain he is in fact looking at a mind, and at everything which a mind consists of. The biologist does not see reality as that dissected brain experienced it when that dissected brain's human was a living perceiving mind, because the biologist is looking at the brain, rather than the biologist actually being that brain looking at reality by means of the concepts and perceptions inside that brain.

Subpart 24. The "misperception" argument: When someone sees a person in the distance, and recognizes them as someone they know, and then when they come closer they realize that he or she is not the person they thought, this is an instance of deceptive senses. All other cases where one thinks that one perceives something and one turns out to be wrong are similar. I argue that this is not a case of deceitful perception. Referring to what I said above, the misidentification of what one is seeing is a mistake in judgment, i.e. in conceptual identification, it is not an error in perception. An error in conceptually identifying a perceived thing is not an error in perceiving the perceived thing itself. One is not deceived about the perceived qualities that one sees. Rather one makes an error in judging what thought-about qualities the perceived thing has. In other words, one makes a mistake in one's awareness of what the thing is that one sees.

This concludes the general trompe l'oeil refutations. Let us turn to the trompe l'oeil arguments of the old masters, Plato and Descartes.

Subpart 25 Plato's "wind" argument: In Plato's "Theaetetus" dialogue Socrates argues against the position that "knowledge is perception," i.e. that perceived things exist, perception is infallible, and everyone perceives the same things, with many different arguments. I have refuted some of these above already. It is pertinent for us to examine two arguments in the "Theaetetus" in detail.

Plato's first argument regards the temperature of wind. He says that a certain wind that blows past two people might feel cold to one and hot to one, or cold to one and very cold to one. But the same wind could not be both hot and cold at the same time, so if perception is knowledge then the wind is cold and hot to whomever perceives it as such, and perception is subjective.

In general this Platonic argument is refuted by the distinction between perception and conceptual identification, i.e. judgment. If two people perceive the same thing but they conceptually identify that one thing in two different ways then they will be aware of two different things even though they both perceive the same thing. However, something about perception and judgment is specific to the case of wind. This is the theory of relative judgment, which I refer as the theory of the "objective relative". When two people feel the same wind, and one person feels cold and one person feels hot, why does what they feel differ? The answer is that it does not actually differ. They are both perceiving the same temperature (temperature being the degree of the activity, i.e. the speed of the action, of the particles of matter at the atomic level, hotter being

215

more or faster and colder being less or slower). The wind has a certain objective temperature. This temperature could be measured in terms of its effect upon a chemical, and, if judged in exact numeric terms in this way, by means of a thermometer.

But that is not how the human brain judges temperature. The human brain judges temperature relatively, which means that when a certain temperature is perceived by the skin, the temperature is judged by means of a comparison to some other temperature. Colder than this standard of measurement is called "cold" and hotter than this is "hot". "Cold" or "hot" are no different from "colder than" or "hotter than" the base temperature in relation to which the temperature is measured. The only reason why people say "hot" or "cold" is because they are ignorant of relative measurement and the temperature they used as the frame of reference, i.e. the temperature that other temperatures are compared to. It is so common that they don't even think about it, and the relative measurement comparison is done unconsciously, by unconscious reasoning.

For human judgment of temperature, except when one is consciously comparing two things, it is usually the average room temperature that one judges in comparison to. When the temperature changes it is with the last temperature that the new temperature is compared. This is why two people can think that the same temperature is hot or cold, because it is hotter than the temperature one person was before and colder than the temperature the other person was before, or else their frame of reference temperature differs similarly. In either case it is the same temperature that they both perceive. This is also why the same bathroom can feel warm when one enters it and cold when one steps out of a warm shower, because the last temperature against which it is compared differs. This explains why things will feel hot or cold differently depending on whether one's hand is in cold water, or just came out of cold water, or was next to an electric heater, etc.

If you are under the influence of modern philosophy, my reader may think that I am acknowledging the subjectivity of relative measurement. Let me deny this and discuss how objective relative measurement works. Relative measurement is measurement made by means of a comparison between two things, in one of two forms, either by comparing two things to each other directly, or by taking one thing, calling it the standard of measurement or frame of reference, and then measuring all other things in terms of it, which consists of measuring all things in relation to it or measuring how many of the standard thing all other things consist of. All quantitative measurement is relative. For example, the number one is the standard for counting all numbers, i.e. one counts how many number ones each other number consists of, e.g. ten is ten ones, and each number is measured by the relationship between itself and one, e.g. 120 is 120/1, etc.

For all other quantities, something is chosen as the standard, and this thing is used to measure all other quantities of that kind. For example, the length of one inch is a standard for measuring length, and counting how many inches are in any other length is how we measure it relative to the inch. A thing that is six inches long is measured by one inch six times, etc. Location is measured relatively, for all location is either closer or farther away from some other location. On the Cartesian Plane, used in math and geometry, the location of all points is measured relative to the origin at (x=0, y=0). Many qualities are also measured relatively, by comparing the quality of one thing to that quality in another thing.

The question at the heart of the Platonic argument is whether relative measurement as a means of judgment yields knowledge of objective quantities and qualities, or yields only subjective qualities and quantities. "Relative" is often used as a synonym for "subjective", and Plato originates that confusion later in the "Theaetetus" when he says that the objects of subjective perception exist relative to the perceiver, only in the relation between perceived thing and perceiver. Plato's semantic equation cannot be taken as proof that Plato is correct. The question is whether the standard of measure creates or alters that which is measured, or whether as a means of judgment relative measurement using an objectively existing standard or frame of reference shows an objectively existing thing or measures an objective property of a thing. If relative measurement is objective, then we can conclude that objective relative measurement is possible.

The answer to our question can be reasoned from perception directly, or from considering the essence of a ratio. Look at one thing with a straight edge, and look at something else. Study what they look like. Hold the first thing up to the second, and count out the number of times that the first thing's length fits into the second thing. If one calls the first thing's length "one unit," then one can say that the second thing's length is however many numbers of units that were counted. If one then looks at them again, they will both look exactly the same as they did before, so the measurement did not change how long the second thing was, and yet that length was reasoned to consist of a certain number of lengths equal to the first thing's length. Thus, the objective quantity of length, the directly perceivable length, was reasoned to be that number of units.

The thing itself and the length are not two different things, rather they exactly the same thing, in that the "thing's length" is the thing considered as the doer of a certain action, being that length, thus the length is an aspect of the thing. The thing is seen, and therefore that length is seen. If a thing is six inches long, being that length, and being six inches, is the same act, but if someone thinks that the means of judgment are subjective and that the act of judgment creates the judged thing, then one will think that the act of being measured creates the measured quality, and that the absolute quantity of length in the thing in itself is interpreted by the means of measurement into a new quantity of length, the subjective relative length. If this were the case, then the absolute length would be impossible to measure, which is what these people believe. But no evidence of this can be found in the perceived world, and the distinction between "absolute" and "relative" as equated to "objective" and "subjective" is deduced from those people's assumption of the subjectivity of judgment.

Let us conclude by considering the essence of a ratio. A ratio is one thing measured in relation to another thing. The two terms of the ratio are both absolute, objective quantities. Therefore the relative quantity is the absolute quantity measured by means of the standard of measurement, which does not alter what is measured but only reveals its being to the mind. The sense in which thirty units is relative to one unit is different from the sense in which the subjective is relative to the perceiver. Being thirty units or being hotter than the last temperature you felt is an act objectively done by the things in themselves. If one inch equals one inch, and a ruler is twelve inches long, then the act of being twelve inches long consists of the ruler doing its act of length and one inch being its act of length. The act of being twelve inches long is an act done by the ruler, which has an objective physical existence as a thing of that length, but it is an act which is known because a thing that is one inch long does the act of being one inch long. Thus a ratio is two things each doing an action with one action relating to the other action. A ratio is a relationship between two things, and a relationship between two things actually consists of the two things themselves, with both things considered together and with one thing judged against the other thing. Therefore if two things both exist objectively, then the relationship between them exists objectively, in which case the ratio exists objectively.

If one thing is hotter than another thing then being hotter than the other temperature is consequential of being that temperature, but being hotter than the other temperature is a part of the act of being that temperature, not a second act that being that temperature causes the thing doing it to do. Being that temperature and being hotter are the same act, although the latter is a part of the former. In the same sense, "12" is the same thing as "more than 8", the "is" in these cases being the same "is" as when one says "a square is a rectangle," which is to say, the "is" of is-plus and not is-equal. Similarly, San Francisco being west of New York is consequential of the essence of being in the location of San Francisco, and New York can be used as a relative location to measure the absolute property of the location of San Francisco. Being on the East Coast of the USA is an objective property of New York City, and being on the West Coast of the USA is an objective property of San Francisco, in that these places exist in physical reality and not merely in people's minds, although east is measured relative to west. To return to the temperature example, the measured temperature is the absolute temperature measured by means of the standard, and both the measured thing and the standard exist objectively and have that property objectively, so the measured property is not a subjective thing dependent upon the relationship or created or altered by the act of being measured.

Subpart 26. Plato's "more by less" argument: Plato argues in both the "Phaedo" and the "Theaetetus" that a perceivable contradictory paradox follows from the increase and decrease of numbers of things. Let's deal with the latter dialogue first. In "Theaetetus" he says that, if one takes 6 dice, and adds four dice to them, the amount grew by one and a half, an instance of things growing by halves. Then he says that if one puts twelve next to the six, they are half as many, and they have become half by increase.

219

The paradox, a contradiction according to him, is that things increase by decrease and vice versa.

Plato says that this is a case of the dice becoming greater in number by a process of reduction, and becoming lesser in number by a process of increase. But nowhere in what he describes does that actually takes place. When one adds four dice to six dice one creates a group of ten dice. The four dice become ten dice by six dice being added, and addition is a process of increase. To say that the four have grown by one and a half is the same as saying that the four have grown by the addition of six and that six is equal to one and a half fours, i.e. is equal to 1.5 x 4. Six is one-point-five fours because one four is four and one half of four is two, and two plus four is six. Thus, four growing by one and a half is a process of increase, not a process of decrease. The appearance of contradiction is created by Plato's intentional lack of clarity, which was obfuscation worthy of any sophist. To say that six is half of twelve is not saying that the six have been reduced, since one is not saying that anything has been removed from them, one is only saying that the twelve is composed of two groups of six, and the group of six is equal to one of these groups. That is what "six is half of twelve" means. Nowhere is there any reduction in the perceived increase, and Plato's example of perceived decrease is not an actual example of decrease.

In the "Phaedo" dialogue Plato argues that for a man to be taller than another man by a head would be a contradiction, first because the tall man would be bigger by a head and the short man shorter by a head, and therefore the tall man would be taller and the short man shorter by the same thing, and secondly because the tall man would be taller by a small head. If ten were more than eight by two then it would be made greater by something smaller. And for two cubits to be longer than one cubit by a half is a case of something being greater by half. Each of these is a case of the process of increase containing decrease, and since increase and decrease are opposites it is a contradiction for the same thing to be both increase and decrease.

For a man to be one head bigger than a shorter man means that the height of the tall man is equal to the height of the short man plus the height of one head. For a man to be one head shorter than a taller man is for the small man's height to be equal to the height of the tall man minus the height of one head. Nowhere in this is any change in the height of either man taking place, and both the things I have just described consist of the tall man being that height that he is, and the short man being that height that he is. For the tall man and the short man to be taller and shorter by the same thing means that the length between the tall man and the short man is the same as the length between the short man and the tall man. Nothing is contradictory about this.

If one thought of the head as if it were something that the two men were both "participating" in and each received a different identity from, i.e. a Platonic Form, then it would be strange for different identities to have the same source, but we have no reason to think that way. The tall man has to be taller than the short man by a length that is smaller than the tall man's total height, because that length is the difference between the tall man's height and the short man's height. Again, if one is not thinking only about the physical objects and their being, if one is thinking about the head as something that the tall man "participates in" to get the identity of tallness from, then it would be strange for something shorter than him to make him taller. But the head does not cause him to be taller. Rather the head is merely the difference between the two men, whereas the cause of their being taller and shorter, the only active actions taking place in that example, is each man's act of being the height that he is.

Similarly, two and eight are ten, eight plus two is ten, and for ten to be more than eight by two is the same as two plus eight being ten. To believe that ten being more than eight is something that has its source in the two is to see the smaller as creating the greater, but the only actions that are the cause of all of this is ten being ten and eight being eight. No contradiction exists in the perceived things. The Platonic mistake is to think of being as something that is bestowed by participation in a Platonic Form. The process of increase that adding two to eight is, does not include any process of decrease, and two being less than ten is not a process of decrease, since nothing is removed, rather being less than ten is an act which two always does. No change is involved. Only addition and subtraction are processes of change, whereas the being of numbers is not change. The active action that being a number consists of is the parts of each thing acting to form a whole of the same kind as the other whole units in the group. For example, of a group of ten books the first book does the act of being a book, the second book does the act of being a book, etc. Each number consists of being one book, which consists of one act of being a book. One does not consider the specifics of that action of being one thing when reasoning about essential numbers.

For two cubits to be more than one cubit by half is for the length of two cubits to be greater than one cubit by an amount equal to half of two cubits. The length of two cubits is greater than one cubit by the length that when removed from two cubits leaves one cubit. Nowhere in one cubit being half of two or two cubits being twice as many as one or two halves is there anything that both increases and decreases simultaneously in the same sense. One cubit is one cubit and two cubits is two cubits. It is sophistry to confuse being greater by half with being greater by half and having a half taken away, to equate containing a half with dividing by two. One should remember Aristotle's principle of non-contradictory identity that "a thing cannot both be and not be at the same time in the same respect," with a precise understanding of the "respect" as the thing which other things are measured in relation to. This clears away Plato's confusion.

Plato is relying here upon a concept of "pure being" vs. "mixed being," a theory that implies that to really be something a thing cannot have anything to do with that something's opposites. For example, to be double, according to Plato, is to partake of the "pure being" of doubleness, and such a thing cannot have anything to do with half, i.e. a thing cannot partake of halfness in any way. However, this concept of "having to do with," "partaking" or "participating" does not exist in reality, and implies a very watery and vague notion of being, based on Platonic Forms. Also, the being of this theory is not the being that exists in reality. In reality a thing that is two doubled is half of eight, and there are many other examples of this, and if something is not the being of real things then we need not be concerned with it. Pure being is not being that does not contain anything but that being. This Platonic error confuses the essential X, which indeed does not do anything but be X, with the act of being X. The act of being X is the "pure being" of X, doing the X act is being X, and a thing can do that act and also do any other acts that the nature of being X does not prevent a thing doing X from doing.

Plato's philosophy is rife with mistakes, such as his theory of Platonic Forms. The idea of Platonic Forms originates from Plato confusing the essential X with the essence of X. The essence of X is the act of being X, whereas the essential X is the object of thought which does only the X action and no others. A thing can both really be an X in physical reality and also be the doer of every other action it does, and its being X is no less pure for being whatever else it is, even if everything else is specific with respect to its being X. Plato's error, which eventually leads to the distinction between the being of the ideal things and the being of the perceived things, i.e. between "pure" being and impure being, should be avoided.

In "Phaedo" Plato also argues that "one plus one equals two" contains contradiction, first because they are one and one but not two when apart and two when they come close together, which is weird, and secondly because when one is split it becomes two, and for one to become two both by a process of uniting and a process of dividing is contradictory.

Plato begins by confusing and equating two very different things, the act of physically bringing two things together and the act of mentally bringing two things together. Plato talks about taking two things in each hand and squishing them together, as well as mentally considering one thing, mentally considering another thing, and then considering both of those things together as a whole, as if they were the same act. Any one thing is one thing so long as it continues to act as a whole, and any two things are two things so long as they continue to be two things each acting as a whole. The only act that each one of the essential two things both do together is acting as a thing as such, which is why they are not one but two things.

To take two things and physically bring them closer together in no way makes them be two things. Those two things were two things before you had them in your hands. Two birds are two birds whether one foot or one mile exists between them in physical space. However, when reasoning about essential things, if you consider an essential one, and then add another thing to what you are considering, then you have changed the object of consideration from one thing to two things. The only way to physically create ones, twos, or any number of things, is to cause matter to act as a certain number of wholes. The only reason why physically grouping things together causes them to be counted is that people have a tendency to count things in the same location, so that when you bring two birds into the same room you consider them as a whole and count them as two birds, because two things do the act as a whole of being the birds in this room.

In response to the second objection, I provide a deeper analysis of this in the later section which discusses the "meaning of math", but I will sketch my reply now. To be added or subtracted is to bring things together or separate them. But to multiply by X is to consider a thing to be doing the act of being X, and to divide by X is to remove the act of being X from a thing. When one mentally divides numbers one ceases to consider the thing being the number that one has divided out. For example, 3 x 5 is three fives, which is the same as three being five (being five threes). So when one divides by 5 one removes the act of being 5 from consideration, leaving only the act of being 3. This can be pictured visually: fifteen is XXXXX + XXXXX + XXXXX. When 15 is divided by five the division removes the act of being five from the act of being three. This leaves behind X + X + X which is three.

When one mentally divides one by two, the thing isn't doing the act of being two, so one removes the act of being two from the act of being a whole and is left with a divided whole. A whole divided by two is the negation of being two, which is why 2 times 1/2 is one, being two as a whole negates the reverse of being two, which is a whole being divided into two, and leaves being a whole. And a whole from which the act of being two is removed is one half, because one half being two, i.e. being two halves, is one whole. When one physically divides a whole into two parts, on the other hand, one causes the thing to stop acting as a whole by cutting it through the middle so that the parts on each side cannot act as a whole, and each severed side then acts as a whole and becomes a one, leaving the net result of two ones. No contradiction or paradox exists in either case.

I shall end this by mentioning that, as Aristotle says in his "Metaphysics," when a thing is both half and equal, this is not a contradiction, because it is both half and equal at the same time but not in relation to the same thing, meaning that it is half of its double and equal to what it is equal to. Being half of that which is twice itself, and being equal to that which it is the same number as, are two different acts that it does. One only thinks that this is paradoxical if one believes the doctrine of pure being, or if one is fooled by the vagueness of the terms "half" and "equal," each of which refers to a relationship without stating the second item in the relationship. Indeed, at heart the "more by less" argument relies upon the lack of precision in mathematical terminology to confuse the reader, and is sheer trickery.

Subpart 27. Descartes' "wax" argument: In the second meditation of his "Meditations on First Philosophy," Descartes considers a piece of wax and tries to reason some conclusions from this wax. Descartes says that his piece of wax retains some of the flavor from its honeycomb origin, that it smells of flowers, that it has a color, shape and size that he sees, that it is hard and cold, and that it emits a sound when hit. Then he melts it with fire, and notes that it has lost its flavor, lost its smell, its color and shape have changed, it size had increased, it is liquid and hot, and it no longer makes a sound when hit. His conclusion is that, since all perceivable properties of the wax have changed, and yet the same wax itself is still believed to be present, that the wax itself is different from the perceivable properties which changed, and that the color, shape, size, etc. never were the wax itself. He ultimately argues that nothing is known by means of the senses and that the wax itself is known only by the pure intelligibility of the mind. Thus, as a "trompe l'oeil" argument, every perceivable property is said to be an illusion which deceives the senses, since all perceivable properties change and yet things in themselves remain the same. This is a modern version of the ancient argument that the perceivable world is always changing and is a world of becoming, of which there can be no knowledge, while the intelligible world is a world of being, of which there can be knowledge.

So let us consider the piece of wax and see if Descartes' conclusions can actually be reasoned from the perceived things. Descartes sees some one thing that has a certain taste, smell, color, shape, size, hardness, and temperature. He could easily reason that it is the same one thing with all of the visible properties because they can all be visually identified as one thing. If he takes the thing with that color, shape, and size, and brings this visible thing to his nose and mouth to smell and taste it, and to his fingers to touch it or hit it, then he could, from perceiving the same thing being that color, shape, size, etc., reason that one thing, the wax, was being all of these properties. Before he melts the wax, as he watches it, he will perceive no change, he will see the same shape, color, size, and all the other properties, and the wax will remain exactly the same with respect to being all that it is perceived to be. The only thing that will change is the moment in which it exists, which is non-essential to all of those qualities.

Then Descartes melts the wax, and turns the cold hard wax into a puddle of hot liquid wax, with a different size, shape, color, taste, smell, hardness, and solidity. Indeed, every perceivable property that was perceived is no longer perceived, and a new set of properties is perceived. Descartes is then faced with two questions: (1) does, in fact, the same wax exist, and (2) if the same wax still exists and all of the perceivable qualities have changed, are the wax and the perceivable properties two different things?

Descartes was ignorant of what being and identity consist of, but we are not, and the reasoning of this essay's discussions of metaphysics and essences can be applied here. To know whether the hot liquid puddle and the hard cold block are both wax, one must know what the essential act of being wax is. To know if they are both the same thing, one must know what the essential act of that thing was. Being the same consists of doing the same action, and not being the same consists of not doing the same action, and being different consists of doing different action. This applies not only to two things being the same or different, but to one thing being the same or different over a period of time. Whereas being a certain shape is a property of the form, being wax is a property done by the substance. You may refer back the section on metaphysics where I noted that the form is the thing as a whole and the substance is the thing as separate parts, so that a property of the substance is the parts as the doer of the action that all the parts do individually, and if all the parts of the whole do the same act individually then it is an act done by the thing as the substance.

The essential act of being wax is the act of the substance being a certain chemical compound of molecules, and any substance that does the act of being that particular compound is doing the act of being wax. Both the solid block and the melted puddle are that substance, their substance does that act, so they are both wax. Also, they are the same thing in that it is the same group of particles, of molecules and atoms, which do the act of being wax both as the hard wax and as the melted wax. This is the sense in which the wax is the same thing when hard or melted. However, as the doer of the act of being either solid or liquid, in that aspect, the wax is not the same thing when it is a solid as it is when it is a liquid.

But the second question is whether the piece of hard wax and the puddle of melted wax are, as a whole, the same piece of wax, or whether the old thing was destroyed and a new thing has taken its place. This is the question that Descartes is asking. Although Descartes is not an Aristotelian, when he asks whether it is the same wax or not and answers that it is the same wax, he is asking a question that Aristotle also grappled with. Descartes is not asking whether the chemical substance is wax both before and after the melting, rather he is asking whether some being is in the wax which is independent of its physical properties. And that is where Descartes makes a mistake which dates back to Aristotle.

Aristotle believed that all things consisted of a substance that had attributes predicated of it, and he defined the attributes as accidental to the substance, and the substance as essential to the substance, and said that the substance was what remained the same when the attributes predicated of it changed. One of his examples was Socrates vs. Socrates sitting. Aristotle said that they were both Socrates but not both sitting, because the substance was Socrates and the attribute was sitting or not sitting. The mistake that Aristotle made, from an imprecise understanding of essence, was that the action that a thing does is specific only in relation to its essential action. A specific action is only specific relative to the essential action, and before one has chosen which action to consider as the essence and use essential reasoning on, all actions are specific and no action is more essential than any other, since every action is a part of the thing's total being action. Essence is relative in the sense that one action in a real thing is the essence relative to all other being actions of that thing as specifics. But essence is objectively relative in the sense that it is objectively true that every action as the essential action is the essence relative to all other actions, and what consequence follows from each essence is objective and absolute. To be specific is relative to the essence.

In reality, every real thing is completely specific and every being action is specific, but a specific act becomes an essence when

227

it is analyzed using essential reasoning. The being of the essence exists objectively in reality, and the consequence also exists objectively and as an absolute, so the essence exists in reality and not in the mind, but which action one chooses to consider as the essence in one's essential reasoning is a human choice which is not in the thing. If a red square exists, then it is objectively square and red, and being square will objectively cause it to possess four right angles as a consequence of its essence, but human choice decides whether to analyze the red square in its aspect as a red thing or in its aspect as a square thing.

Aristotle's substance is the thing as the doer of the essential action, but he believes this to be absolute, i.e. he believes that being the essence is a quality of the essential action rather than arising from how the action is considered, and so he believes that one can ask whether the same substance exists before and after the change of predicates, and answer this just by examining objects, without first considering which action one is going to consider as the essence. But this is impossible. Because a thing only remains the same as the doer of a certain action, it is impossible to ask whether a thing has remained the same or has changed without first knowing which action one is considering to see whether the thing still does the essential action or has ceased to do it. The only way to do this would be to ask whether the thing's total act of being had changed, but since being at a certain moment in time is an act of being, everything's total act of being is continually changing.

When Descartes asks "does the wax remain?" he is asking "does the substance of the wax, that which was not its attributes, remain?" and he gives Aristotle's answer, that the substance does remain, and draws Aristotle's conclusion, that the substance of a thing is different from its attributes, meaning that the substance is some sort of substratum upon which the perceivable properties hang. Descartes merely took Aristotle's error to its logical conclusion, by concluding that if the substance is different from the attributes, and the attributes are all that one perceives in the physical world, then the substance cannot be perceived and it is not physical, hence the true substance of a thing's being is some sort of spiritual or magical entity. But that is not what one reasons if one studies the being of perceived things. A person who understood essence would only ever ask "has the thing changed as the doer of such-and-such act?"

Thus, in the wax example, we can ask, is the thing the same as the doer of the act of being wax, in other words, is the wax still wax, and we can answer yes. Or we can ask, is the thing still cold, is the thing still solid, is the thing still that color, that taste, etc., and we must answer no, because the thing that is still there, whose only continuing action is being wax, is no longer doing any of those actions. If it were no longer doing any of the same actions, then it would not be the same thing at all, and we could not ask "is the same thing still so-and-so," because there would be no thing the same as the thing that used to exist. But the thing that does still exist is still wax, and it still exists as wax, not because it has a substance separate from perceivable properties, but because it still does the essential act of being wax, as could be empirically reasoned if one studied its molecular substance by a chemical analysis.

Things only stay the same as the doers of actions, and for a thing to change, for a thing that is the same both before the change and after the change to change, is for it to be the same as the doer of a certain action it continues to do, and to simultaneously be different as the doer of the other nonessential actions that it ceases or begins to do when it changes. To answer Aristotle's question, Socrates and Socrates sitting are both Socrates because each is a thing that does the act of being Socrates, but only Socrates sitting sits because Socrates sitting is a thing doing the act of being Socrates and doing the act of sitting, in other words, is Socrates sitting. One cannot say that they are "the same thing" without qualification. One can only say that they are both Socrates, meaning that they are the same thing as the doer of the act of being Socrates, and they are the same thing only as the doer of that act, but they are different as the doer of the act of being in a position. As a sitting thing, Socrates sitting is different from Socrates as such. The difference arises from what one defines as the essence, and a more essential essence is different from a more specific essence in the sense of not being "is-equal", although a more specific essence "is-plus" the more variable essence.

The "being vs. becoming" argument is to my knowledge the oldest and most influential argument against sense perception. It was given its first popular statement by the ancient Greek philosopher Heraclitus, who argued that everything changes and no stable identity exists. It was used by Parmenides, who argued that perception is illusion and reality is all one single unchanging entity.

And it was embraced and extensively relied upon by Plato, who introduced it into the subsequent development of ancient philosophy. It appealed to religious thinkers, since it implies that the spiritual is eternal while the physical is transient. And it was used by Descartes, who made clever use of it with his example of the melted wax. Descartes introduced being vs. becoming into the development of modern philosophy, and it lives on in the thinking of contemporary philosophers. The being vs. becoming argument is that the perceivable physical world is always changing, that perceived things continually cease to be what they were and become new things, that they are always in a state of flux and in a process of "coming to be," i.e. changing into something, without ever reaching the state of "being," i.e. permanently being something. Because physical objects are always changing one can never perceive the same things and therefore one can never get any stable knowledge of anything perceivable or of physical objects.

The ancient world saw the best example of this argument from Heraclitus, when he said that you cannot step into the same river twice. In light of what I said above, this is easy to refute. If the essential act of being a river is being a flowing body of water that runs through a specified location, then when one steps into a flowing body of water at that place then one is stepping into a river, and if one steps into a flowing body of water at that place again one is stepping into the same river again, and if it continued to be a flowing body of water at that place then it continued to be that river. If one further specifies the river's act of being at a certain place to being a real river at a real location on planet Earth, for example, the act of being the Hudson River, and one steps into a flowing body of water at that place, and then steps into a flowing body of water at that place again, then one has stepped into the Hudson River twice.

It is true that one does not step into the same molecules of water twice, for the water one first steps into has washed downstream when one steps into the river again, and indeed the thing one steps into differs with respect to many other things, in any way that its action changed between the two moments when one steps into it. But no matter the different ways in which it changes, the Hudson River stayed the same as the doer of the act of being a flowing body of water at that place. Things stay the same as the doer of actions, and if a thing continues to do the essential action, then it continues to be that something. Permanent identity exists as the doer of an action the thing continues to do. When it continues to do an act it is the same thing as the doer of that act, and when acts specific to that act change, it is a new total thing, but still the exact same something that does that act, i.e. it is the same thing as the doer of that act.

That concludes my refutation of the trompe l'oeil arguments.

Chapter Thirty Five: Perception: Refuting the Circular Argument

I will conclude my discussion of perception with my refutation of the final kind of argument against certainty in the senses that I have encountered. This is the argument from the hypothesis that all perceived things are illusions. I call this the "infinite doubt loop" argument, for reasons that shall become apparent. This argument was given its first famous statement by Descartes, so let us consider his formulation of it. According to Descartes, it is possible that the whole world is your dream. If it was your dream then you would have no way of distinguishing it from the waking world. According to him it is also possible that every perceived thing is an illusion put before your mind by a demon intent on deceiving you. If this was the case then you would have no way of distinguishing perceived reality from a perceived illusion. For both of these reasons it is possible that all perception is deception and all perceived things are illusions, and one can never be certain of the reality of perceived things. One could even extend this beyond what Descartes presents, and say that if it were possible to reason from perceived things that perceived things are real and exist objectively, that one could not trust such reasoning, because the perceived things from which one reasoned could have been illusions, and one could never distinguish these original perceptions from illusions by means of reasoning from perceived things.

Both Descartes' original argument and my extension are loop arguments. A loop argument, which is also called a circular argument, is an argument that begins with an unreasoned assumption (usually about possibility, but not always), reasons a conclusion from that assumed premise, and then uses the conclusion to justify belief in the initial assumed premise. Thus it is a sophisticated technique that gives circular reasoning the appearance of rational deduction. A loop argument is a trick which, if successful, leaves the person convinced that the original assumption has been proved to be true, so that the assumption is never questioned. The only way to refute a loop argument is to first identify its structure as a loop, and to then refute its original assumption. If the assumed premise is refuted, then the conclusion is proved to be invalid.

Let us look at the structure of Descartes' argument. Descartes' first argument is that all life could be a waking dream. Descartes' second argument is the possibility of a demon placing illusions before the senses. The structure of this argument can be analyzed by identifying the line which the points of the argument seem to form, in which a valid conclusion is reasoned from an acceptable given premise, and then identifying the circle which the points of the argument really form, a circle that begins with the first premise (the assumption), proceeds to the conclusion, and then jumps back to the first premise to justify its origin with its conclusion. The argument, in its own terms, could be described as a series of points in this way:

0. We may begin with one definition, that an illusion is something that is perceived but does not exist.

1. What if all perceived things were illusions placed before the senses by a demon or by a dream?

2. Then all perceived things would be illusions.

3. But many people believe that the things they perceive are real.

4. Then those people would believe that perceived illusions are reality.

5. Then those people could not tell the difference between illusion and reality from perception.

6. But those people were defined as people who believe that what they perceive is reality, and anyone could do that, so this must be true of all people.

7. Then one could never have any certain knowledge of perceived things, since one could never know whether the perceived thing was real or an illusion.

8. Then one could never deduce any certain knowledge of reality from perception, since one could never tell whether any perceived thing that one reasoned from is real or an illusion.

9. The conclusion: One can never be certain that what one perceives is not illusion, nor reason certain knowledge of reality from perception.

Thus, by reasoning from a hypothesis that one can easily imagine as a realistic scenario, one can reason that it is impossible to derive knowledge of reality from the senses, that one can never be certain that what one perceives exists in reality, that one cannot distinguish reality from illusion in perception, and that the possibility of a dream or a demon deceiving the senses is possible and realistic. Doubt of this kind seems to be a very rational position as a result of the supposed logic of the deduction.

Now, let us examine the structure of the argument as a circle, by accurately identifying the premises of each point:

1. It is possible for the senses to deceive the perceiver.

2. It is possible for a human to perceive an illusion.

3. It is possible for a demon or a dream to cause the senses to deceive the perceiver.

4. If one perceived thing could be an illusion, then everything that is perceived could be an illusion.

5. If a demon or dream were causing the senses to deceive the perceiver about all things, then all perceived things would be illusions.

6. Many people believe that the things they perceive are real.

7. If perceptions were illusions, then those people would perceive illusions and believe them to be real.

8. Then those people could not tell the difference between illusion and reality from perception.

9. But those people were defined as people who believe that what they perceive is reality, and anyone could do that, so this must be true of all people.

10. If all perceived things were illusions put before the senses by a demon, then one could not tell the difference between reality and illusion from perception.

11. If all perceived things were illusions put before the senses by a demon or a dream, one could never know whether perceived things are reality or illusions (except on the basis of a priori "pure reason" based on a world of spirit or non-physical intuition, as distinguished from a posteriori empirical reason).

12. It is impossible to ever know whether perceived things are illusions or not, and if one does not know whether something is true or not then both options are possibilities, therefore it is possible for all perceived things to be illusions put before the senses by a demon.

13. Therefore it is possible for a demon to cause the senses to deceive the perceiver.

14. If this is possible then it is possible for humans to perceive illusions.

15. Therefore it is possible for the senses to deceive the perceiver.

And this goes back to the beginning and justifies the original premises.

A great deal must be discussed before we can return to this structural analysis and use it to refute the argument. All reasoning consists of a series of reason-computations. Each reason-computation consists of taking one or more things which are known and reasoning something from those one or more things. The thing that is reasoned is said to be deduced from the one or more things it was reasoned from. Knowledge of the deduced thing is based solely upon knowledge of the things from which it was deduced. (I shall explain how the brain does this in detail later). The things that one reasons from are called premises, and the thing that is reasoned from the premises is called the conclusion. If the premises are true and the reasoning is valid then the conclusion is true. A series of reason-computations forms a line of reasoning, with each individual reason-computation being one point in the line. A reason-computation, which deduces a conclusion from a set of premises, can have as its premises three things: (1) one or more perceptions, or (2) one or more concepts, or (3) both perceptions and concepts. Each concept can be either a reasoned conclusion of prior reasoning or an assumption if it was not the result of prior reasoning. Note that when I say perceptions and concepts, this actually refers to unconscious reasoning, whereas for conscious reasoning the equivalent of a perception is a perceived thing, and the equivalent of a concept is a thing that one is aware of or thinks about or judges.

Note also that I do not posit that reasoning works like the so-called "formal symbolic logic" of the Analytic philosophers, who fill pages with written symbols that have no meaning and assert that logic should process such symbols using tools similar to computer software. Here I am concerned with thinking as it is done in reality by human brains, and not with abstract disembodied logic nor with computer software based on Boolean operators that compute "and" "or" "if then" and "not". As I shall discuss later, unconscious reasoning might work similarly to computer software, but conscious reasoning is concerned with things, that is, with the things that we perceive and are aware of and to which our words refer, rather than being merely software processing a set of words, i.e. symbols, according to linguistic or "logical" rules. The range of the ways that a reason-computation can deduce a conclusion from a set of premises includes every valid manner of thinking, such as every deduction that applies empirical essential reasoning to deduce the consequence of the essence.

For example, a reason-computation could be addition or multiplication or another mathematical function in math, or reasoning from empirical data and the results of lab experiments in science, or figuring out how to open a can of food with a can opener by studying the can and the can opener and experimenting with different ways to try to use it to get the can open. Abstract intellectual reasoning is a real-world practical tool, in that the difference between theoretical reason and practical reason lies not in the nature of the reason but only in to what end it is used. Reason as it exists in real human brains is our concern in this philosophical exploration.

An assumption is something that is reasoned from that was itself not reasoned, which is to say, an assumption is an unreasoned premise. If the assumption is true, i.e. if the thing assumed to exist does in fact exist, then the conclusion is true, i.e. the thing that was reasoned exists. And if the assumption is false, then the conclusion is false. Because the assumption is not reasoned, one has no way of knowing whether it is true or false and therefore one has no way of knowing whether the conclusion is true or false, and so one cannot be rationally certain of reasoning from assumption. On account of the fact that it can produce a belief in things that do not exist, and that its conclusions admit of no certainty, I say that reasoning from an assumption corrupts that line of reasoning, and the name I have given to reasoning from assumptions is "corrupted reasoning."

A distinction can be drawn between reasoning that reasons from assumptions and reasoning that does not reason from assumptions. All reasoning is one or the other of these two kinds. As I shall subsequently explain, all reasoning that is not from assumptions is purely from empirical experience alone. Unconscious reasoning consists of reasoning concepts from sensory perceptions without the use of any concepts, and then reasoning other concepts from those concepts reasoned purely from perceptions, reasoning concepts from a combination of perceptions and concepts, and, ultimately, it ends with reasoned concepts being used to identify the objects of empirical experience, i.e. perceived things. Conscious reasoning consists of reasoning things from perceived things to conclusions consisting of things that one is aware, and then further reasoning other things that one is aware of using things that one is aware of and perceived things as premises. Because reasoning from perceptions and perceived things is purely from empirical experience and not from any other sources, I call it "pure reasoning," although it can also be called "pure empirical reasoning." One can begin to see that reasoning from perception and reasoning from assumption are different in kind, for the former admits of certainty and cannot be wrong, but the later cannot possibly admit of certainty and must always have the possibility of error.

I must discuss a few other points before I can give the full line of reasoning to certainty in empirical essential reasoning. Philosophers typically draw a distinction between "deduction", which deduces a conclusion from premises by means of reason and logic, and "induction", which analyzes empirical experience to determine whether an initial premise is true or false. Philosophers, especially Rationalists like Descartes and Analytics like Wittgenstein, focus their attention on deduction, and they don't care about induction, nor do they have a good theory to explain it, because they don't really believe that knowledge can come from empirical experience. My theory does not draw a sharp distinction between deduction and induction, because all logical deduction is based on premises that were ultimately reasoned from perceived things, so the only difference between induction and deduction is that induction reasons conclusions directly from perceived things, and deduction reasons things from premises that were themselves reasoned from perceived things, or reasoned from premises reasoned from other premises reasoned from perceived things, and so on.

Regarding how induction works, I will discuss unconscious reasoning elsewhere, but for conscious reasoning, one sees the perceived thing, and one reasons from the thing that one perceives. For example, if one sees a red apple, one can reason that it is red from seeing its color, and one can reason that it is an apple from its color and shape. Elsewhere in this book I have explained how science can reason from perceived things, by using the scientific method, and also by applying essential reasoning to perceived things. I will not repeat that discussion here. I will however note the example that if one cuts the apple open and looks at its seeds and studies its flesh one can reason that the apple is a fruit, and one could then reason from the perceived thing that one can plant it and grow an apple tree.

If one begins with an assumed premise and later reasons from perception that the assumption is true, then the assumption is no longer unreasoned. Once the assumption is reasoned, it becomes a reasoned premise and the line of reasoning from the original perceptions to the conclusion reasoned from the assumption becomes pure, so that the conclusion is ultimately reasoned from perception and becomes certain. If something is reasoned from an assumption, and is not based in empirical experience and deduced from perceived things, then the conclusion should not be believed. When faced with corrupted reasoning, the only thing to do is to reason from sense-perceived things whether the assumption is true or false, so that one can have certain knowledge of whether the conclusion is true or false. The only rational response to an assumption is to use pure empirical essential reasoning (PEER) to judge its truth.

I have distinguished corrupt reasoning, which reasons from an assumption, with pure reason, which reasons entirely from perceived things, i.e. empirical experience. In the prior two sections on perception, I have already offered the lines of reasoning that enable a thinker to reason from the perceived things as premises to the conclusion that perception provides accurate truthful knowledge of reality and that perceived things exist objectively. As I will discuss below, if one begins with perceived things, and one reasons from the perceived things that perception cannot lie or deceive, then one will have reasoned that perception is truthful. This line of reasoning then knocks out the assumptions at the beginning of the Cartesian demon/dream infinite doubt loop argument, which refutes the loop argument by refuting its assumed first premises. If one can reason from sensory perceptions, without assumption, that all perceived things exist "as a consequence of their essence" (which is the more precise term for what people mean when they say "by definition"), then one can reason that all things reasoned directly from perceived things are true, since what they are reasoned from must be true as a consequence of their essence.

I deal with two loop arguments in this essay, of which the first is the "infinite doubt" loop, which reasons from the possibility of the senses to deceive the perceiver to the conclusion that perception cannot be trusted. The other is the "infinite ignorance" loop, which similarly reasons from the possibility that reason could be wrong to the conclusion that reason cannot achieve certainty and knowledge. Both loop arguments rely upon the implicit view that they are questions which must be answered before one can trust in the senses or reason, and that, for this reason, they both deserve pre-rational consideration.

In other words, the advocate of the loop arguments asserts that his challenge to perception (and reason) must be answered before one may be allowed to reason from perception, since the validity of perception rests upon the answer to his challenge. But this is merely the same loop argument since his view assumes that perception is uncertain until his questions are answered. He reasons his argument from the possibility of the senses to deceive, but the possibility of the senses to deceive is the very assumption in question. One must reason that perception is capable of making mistakes before this can be believed. The demand for pre-rational consideration is merely a loop argument. To begin with pre-rational consideration of the question of perception is to make an assumption. In contrast, to begin with reasoning from perception and then use that reasoning from perceived things to answer such questions is to begin with no assumptions because a perceived thing is not an assumption. Perceived things are not reasoned, they are perceived. Perceived things are self-evident in the sense that a perceived thing is itself the evidence that it is what it looks like, i.e. that it is what it is known to be through perception. Reasoning purely from perceptions without any assumptions is purely free of assumption.

Let me note here my response to the objection that to reason from perceived things implies that reasoning from perceived things is valid, and therefore even the original reasoning to the validity of perception from perceived things has assumptions. This counterattack relies upon an ambiguity between the premises of reasoning and the premises of actions. The premise of an action is that thing which must be true for a certain action to be right, while the premise of reasoning is a thing that is reasoned from. The act of reasoning from perception has premises as an action, but these are not the premises that it reasons from and so these premises of the action are not assumptions.

When one says that a person who reasons something from perception, without reasoning from anything besides perceived things, is assuming that perception is correct, this is not true, since the correctness of perception is not being reasoned from, only the perceived things themselves are being reasoned from as premises. The reasoning would not be correct unless it is true that reason and perception are valid, but the premises of the reasoning itself are the perceived things, since these are what the conclusion is deduced from, and the thought-about things consisting of the essence of reason with a consequence of being correct and the essence of a perceived thing with a consequence of being a real thing are nowhere in the premises of this line of reasoning and are not what was reasoned from. Indeed, those things are the conclusions reasoned from the perceived things, and they are not the premises which are reasoned from. The accuracy of perception and reason are premises of this reasoning as an action, i.e. the reasoning would be wrong unless reason and perception are accurate, but they are not premises of this line of reasoning as reasoning, i.e. we reasoned from an apple, a rock, dominoes, and other perceived things, and not from the abstract ideas of reason and perception.

Here I should mention the big difference between my epistemology as compared to the epistemology of Aristotle, and his intellectual heir, Ayn Rand. They believe that reason begins with first principles called axioms, that the axioms are "self-evident," and that everything can be deduced as a matter of logic from the axioms. I do not believe in axioms. Instead, I believe that perceived things (or generally, empirical experience) are the only things that can properly be called "self-evident." A perceived thing is self-evident because you can directly perceive that it is what it is. For example, if you look at a red apple, you can see that it is red, therefore its being red is self-evident. You can also see that it is an apple, so that is self-evident. And, using the lines of reasoning offered above, you can see from the red apple's perceivable properties that it exists objectively, so the red apple as an existing thing is self-evident. Thus, my philosophy is deduced entirely from perceived things.

As I have said elsewhere, it is possible to reason from perceived things to the principle that perceived things exist and to the infallibility of perception. My reasoning begins, not with an assumption, but with perceived things. Therefore my reasoning is pure. In fact, my reasoning is pure empirical essential reasoning (PEER). On the other hand, an axiom is an assumption, and the axiomatic propositions are not actually self-evident, so Aristotle's and Rand's lines of reasoning are corrupt, despite their good intentions of seeking to give a philosophical foundation for science and empirical knowledge. Rand's axiom of "existence exists" is not exactly the same thing as seeing the things in the physical world and seeing that they exist, because one can only see specific perceivable existing things, whereas the general principle that existence exists is an abstract idea (or, to speak precisely, it is a thing one is aware of which is the object of an idea, or it is an essential thing). I do not dispute that Aristotle's and Rand's axioms are true; for example, I believe that it is true to say "existence exists" (or, as I like to phrase it, "existence exists objectively"), and it is true to say that "a thing cannot both be and not be something at the same time in the same respect," which Rand stated as the axiom "A is A." What I dispute is whether these statements asserted to be axioms are self-evident. The axiomatic statements are not self-evident, and one does not know them unless and until they are proven by reasoning from perceived things.

My reader may find it strange to think that one would engage in my program of reasoning from perceived things without already having some idea that reason and perception are valid, since one would not reason from perception without some basis for thinking this is the right thing to do. This confusion is cleared up by upcoming sections, where I explain that the reasoning of first concepts from perceptions, which I call "first reasoning," happens in reality in the brains of infants and young children, during a phase which I call "the birth of reasoning." The brain of an infant is empty, i.e. is a blank slate, but you, my reader, already have a brain full of knowledge reasoned from experience, and when I try to persuade you I seek to appeal to your reasoned knowledge, not to ask you to ignore it. I am not asking you as an adult to start fresh from scratch with an empty mind and choose to reason from perceived things, since the conscious reasoning to understand reality inevitably builds upon the unconscious reasoning which you did as a child, and hopefully you are already a smart person with some concept of reality capable of seeing the truth in my theory. Let me repeat, mine is a philosophy based on reality as it really exists, and I do not bend or twist your knowledge of the world or your grasp of reason which you already understand in order to fit my ideas. My concern is always with avoiding external contradictions, and I think that my theory of how one reasons that perception is truthful and how one can refute the demon/dream argument by means of using pure reason to refute the loop argument's assumption is an accurate reflection of how people can really think and what exists in reality.

We need to discuss only one other topic here, which is the question of what is possibility. When one says "possibility" one refers to one of three things. These three have only their names in common, and different words for each should be created so that we can speak with greater precision. I have taken to calling them "metaphysical possibility", "mechanical possibility", and "ignorant possibility". Generally, "possibility" is potentiality. For a thing to be possible is for it to be able to be, and for a thing to be impossible is for it to be unable to be, and ability is that which a thing can do. One can say this of all three kinds of possibility, but the words will mean different things in each case, so that nothing I have just said has a distinct meaning without reference to which kind of possibility one is talking about.

It is metaphysically impossible for X to be Y if a consequence of the essential X action is that a thing cannot both do the X action and the Y action. It is metaphysically possible for X to be Y if it is not metaphysically impossible, which means that it is metaphysically possible for X to be Y if the essence of X allows or permits X to also be Y. It is metaphysically possible for X to exist if the act of existing does not have as its consequence that a thing cannot both do it and the X action. Likewise it is metaphysically possible or impossible for a thing to both do an act and any other depending on the consequences of the two essences, e.g. the metaphysical possibility of X being Y depends on $X(e)$ and $Y(e)$ and $XY(e)$. Whatever the consequence of the essence leaves as possible is metaphysically possible. An example of this would be that it is impossible for a physical thing to be in two different places at the same time.

Mechanical possibility is what it is actually physically possible for a thing to do, how it is possible for it to respond, and what it is possible for it to become, as a consequence of the actions that it is doing in the immediate present. One example of this is that it is possible for you to read these words, since the actions that you are doing give you that power. An example to distinguish these would be if, back in ancient Rome, one had said that it is impossible for humans to fly. For humans to fly was, at that time, mechanically impossible, because the Romans had not invented airplanes, but it was not metaphysically impossible at the time, since the essences of being human and flying are not mutually exclusive, the Romans could have flown if they had invented an airplane, and it later became mechanically possible for humans to fly with the creation of the airplane. What is mechanically possible is always metaphysically possible, because if the thing that is mechanically possible had contradictory essences rendering it metaphysically impossible then it could not be what it is. In other words, metaphysical impossibility causes mechanical impossibility. But what is metaphysically possible is not always mechanically possible, because for a mechanical possibility to exist the thing with that power to act must actually physically exist at that place and time, and not all things necessarily exist at all places and times (in fact, only space and time necessarily exist at all places and times).

The third kind of possibility is quite different from the last two. For an example of ignorant possibility, if you ask someone "Will it rain?" and they answer "It's possible," they do not mean that rain is a metaphysical possibility, or that the clouds are capable or raining, rather they mean that they do not know whether it will rain or not. Ignorant possibility is what one refers to when one says that "it is possible" and one means that one does not know whether the fact exists or not. If it is mechanically possible for a thing to do X, but it has not done it yet, but it could do it at any time, then you don't know whether it will do X in the future or not, and so ignorant possibility in the future follows from mechanical possibility in the present, but not vice versa. Therefore one can reason ignorant possibility from mechanical possibility, but one cannot reason mechanical possibility from ignorant possibility.

If one doesn't know whether a thing is X or not, and if it is metaphysically possible that it is mechanically impossible for the thing to be X, and one is ignorant of whether X is mechanically possible, and one does not know whether a thing is X or not as a result of ignorance, then ignorant possibility would follow from sheer ignorance, and not from knowledge of mechanical possibility. When one says that something is ignorantly possible, one implies that it is mechanically possible, since one is saying that the thing could mechanically do what is possible, but one does not know whether it will or won't do it. But this implication does not make ignorance a basis for knowledge of mechanical possibility.

To return to the rain example, if a person is asked "Will it rain?" and he replies "It's possible", and at the time he believed that the skies were mechanically capable of raining, but he was actually wrong and the air was too dry to rain, then it will be true that the man does not know whether it will rain or not, so that he statement will be an accurate expression of his ignorance regarding whether it will rain, which was the meaning that he intended to communicate, even though it will be false that the mechanical possibility of rain exists. To be precise, when a person is asked "Will it rain?" and he replies "It's possible", he may actually have one of two meanings, either (1) "It is mechanically possible for rain to fall from the clouds in the sky, although I don't know whether it actually will rain or not," or (2) "I don't know whether it will rain or not." The latter is a common meaning which probably evolved from the former meaning. This latter meaning is what I mean by ignorant possibility.

So one can reason metaphysical possibility from mechanical possibility, but not mechanical possibility from metaphysical possibility, and one can reason ignorant possibility from mechanical possibility, but one cannot reason mechanical possibility from ignorant possibility. What was said of possibility goes for impossibility also, since impossibility is the absence of possibility. Regarding the ignorant possibility of X or not X, if one learns which it is, for example if one learns that it is X, the ignorant possibility ceases to exist, since it was merely one's ignorance of which it was. If a thing is X then it is only possible for it to be X and it is not possible for it to not be X, both metaphysically and mechanically, for a thing cannot both do and not do an action at the same time in the same respect, because doing the action would cause the thing doing the action to not not do the action as a consequence of doing the action.

Now I must explain how this applies to the infinite doubt loop. Note that the strategy to refute a loop argument, i.e. an argument which justifies its assumed first premise by means of its conclusion, it to identify the first premise as an assumption, and then refute the assumption to invalidate the conclusion. Look back at my diagrams of the infinite doubt loop argument. At point 1 in my second diagram, and also at point 1 in my first diagram, the argument begins with the trick of confusing the mechanical possibility of deceiving the senses with the ignorant possibility of deceiving the senses, and seeking to deduce the mechanical possibility from the ignorant possibility. It is plausible to begin an inquiry such as Descartes', which philosophically explores whether the senses tell the truth or lie, by stating that one does not know whether the senses tell the truth or lie. But to equate this initial ignorance with the knowledge that the senses are actually physically capable of deception is an irrational inference. In such a philosophical exploration, one should accept nothing without reasoned proof in support of that which one accepts. What is at stake is whether it is mechanically possible for the senses to deceive the perceiver or not, but the argument makes no distinction between the three kinds of possibility and ignores the question of mechanical possibility. It implicitly equates mechanical possibility with ignorant possibility from the very beginning.

With respect to the question of the mechanical possibility of the senses deceiving the perceiver, if one can reason that perceived things exist from the perceived things themselves, and if the perceived things are self-evident, which means that they are their own proof that they are what is perceived and they are the evidence that they have the properties from which one deduced that they objectively exist, then one can reason that it is impossible for perceived things to be illusions. One can reason from perceived things that they are what they are perceived to be directly from perception, e.g. if one sees a red apple one can reason from the red apple that it is red. One can reason that perceived things exist objectively in physical reality from the perceived things themselves using the various methods which I described earlier, e.g. the closed eye argument that not seeing the apple when one's eyes were closed did not destroy it, therefore it exists outside of the eyes. One can reason that perceived things exist consequentially of the essence of being perceived from the means argument, e.g. the apple is what you see and you see it by perceiving it so it must be the thing shown to you by your means of perception, hence it must be that object which you see and could not be the means which showed it to you, therefore what you see is what you see and the seen thing cannot be composed of the act of seeing or the means of sight.

In other words, when you look at an apple you see an apple, you do not see your own eyes or your own brain, and this is proved because the apple looks like an apple not like eyes or brains. One can also reason using the argument from the act of perception, e.g. your finger tracing a line in space from the apple to your eye shows that the apple exists outside of your head. And one can reason that perception is infallible from perceived things existing consequentially of being perceived, because if the act of being perceived requires a thing to exist then all perceived things must exist, and it is therefore impossible for a perceived thing to not be real, i.e. to not exist. One can reason that to be perceived requires the thing to do the act of existing from the means argument and the argument from the act of perception and from the closed eye argument, because to be objectively in space and time is to exist, and a thing must be objectively in space and time to be perceived, in other words a thing's existence causes it to be perceived, it is perceived because it exists, so existing is a consequence of the essence of being perceived.

Descartes' argument begins with an assumption, that it is possible for the senses to deceive the perceiver, which he derives from mere ignorant possibility and not from any reasoned proof. Once one has reasoned that the original assumption that was made by Descartes, the first premise of his argument, is false, then one can reason that the rest of the argument is false on the basis that the truth of every later conclusion was reasoned from the first premise.

The inability to tell the difference between reality and illusion from perception is meant to prove that rational 100% certainty, i.e. knowledge, in the existence of perceived things is impossible, and this in turn justifies doubt in the senses. Thus a conclusion drawn from the original assumption, that it is possible for the senses to deceive, is meant to prove that one can never know whether the senses deceive or not, from which one can conclude that it is possible for the senses to deceive. But this does not prove that it is possible for the senses to deceive, since there would need to be a basis for knowing that prior to the first assumption, to justify what is reasoned from the first assumption, otherwise what the argument deduces loops back to the first assumption. And, since the conclusion constitutes its only proof of the first assumption, if the first assumption were wrong then the whole circle of points would all be wrong.

In the physical world as known by science, the rod and cone nerve cells in the eyes' retinas are sensitive to rays of light reflected off of objects that physically exist. As known by science, perception brings to the mind's eye a world that objectively physically exists in space and time external to the subjective mind. The rejection of knowledge from empirical experience made sense in ancient eras, but not in our modern age where we benefit from the study of physics, chemistry, biology, etc. Science, having understood the means of perception of the five senses such that it is known that sensory experience shows an objectively existing world, should be free to rely on sensory experience in order to discover truths and achieve knowledge.

It may seem counterintuitive to say that the question of whether perception is true or false assumes that it is possible for perception to be false, and that attacking this assumption is the basis for proving that perception is true. If one feels this way, one should know that my argument can be phrased in one of two ways. I could say that Descartes uses an assumption that is false, or I could equally say that I do not use any assumptions. I have deduced from perceived things that perception is true, and the perceived things are self-evident, therefore I have not assumed that perception is true when reasoning that perception is true. In that sense I cannot be challenged for having made a faulty assumption in my premises. Or I might say that it is perfectly proper to begin with the ignorant possibility of perception, but to then say that sensory deception is mechanically possible or metaphysically possible assumes the conclusion, as I have argued.

Note that it would be a misinterpretation of my argument to say that I do not allow people to wonder whether the senses deceive the perceiver or not. One is allowed to wonder, but one's wonder should be satisfied by a strictly rational analysis. One should be allowed to say that one does not know whether perception provides truth or lies at the outset of one's philosophical inquiry. But if my critics want to wonder about perception, they still could not make any claim to reason and logic if they attack my deduction of knowledge from perception by saying that I assumed that perceived things exist. I make no assumption, I assume nothing, and everything that I deduce is reasoned from the self-evident, immediately perceivable properties of perceived things, which is an empirical experience that all humans share, so everyone can see the same things that I see and make the same inferences that I make.

If my critics, the Cartesians, Kantians, and other Rationalists, begin by assuming an evil demon who places lies before the eyes, and then use this assumption as a basis to doubt all of perception, then they never gave perception a chance. This idea of a deceptive spirit or an all-encompassing dream that attacks empirical reasoning comes from a bizarre world of magic that exists outside the realm of the objective physical world as known by science and as seen by perception. Science does not accept evil spirits, and dreams happen when the brain is asleep, not when it is awake and receiving nerve signals from the sensory organs. If one chooses to believe in such fantasy imaginations as the basis of one's philosophy, then one can make no claim to being rational or scientific.

The reason why I call Descartes' argument the "infinite doubt loop," and the strategy for refuting loop arguments that I have developed, are both evident from the above. Before I move on, I must note that Descartes' whole grand mission of establishing a foundation of certain knowledge, which influenced every modern (and Postmodern) philosopher who came after him, consisted of him beginning by doubting the senses and every other proposition that he was capable of doubting, and then trying to find those propositions which it was logically impossible to doubt. Ultimately he concluded that he could not doubt "I exist", because he exists to think thoughts, and he could not doubt "God exists," because God created him to doubt. But why does he begin by doubting? What rational justification does he give for believing that it is possible for what he doubts to be false? The few arguments he gives for uncertainty in the things that he doubts are all flimsy. It is implicit in what he says that he does not believe that one must reason that something is possible before one reasons from that possibility. This enables him to reason from the hypothesis that nothing is certain and everything is doubtful without justifying this premise. But what this means is that the entire Cartesian philosophy is reasoned from unjustified assumptions regarding possibility. The Cartesian plan for the attainment of reasoned certainty by means of the methodology of doubt is founded upon the shaky foundation of assumptions and cannot possibly produce certain knowledge of anything. Doubt, like any other idea in a rational philosophy, must be reasoned in order to be valid. In other words, the so-called Rationalist philosophy is actually quite irrational.

Indeed, the Rationalist Cartesian arguments for the knowledge of self and God, e.g. "cogito ergo sum," all stem from a mind turned in upon itself, a mind looking only at its own thoughts, rather than a mind turned outward and looking out at the external reality seen in our experience of the world. The Solipsist, Subjectivist Descartes' so-called "reason" is completely different from how reason really works, e.g. true scientific reason would see that "I" refers to the body I see when I look in the mirror or examine my hands, and would conclude "I exist" from observing myself as a body and a brain acting in physical reality. The mind which looks only at itself will be insulated, perhaps immune, from testing any of its ideas against empirical experience of the external world by means of the scientific method, so one would expect an anti-science philosophy to result from the Cartesian Rationalist method.

A similar critique, namely, Solipsist Subjectivism, can be leveled against Kant and Kant's desire to derive the principles of science from the mind's internal subjective thoughts, i.e. Kant's "pure reason" is what the mind creates and imposes upon experience, instead of looking at the external world outside the mind to discover the sources of scientific knowledge. Descartes and Kant may perhaps be forgiven, because they lacked the work done by Mortimer Adler, and only the Objectivity Proviso solves the Solipsism Subjectivism problem. If the means of perception are assumed to alter or create what the mind's eye perceives, then perception could offer only subjective knowledge, and some source of knowledge other than perception, i.e. mystical insight or intuition, would be necessary to access objective reality. Only if empirical perception directly experiences objective reality and things in themselves do we become able to say that we, as physical human beings, can look at objective existence and know reality.

This concludes my sections on perception. The conclusion of the reasoning which I have presented above is that perceived things exist objectively in physical reality, that sense-perception is a means of direct knowledge of reality, that the senses are infallible and incapable of deception, and also that the act of perception and the means of perception act in the way I have described. The question of the subjectivity and deceptiveness of the senses has been answered. And the answer that the senses yield knowledge of perceived things which exist and are objective consequentially of their essence has been reached. Yet the questions of how to reason concepts from perceptions, what are "concepts", what is the entire scope of what we mean by "reason" in the sense of deducing things that are thought about from perceived things, and whether or not conceptual reasoning admits of certainty, remain for us to think about.

Chapter Thirty Six: The Birth of Reasoning: Words and Definitions

Aristotle developed a popular theory that has dominated philosophy on the topic of how reasoning works for the past 2000 years. His theory is that all demonstrative reasoning begins with axioms and definitions. In this model of how reasoning works, a set of facts are "given," that is, assumed to be true, things are reasoned from the given facts using the rules of logical syllogisms, and all conclusions are ultimately reasoned from a set of self-evident first principles called "axioms" and a set of definitions which are true by definition. Examples of this theory can be seen in the methodology of mathematical and geometric proofs, as well as modern theories of logic. What this theory means is that all reasoning begins with assumptions. I argue that Aristotle's theory of reason is incorrect, because nothing reasoned from an assumption can be certain or constitute knowledge. The error of this theory becomes clearer as one examines it in detail. Let us first look at this theory's understanding of what propositions and definitions are, and then turn to the study of axioms.

According to Aristotle's theory, a proposition is a sentence that affirms or denies a predicate of a subject. Having already dealt with the flawed theory of predicates and subjects, I could rephrase this and say that a proposition is a sentence that says that a thing is or is not doing something. But, in light of what people are actually talking about when they talk about propositions (and they talk the same way about "facts"), that would still be wrong. For example, when one says "lead is heavier than water," or "every point in a circle is equidistant from its center," those are what one would call propositions. But if asked what a sentence is, one would say that it is a set of words. Yet when one reasons from the so-called propositions, one is reasoning from lead being heavier than water, or from a circle being every point equidistant from a point. One is not reasoning from or about a set of words, one is reasoning about things doing actions.

Whether one is reasoning about essential things or real things makes no difference, since if one considers the proposition "Mary wore a red dress," it is still the girl Mary doing an action that you are considering, not the word that names the girl. This mistake is dependent upon a misunderstanding of the nature of words which fails to understand the difference between a word and that thing which a word shows to the listener's or reader's mind. The mistake arises from the general ignorance of the act of showing prevalent in philosophy. A word is that which shows a thing, and the thing is what the word shows. Similarly, a sentence is a group of words which show something, and the thing is what the sentence shows. Words show by means of an association between the word and the shown thing such that the word causes the brain to bring awareness or memory of the thing to the mind's eye. The association between a word and a thing is arbitrary in its origin but makes sense in the context of the rules and sets of words that form an entire language.

The reasoning which we do has things, not words, as its objects (unless we are specifically reasoning about words, such as when interpreting a foreign language). Things, and not words, are the objects of essential reasoning as I described it, and empirical essential reasoning is the kind of reason with which this essay is concerned. Essential reasoning is the kind of reasoning that demonstrative reasoning consists of. According to Aristotle's theory of how reasoning works, we would be reduced to the absurdity that reason is dealing only with sentences, rather than with things. Some logicians, such as Wittgenstein and his followers, have exploited Aristotle's erroneous belief and come up with the theory that the necessity of the conclusion, which is to say, that which draws the conclusion from the premises, is merely the manipulation of symbols, i.e. reasoning has words, not things, as its objects. In other words, "Mary wore a red dress" is necessary not if Mary actually wore a red dress in physical reality. Rather it is necessary only if "Mary" is a word which we defined to mean "a girl wearing a red dress." For Analytics such as Wittgenstein, and also according to a certain interpretation of Kant, necessity is always "by definition", and comes from language, instead of necessity as a consequence of the essence, from things in themselves. That mistake is the foundational premise of the theory which is called "symbolic formal logic," which was developed by the philosophers Frege, Russell, and Wittgenstein.

Wittgenstein's exploitation of Aristotle's mistake is popular among logical positivists and contemporary analytic philosophers and philosophers of language. Some of these people even believe that language can only refer to words, e.g. "Sun" is a word and not a star in the sky. They think so because of their ignorance of the function of words as things that show other things to the mind. But it becomes clear to anyone who understands essences that essential reasoning draws a necessary conclusion from the premises, and necessity comes from things, not definitions of words. All things, including premises, i.e. the things reasoned from, and the conclusion, i.e. the thing reasoned, are things doing actions, not sets of words. In order to use the words consistently (a matter which will be addressed shortly) one must either talk about the things shown by propositions, or consider propositions to be things shown by sentences. Given the way the words are used now, the current meaning is that a proposition is the thing shown by a sentence. Thus a sentence is a group of words which show a thing, and a proposition or fact is a thing shown by a sentence.

An extension of the mistaken theory of propositions is this theory's ignorance of what truth and falsity consist of. To be true or false are not actions that things do as things. Instead, being true or false are actions that can be done only by things that show other things which are not themselves, e.g. by words. Being true and false are things that only words and sentences (and concepts, perceptions, and perhaps works of art) do, because the act of being true is the act of showing a thing that exists to the mind, and the act of being false is the act of showing something that does not exist to the mind. Thus, the truth and falsehood of sentences that one supposedly is reasoning from is actually the existence or non-existence of the things that one is thinking about.

This is the understanding of truth based on the correspondence theory of language, which states that the meaning of a word is the thing which the word corresponds to. Words correspond to their meanings because a word shows something other than itself to the mind, and words in a language show something to the mind of the listener or reader either because the force of habit has associated the word with its meaning in the person's mind, or, as a result of learning and speaking a language, a person knows that the speaker or writer intended this set of symbols, for example "d" and "o" and "g", to show a dog, so the listener or reader interprets the meaning from the symbols to discover the intent of the author using his knowledge of the set of meanings and associations which constitutes the language. Just as a word can show a thing, like "dog" showing a dog, a sentence is a set of words organized according to the rules of grammar, and the grammar unites the words to show one thing. The sentence "a person is at the front door of the house" shows a person being at the front door of the house. It is true if a person really is at the front door, and it is false if nobody is at the door.

The sentence "the red apple is on the table" shows a thing which does the act of existing, of being red, of being an apple, and of being on the table. As we would expect if essential reason is the primary tool of human thought, words and sentences often pick out an object by reference to a specific property as the essence of the thing. Thus, "the original capitol of the United States" and "New York City" refer to the same object, but they pick it out by means of two different properties which it has. This explains what is meant by the "sense" of a word, as a solution to the paradox articulated by Frege, namely that two words can be different yet mean the same thing. The sense of a word is which property it uses to identify a thing to which it refers. The reason why "New York City is New York City" means something different from "New York City is the original capitol of the United States," is that, although they both refer to one object in reality, the two sentences identify the same thing by two different essential properties. The second sentence asserts that the thing with one property also has the other property, while the first sentence merely states that the thing which has one property does in fact have that one property. "A is A" states that the thing being A is A, whereas "A is B" states that the thing being A is also B.

A sentence which is true shows a thing that exists to the mind, but a sentence shows a thing as being the entire thing which the sentence asserts, doing the total being act which is shown. So if the total thing shown does not exist, then a word in a sentence might be true even though the sentence is false. For example, "the sky is blue and is made of ice cream" shows the sky being blue and being made of ice cream and being an existing thing. Something that exists, the sky, is pointed out, and the sentence shows it doing something that it really does, being blue, so "the sky is blue" is true. But the sentence also shows the existing thing as having a property which it does not have in existence, being made of ice cream. Thus, "the sky is blue and made of ice cream" is false, even though the sky exists and is blue, because, although the sentence picks out a thing which exists by reference to its essence, it shows a thing which in its entirety does not exist.

The logical positivists, and logicians like Russell and Wittgenstein, tried to translate the English language into some whacky new language of logic to force statements to correspond to empirical observations so that statements could be empirically verified. But no new language of logic is needed, because the common languages like English as spoken and written already correspond to things in observable reality. The sentence "a person is at the front door of the house" can be verified by looking out the window and seeing whether someone is there.

The correspondence theory of language and truth also solves the supposed paradox of the sentence "this sentence is false," which is true if false and false if true. The sentence actually has no meaning and is neither true nor false, because a thing is only true or false if it shows something other than itself to the mind, and "this sentence" refers to itself and not to something else, so the meaning can neither exist to cause its truth nor not exist to constitute falsity. A person can perform the act of self-reference by saying "I" or "me," but a word or sentence as an act of speech cannot itself do self-reference and yet still have a meaning.

I refer to words (and concepts and perceptions, and works of art) as showing things, and I refer to meanings as shown things. A word or sentence is true if the shown thing, i.e. the thing that it shows, exists, and it is false if the shown thing does not exist. Truth is the words that are true, although, as a result of a mistake in the philosophy of language, people often say "the truth" when they really mean "reality," i.e., they confuse the shown thing with the showing thing because they are ignorant of the correspondence theory of language. A word can refer to a real thing, an essential thing, or a non-existing thing, although it will only be true if it either refers to a thing that exists, or says of an essential thing that it has the consequence which it really has, because, for example, when one says "a square has four sides" one is saying "every square has four sides" and so the meaning of the sentence reduces to all of the existing squares in reality, which exist as described to make the sentence true. When one speaks the truth of that which does not exist, it is true because what exists is not it, i.e. the truth of "God does not exist" really consists of the physical reality which exists which makes the sentence true.

The theory of symbolic formal logic offers a software-based system for deducing the truth or falsity of sets of symbols, but it has no answer for the question of how to discover whether a symbol is actually true or false in the first instance. My analysis explains why the theory of logic cannot say what truth or falsehood are, namely, it does not know true or false because it sees only words and is blind to things that words refer to. My theory explains what it means when one says that a proposition is true or false. And, because of this explanation, my theory adds depth to our understanding of how reasoning is actually working.

For example, in the case of the syllogism "1. All men are mortal, 2. Socrates is a man, 3. Therefore, Socrates is mortal," the truth of the first sentence consists of the thing it shows existing, e.g. the essential human man being mortal, which itself consists of the act of being mortal being consequential of the act of being a human man (which would have been the result of prior reasoning on the essence of being human). The truth of the second sentence consists of the thing it shows existing, e.g. Socrates being a man. And the conclusion is reasoned from the premises by essential identification, e.g. first identifying Socrates as a man, reasoning that he must do the act of being a man, and then reasoning that he must do the act of being mortal, which is consequential of the essential act of being a man. So my theory gives a meaning-based account of logic, in contrast to the theory of logic implied by Aristotle and embraced by Wittgenstein that logic is merely a set of rules to manipulate symbols and words and sentences. The theory of the logicians would have the truth of the sentences have nothing to do with what they show, i.e. the meanings of the words, and those philosophers would have the conclusion be a matter of the manipulation of the words rather than reasoning about the objects.

Back to the topic of axioms and definitions, let us first consider what a definition is, and then consider what it is for a definition to be "given." A definition is actually two different things, and they have been equated because of the confusion between words and shown things. A word is a set of sounds or letters used by one human to show one thing to another human. When a group of humans gets together and uses the same words to show the same things to each other, they call this a language. It is very useful for communication to use a language, because, since everyone consistently uses the same words, one can communicate without having to try to figure out which thing is being shown by which word. One can be taught language and then know what thing anyone using that language means when they use a certain word. Without consistent usage, everyone would use different words for the same things, and to communicate one would have to learn each person's private language before one could communicate with that person. Also, if one changes which thing is meant to be shown by a certain word in the middle of communication, i.e. if one changes the meaning of a word while communicating, then the person whom one is trying to show things to will have to first figure out what the word now means before the word will show that thing to the person. Otherwise, the person will think that it is still being used to show the old meaning, and that is the thing it will show to his mind. Thus, for a variety of practical reasons it is necessary that the usage of words be as consistent as is possible.

The first kind of definition is a statement of what thing is shown by what word. This definition is necessary so that everyone can use their words consistently. That is the kind of definition that dictionaries do. The second kind of definition, on the other hand, consists of stating what the essential action of a thing consists of. For example, if one has defined that the word "X" means X, then one will not know which things that word refers to unless what knows which things are X, for which one must know what is the essence of X. If one is aware of what the parts of the total act of being of an essential thing are and one can find words to name them, or if one can find either one single word or some group of words which shows the essence of a thing, then one can state what the essence of a thing is, showing the essence of the thing to another person by means of that sentence. Even in one's own reasoning, the decision to consider a certain act as the essence and to reason essentially about that act may be considered a definition of that act as the essence, in the sense that you tell yourself what the essence is.

The sentence used for the two kinds of definitions may look the same, e.g. one could say "a square has four sides" either to say that the thing one is committed to showing by the word with those letters s, q, u, a, r, and e, is a thing with four sides(e), or one may be trying to show that the essence of a square is that it has four sides. One could use the same sentence for both purposes. One sentence can both define a word and specify the essence of the meaning whenever one defines, because when one says what thing one is going to show consistently with a word one will want that thing to be known by its essence. So defining a word and specifying an essence go together. But the two different kinds of definition have different purposes as they pertain to reasoning, because the only use of defining a word is to enable consistent word usage so that multiple people can verbally discuss or argue about the same things, while the purpose of the second is to actually show which essences are being reasoned from.

Considering that the definitional sentences show what is reasoned from rather than being reasoned from themselves, it is clear that the "definitions" at the beginning of a proof, while they may function as the definition of words, have as their primary purpose the specifying of essences, to show what are the essences that are being reasoned from are. If the thing shown rather than the sentence itself is reasoned from, it is the essence itself, not its definition, that one reasons from. Thus, to say that a proof begins with a definition is to say that it begins with an identification of the essence.

Chapter Thirty Seven: The Birth of Reasoning: Axioms

If a line of reasoning, e.g. a math or geometry proof, begins with specifying essences, then where do these essences come from? Or, to be precise, where does one's knowledge of these essences come from? To say that the definitional statements are true "by definition," if that has any meaning, does not answer the question of what the source of the knowledge of essences is. The theory that demonstrative reasoning begins with definitions means that demonstrative reasoning begins with knowledge of essences, and since the Aristotle theory has no idea where this knowledge comes from or what basis exists for certainty in this knowledge, the theory says that demonstrative reasoning begins with assumptions about essences. Reasoning from an assumption is only as certain or uncertain as the assumption itself. Unless the way in which the mind comes to know essences can be discovered, then demonstrative reasoning will be totally uncertain. We are left, then, with the question of where the knowledge of essences comes from.

Now let us tackle the question of axioms. An "axiom" is a self-evident proposition. One or more of these axiomatic propositions are assumed to be true at the beginning of every line of demonstrative reasoning (i.e. proof), and the conclusions of the proof are reasoned demonstratively from these axioms. The truth of the conclusion rests upon the truth of the axioms, and yet the conclusions are believed to be necessarily true, so the axioms must also be necessarily true. And yet how does the person doing the demonstration know whether the axioms are true or not? Either the person has no reason to believe that the axioms are true and believes them for no reason, or they are believed because of some cause other than reason, or else some kind of reasoning produces knowledge of these, but such reason could not be "demonstrative" reasoning because the axioms are prior to demonstration, so such reason would be "inductive" reasoning.

If the axioms are believed for no reason and are believed even though the person knows that they were not reasoned, then two things would be true. First, there would be no basis for rational certainty, i.e. certainty that is entirely the product of reasoning, in the axioms. If one could not reason the axioms then one would have no

259

basis for rational certainty in the conclusions of reason because reason deduces conclusions from the axioms. While one could be rationally certain that the conclusions would be true if the axioms were true, one could never be rationally certain that the axioms were true because one could never reason that they were true. Second, if things, and not only things but the most universal of truths, can be known without the use of reason, then there is no point in using reason. If the axioms are known by means of some other source of knowledge than reason, like intuition or revelation, then it is pointless to use reason as our tool for understanding reality.

If the axioms are reasoned from inductive reasoning, then they would be reasoned, and rationally certain conclusions could be reasoned from them. But this theory gives no account of what such inductive reasoning could consist of. Aristotle's epistemological theory does not believe that the axioms need to be reasoned, and therefore it is implicit in this theory that all reasoning begins with assumptions. Furthermore, the way that inductive reasoning is believed to work, by going from something true of the individual to something true of the general, would not produce certainty (since it would be a kind of associative reasoning, discussed above) and thus would not achieve rational certainty.

But Aristotle does not believe that any rational or irrational source for knowledge is necessary to know the axioms. He believes that one should believe in their truth without having reasoned them. Although he says that one needs no reason to believe in the axioms, still he gives several reasons in his "Metaphysics," and these can all be refuted. First, he says that the axioms are self-evident, which means that if one grasps what the axiom means and fully understands what an axiom says then its truth is evident merely from what it means and one cannot doubt it. But in reality the axioms can be doubted. People have been doubting and debating the so-called axiomatic propositions since before Aristotle's time and we continue to do so up to today's era. Indeed, my (and your) ability to question the axioms is proof that they can be both understood and doubted. And if the truth is evident to someone who grasps it just from knowing what it means then by what means is the truth grasped if not by reason? And how does one come to know what the axioms are in the first place?

If the truth of the axioms is grasped by means of reason then they are not the beginning of reason. If they are grasped by some other means then the same thing follows about an irrational means of knowledge being the foundation of reason that I said above. If when one is aware of what they mean one grasps their truth without any means at all, then one would still have no justification for rational trust in axiomatic propositions, since the origin of the knowledge is not only unreasoned but unknowable. And none of these possibilities explains how the knowledge of the axioms enters one's mind to begin with. Aristotle has no reason not to believe that a deceitful demon has placed the axioms into his mind, since he has no line of reasoning to certainty in the axioms. Aristotle's theory does not even account for where the concepts of the things that the propositions talks about come from or where knowledge of what propositions are and how to judge their truth comes from. Yet knowing the truth of these propositions is supposed to be the beginning of all reasoning that produces certain knowledge. The theory not only demands that one reason from assumptions. It claims that the origin of these assumptions cannot be known while claiming that one can be certain of them.

Aristotle says that an ignorance of knowing what can and cannot be demonstrated shows a lack of education, but that is more of an insult than an argument. He also says that the axiom is not a hypothesis because it is necessary for anything to be known. But this does not make its truth any less of a hypothesis, because if it were false then nothing would be knowable, and he does nothing to prove that things are knowable, other than asserting the truth of the axioms. He also says that an axiom must be believed because one must know it before one studies anything. But this is no basis for certainty, since he never disproves the possibility of the axioms being wrong and of all knowledge being wrong, nor can he, because his axioms are unreasoned assumptions.

Aristotle's "vegetable" argument and his "infinite regress" arguments are better than those above, so let us look at those. Aristotle's "vegetable" argument is a defense of his primary axiom, that the same attribute cannot both belong and not belong to the same subject in the same respect, or in other words that a thing cannot both be and not be the same thing at the same time. (Let me stress that I am not arguing that this is untrue. I believe that it is true.

What is at stake here is whether the truth of the axiom can be reasoned from something else, or whether its truth must be accepted as an unreasoned assumption, i.e. as an axiom). His "vegetable" argument comes in many forms, which all boil down to his belief that a man cannot consistently believe that the same thing can both be and not be. This argument is flawed in general because, first, an inability to disbelieve something is no reason to believe that it is true, as for example some people say that a belief in God is necessary for humans to have hope in order to survive, but this does not rationally prove the existence of God. Secondly, doubt in that proposition does not take the forms which he describes. Instead, the form it takes is to debate the topic of whether or not the axiom is true, which is what many thinkers and philosophers do.

Dealing with the specifics, first, Aristotle says that a man who disbelieved his axiom would have contradictory beliefs about the same thing, believing it to both be and not be so, but he does not disprove the possibility of this. Second, he claims that when a person speaks, he is implicitly accepting as a premise that the same thing cannot both be and not be. Even if this were true, what is necessary for language is not necessarily true, and his arguments about this implicit acceptance are also faulty. He says that to call "X" and "not X" by the same name is contradictory, because for a word to have a meaning is for it to mean one definite thing and to not mean anything else, such that to use a word implies that the meaning is what it is and is not not what it is, so that anyone who speaks implicitly accepts his axiom. He argues one can say a thing is X and is not X only by changing the meaning of the word "X" halfway through. Here he is relying upon a lack of distinction between a thing as that named by the word that shows it to the mind, and a thing as the doer of the essential action it does. If I were to say "that thing which is X is not that thing which is named by the word X" then I would be using the word X in contradictory ways, and my language would ultimately have no meaning since my words would have no consistent usage. And that would be one way to apply a disbelief in the axiom and destroy communication. But if I say "a thing both does the act of being X and does not do the act of being X," then, regardless of whether what I have said is true or even possible, nonetheless the words I used do have a consistent meaning, one which is obviously comprehensible, for the essence of X is not that it is named by two crossed lines, but rather that it does the X action.

For an example, I could say "a pencil both writes and does not write at the same time." I had obviously said "and does not writes" in order to say that the pencil both wrote and did not write at the same time, and not in order to change the meaning of the letters "p" plus "e" plus "n" plus "c" plus "i" plus "l" from "a writing thing" to "a non-writing thing." I would be using words consistently and expressing an evident meaning. And in any case, for language and knowledge to depend upon knowing that a thing is itself are not proof that a thing is itself, and only make the need for that proof all the more important. In a very real sense, Aristotle's axiom argument is a loop argument, in other words it is a circular argument. His assumed first premise is the axiom. He then deduces that knowledge is possible from the axiom. And he then uses the fact that knowledge is possible to prove the axiom.

Aristotle also says that belief in the axiom is implicit not only in the use of language but in every human action, since it follows from a thing not being itself that everything is nothing and that the person in question would not exist, but that he walks and talks implies that he believes that he exists and does not not exist. Also if he does anything it implies that he exists, and if he prefers anything it implies that what he likes is not what he dislikes, so that to consistently live a life of disbelief in the axiom would be to live as a vegetable. This argument is faulty for the following reasons, first because doubt would manifest itself in debate rather than living as a vegetable, second because if a man believed that everything was nothing he might believe that he walked and talked as an existing thing and did not walk or talk as a non-existing thing, and also that the good aspect of what he preferred was visible while the opposite aspect was invisible.

Here Aristotle confuses the premises of reasoning with the premises of actions. The premise of an action is that thing which must be true for an action to be right or correct. A premise of an action may be implied by the fact that one does the action, but it is not always the premise of the reasoning or thinking that actually led one to do the action. For example, if a computer breaks down and a repairman replaces its circuit board, the premise of the action is that the circuit board caused the computer to breakdown, and that replacing the circuit board will repair the computer. But the premise of his action could not possibly have been a premise of the reasoning involved, since it was the conclusion of the reasoning, not the origin of the reasoning. In that example, the premise of the reasoning would have been the observations of how the broken computer was operating from which the repairman would infer that the circuit board was the root of the problem. For another example, if you are hungry and you prepare a chicken sandwich to eat, the premise of the action is that the sandwich will make you full and satisfied, whereas the premises of the thinking leading you to make it might have been (1) that you are hungry and (2) that you have fresh chicken in your fridge, from which you reasoned the conclusion that you should make a chicken sandwich.

The premises of reasoning are those things that you reason from. For example, one reasons that a square has four 90 degree angles from the premise that a square has four equal sides. Reasoning is an action done by people, and it has premises as an action, but its premises as an action are not necessarily the same as its premises as reasoning. What must be so for reasoning to be the right action to do are not necessarily those things it reasoned conclusions from. And the premises of a person's actions are not necessarily the premises of a person's reasoning. If a person eats and drinks this does not prove that his existence is a premise of his reasoning, although it is a premise of his action. A crazy person is a good example of this. A crazy person could eat and drink but not think that he exists.

Even if the "vegetable" argument were granted in full, the fact that disbelief in the axiom cannot be carried out consistently in behavior is no reason to believe that the axiom is true and it in no way makes knowledge of the axiom certain. As Aristotle notes, it is only a negative demonstration and gives no basis for certainty or knowledge of the axioms. And it does not explain how one comes by knowledge of the axioms. The argument merely states that everyone has already accepted the axioms so there is no point in denying them or arguing about them. But this ignores the possibility of believing the axiomatic proposition but not as an axiom, which is to say, believing that a thing is itself and contradictions are impossible as a result of having reasoned this conclusion from some other source of knowledge (in my philosophy, from the perceived things of empirical experience).

Aristotle's "infinite regress" argument is a simple reductio ad infinitum argument, in which he argues that if a conclusion is reasoned from premises, which were reasoned as a conclusion from prior premises, which were reasoned as a conclusion from prior premises, etc., and if there were no first premises to reason all the other conclusions from, then the line of reasoning would have no beginning and would extend back to prior sets of premises infinitely, such that no demonstrative reasoning would exist at all. He says of people who question the axioms that they "seek a reason for things for which no reason can be given; for the starting-point of demonstration is not demonstration" (Metaphysics Book IV, 6). This argument is asserted to prove that demonstration must depend upon things that are not demonstrated. Given the dependence of the conclusion upon the premises, Aristotle's argument claims that either one must be certain of the first premises, or else one could not be certain of the conclusions. His argument concludes that the only possible foundation for reasoning with certainty is axioms.

Aristotle's argument is cunning because in its first part it is completely true, but it does not consider a possibility in its later reasoning which leads it to a false conclusion. It is true in the sense that, if every conclusion is reasoned from one or more premises, and the premises themselves are reasoned from other premises, that some original things that were reasoned from would need to exist for anything to have been reasoned at all, because if there were not a first thing to reason from then none of the later reasoning could have taken place. Also, since the certainty of the conclusion depends upon the truth of the premises, for demonstrative reasoning to produce conclusions of which one could possess certain knowledge, the original things that were reasoned from must be known with certainty, and as such these originals would be the foundation of certainty in reasoning. Aristotle is correct up to this point. But he goes awry when he considers the two possible options for the originals of reasoning to be his axioms or loops of self-proving reasoning. He is wrong when he says that, given that the originals cannot be reasoned and therefore must be self-evident, those propositions which are most self-evident must be the originals. When he says this he believes that his axioms are self-evident, when they are not, and he also assumes that the two options he considers are the only two options, when they are not.

Aristotle is reasoning from the assumption that all demonstrative reasoning consists of deducing propositions from propositions, so that the originals must be some kind of proposition which is self-proving. But this cannot be the case, since if it was then all demonstrative reasoning would reason from assumptions, and the uncertainty of the assumptions would prevent rational certainty in all of the conclusions. I argue that propositions can be reasoned directly from perceived things, and the first propositions of reasoning do not appear out of nowhere, but are reasoned from perceived things. Aristotle's axioms are not self-evident, but perceived things are, e.g. if you look at a red apple you see that it is a red apple, and if empirical essential reasoning can be used to reason the first propositions from perceived things, then those propositions would be the products of essential reasoning and could not be wrong, so the certainty of the entire line of reasoning of philosophical inquiry would have its foundation in perceived things. If the so-called axiomatic propositions could be reasoned from perceived things then everything else reasoned from those propositions would not be reasoned from assumptions and one could have certain knowledge of them.

I can describe how the things that the classic model of reasoning calls "axioms" and "definitions" are reasoned from sensory perceptions. The classic model of logical deduction begins with axioms and definitions and then deduces truths. I argue that the so-called axioms and definitions are not what one begins with. Instead, one begins with sensory experience, one reasons the definitions and axioms from experience via inductive reasoning, and one then deduces truths via deductive reasoning. Here I will show that what are called "definitions" and "axioms" are actually concepts reasoned from perception. First I will discuss definitions, and then axioms.

The so-called "definitions" that are reasoned from are actually the essences that are reasoned from, and the essential actions are reasoned from perceptions either when the concept of the essence is reasoned directly from the perceptions, or when the concept of that essence is reasoned from other concepts reasoned from perceptions, or reasoned from concepts reasoned from concepts reasoned from concepts reasoned from concepts reasoned directly from sensory perceptions. All of the concepts are ultimately reasoned from perceptions, and the concept of the essence is the mind's means of

knowing the essence that it reasons from. I shall demonstrate this
with one example of how a definition is reasoned from perception,
and one example of how an axiomatic proposition is reasoned from
perception. Although first reasoning is unconscious reasoning, these
examples will be lines of conscious reasoning to make it easier for
you to understand. It will be seen later that every line of conscious
reasoning that reasons from perceived things and thought-about
essences has a parallel line of unconscious reasoning that reasons
perceptions and concepts in the brain.

Let us consider the example of this so-called definition: a square
is a thing with four equal sides. What this actually means is that the
essence of a square thing is the act of having four equal sides. Thus,
reasoning from this is reasoning from the act of having four equal
sides, and what we are looking for is how a concept of the act of
having four equal sides is reasoned from perception. This is easy to
find, since there are directly perceivable things with four equal sides
and one could reason a concept of the act of having four equal sides
from looking at any of them. When one sees a thing, one is seeing
the thing do its total visible act of being, and the shape that one sees
is the thing doing the act of being that shape. Thus one can see the
act of being square, and one can reason the concept of being square
from the square perceived thing, by isolating its shape in one's
awareness, which consists of staying aware of or thinking about the
perceived shape and then ceasing to consider or think about any of
the perceived thing's properties besides its shape.

This is what one does when one reasons about an essential
square in a proof from a drawing of a square figure. What one
reasons is not specific to the figure because one is reasoning from
the act of being square that the figure is doing, and nothing else. In
other words, essential reasoning reasons from the essence and lets all
specifics be variable in the line of reasoning. By considering any
perceived action and essentially reasoning from it, one can form a
concept of the essence from the perception, and in this way
knowledge of the essential thing is reasoned from the perceived
things that one reasons from in order to become aware of the
essence.

This is also how any mathematical essence is reasoned. For
example, that seven is the number after six can be reasoned from
perception. First, one looks at seven things and from that becomes

aware of the act of being seven. A lot of confusion exists among philosophers about what numbers actually are, because people have a tendency to confuse numbers and their names, and also some anti-science philosophers like to think that math is a spiritual world separate from the physical world. But the number itself is the group of those things which are that number, and so by considering a group of seven things, e.g. seven apples sitting on your kitchen table, one can form a concept of the number seven from perception. By ceasing to consider one unit in the group of seven, one is left considering six things, and in this way one can reason that six is one less than seven and, considering the first six things and then another thing with it, one can reason that seven is the number after six. The first concepts of numbers are reasoned from perceiving groups of things, and looking at the number of things in the group, and counting the things, and then after the first concepts of numbers are reasoned everything else about numbers can be reasoned from the essential numbers that one is aware of by means of those concepts.

My theory allows for an understanding which accurately describes the progression that children and teens go through in the gradual progression of learning math, beginning with counting, then addition and subtraction, then multiplication tables, to geometry, all the way to calculus. In contrast, if math were a spiritual world known by the soul then one would expect the knowledge of math to be in the mind at birth, and one would not expect to need to teach it to children at all, since they would either remember it or know it through mystical intuition and would not need a classroom and teacher existing in physical reality to learn it. Knowledge of the so-called definitions, the essences from which one reasons, is all ultimately based on reasoning from perceptions.

In the case of the so-called "first principles" or "axiomatic propositions," in light of what I have said a principle and an essential thing are identical, and the axiomatic proposition, the axiom itself, is the essential thing that the sentence which states it shows to the mind. For example, the axiom that "A is A", means "a thing cannot both be and not be at the same time in the same respect," but this really reduces to "a thing is itself consequentially of the essence of being itself". Thus the axioms are also essential things that the subsequent proof reasons from, and these too are all ultimately reasoned from sense perception.

One can reason "A is A" from the essence of a thing as such. And the concept of a thing can be abstracted from the things that one perceives, because the perceived things are things so the information of what it is to be a thing is in the perception of the things that one receives from the sense organs. For example, one can consider an apple, form a concept of it from perceiving it, and then reason from the apple's being that it cannot be both ripe and rotten at the same time because being ripe means it is not yet rotten and being rotten means it is past being ripe. One can then further deduce the principle that there exist sets of attributes that a thing can be one or the other but not both. One can then reason, again from one specific apple in experience, that to be something and not be something are like that, because to be something will cause a thing to not be not it, and to not be something will prevent a thing from being that something at the same time.

Having knocked out the theory of axioms in general, let me dissect Aristotle's axiom in detail. Aristotle's axiom is the proposition that a subject cannot both have and not have the same attribute at the same time in the same respect. Before we can examine how this is reasoned from perception we will have to decipher what precisely is the thing that the sentence shows to the mind. If we filter out Aristotle's flawed theory of subjects and predicates we are left with a sentence which shows "a thing cannot both be and not be the same something at the same time in the same respect." If we reason from a thing being something by doing an action, this becomes "a thing cannot both do and not do an action at the same time and in the same respect."

What does it mean for a thing to be in some respect? This can be reasoned from the example that two is both double and equal at the same time but not in the same respect, since two is double to one and equal to two. "In the same respect" means "in relation to the same thing," and this could be more precisely stated as "as measured in comparison to the same thing." This can be seen in other examples, for example the same meal can be both the right amount and too much, and it is both at the same time but as measured in comparison to different things, in this case when measured in comparison to two different amounts, the right amounts for two different people. Or the same meal can taste both good and bad, at the same time but not in the same respect, in that case as measured by two different objects of the concept of good-tasting, the objects of two different people's concept of good-tasting.

So we have "a thing cannot both do and not do the same action at the same time, as measured in comparison to the same thing." The sentence is showing an essential thing, and the possibility referred to is obviously metaphysical impossibility, so the sentence says that a thing does not do something consequentially of its essential action: "a thing consequentially of doing an action does not not do that same action at the same time, as measured in comparison to the same thing." If a thing does not not do something, then it does it, so the thing that the sentence is really showing is "a thing doing an action(e) does the action it does at the same time, as measured in comparison to the same thing, consequentially of doing that action." And all this means is that a thing does the action it does consequentially of doing the action it does, or, in other words, a thing cannot both be itself and not itself.

What the sentence shows is a thing being itself consequentially of being itself. In order to reason this one would first need to reason the concept of a thing, the concept of action, the concept of essence and consequence, the concept of "not," the concept of the same time, for which the concept of the same and the different and the concept of time would be needed, and the concept of relational measurement. If one could reason these concepts from perception, and then reason that something cannot not be what it is consequentially of its essence, and then reason that a thing is itself consequentially of its essence, then one would have reasoned the thing shown by the sentence. First the concepts must be reasoned, and then the essential reasoning can be done. Once these essences are reasoned, the proposition can be deduced from thinking about them. Specifically, a thing doing an act at a time in a certain way and not doing that act at that time in that way cannot happen, because doing it would cause the thing to not not do it.

The concepts of essential things are reasoned from the perceptions of specific things. For example, one can reason the essence of an apple from a perceived apple. Given this, one can reason the concept of a thing from perceiving things, and one can reason the concept of action from perceiving things acting. One can likewise reason the concept of being from perceiving things being, and then reason that doing and being are the same from the reasoning described earlier. From the concepts of things and action one can reason the concept of a thing as the doer of one action and no others, and then get from this the concept of the action whose being done causes a thing to be the doer of that action, which is the concept of essence. And with the concept of thing, action, and essence one can reason the concept of consequence, the concept of the actions that doing the essence causes or requires the thing to do. Along the same example, one can reason that the consequence of being an apple is being a fruit.

One can reason the concept of the same and different from perceiving things that are the same and different and identifying them using the concept of essence, because things are the same and different as the doers of the actions that they continue to or do not continue to do, or as the doer that action that both do or that one does and one does not do. Then from the concept of the different one can reason the concept of a thing not doing an action, and reason the concept of "not," and from the perception of things changing one can reason the concept of time. From the observation of the actions that things do which can only be measured in relation to the actions of other things, and an examination of this process, one can reason the concept of relational measurement. From my example, one can see that a ripe apple cannot be rotten, that the apple exists at a certain place and moment in time, and that the apple is small compared to the table that it rests upon, meaning that its size is measured in relation to some other object. With these concepts reasoned from perception, one can do the essential reasoning necessary to reason "A is A".

So one's knowledge that "A is A" is not reasoned from an axiom that comes out of nowhere. Rather it is reasoned from the objects of the concepts of thing and being and essence and consequence. Those concepts are reasoned from perception, and once those concepts are reasoned, one can use them and this line of reasoning to reason from perception to the axiomatic proposition, without prior reasoning from the axiomatic proposition in any way. In other words, reason is based on empirical sensory experience, for example one reasons inductively from an apple in order to reach the premises that can then be applied deductively to all apples or to all things generally. The proof of reason is not the axioms, rather the proof that one can point to is the things in experience from which one reasons.

The axioms and definitions that the logicians, mathematicians and geometers believe to be the foundation of reasoning are themselves the product of reasoning from perceived things, and perception is the ultimate foundation of all demonstrative reasoning. All of the facts that thinkers in the past have called axioms are actually the objects of concepts ultimately reasoned from sensory perceptions. The axioms are not self-evident, since their evidence is the perceived things that they are reasoned from. The concepts that all further reasoning requires and is based upon, i.e. the first concepts, are not assumptions and they do not exist in the brain at birth. They are reasoned from sensory perceptions in the process of first reasoning. All first concepts are reasoned from perception, and their truth is not assumed, it is reasoned. All reasoning begins not with axioms and definitions, but with first reasoning. All demonstration begins with perceived things, not self-evident axioms.

Only perceived things are self-evident. One should not begin with a belief in the objective existence of the perceived, since this too must be reasoned from perception. The act of reasoning the existence of the perceived from perception does not assume that perception is infallible, nor does reasoning that a thing is itself assume that a thing is itself in the process of reasoning it. Reason can make no assumptions because it begins only with perceived things and reasons from them without the use of any prior knowledge, i.e. any prior concepts, whatsoever, and an assumption would be a prior concept not reasoned from perception. My reasoning begins with an apple, not with "a thing is itself", so I have not assumed that a thing is itself, and the only premise at the start of the line of reasoning is the apple, which is self-evidently what it is, e.g. I can see that the apple is red and ripe. I do not begin with axioms, and it is worth observing that philosophers believe in axioms only because they lack a principle of induction.

However, for the pure assumption-free nature of first reasoning to be understood, we must investigate how first concepts are reasoned from perceptions in greater detail. My theory does have one problem, which is that those first propositions would have to be reasoned from perceived things without the reasoning using any assumptions in any way. We must inquire into the beginnings of human reason in order to learn whether or not this is possible.

Chapter Thirty Eight: The Birth of Reasoning: Early Childhood

I am now going to discuss the period of time between the beginning of your body's existence, i.e. birth, and the beginning of your awareness, a period which I call the birth of reasoning. The first memories that anyone can recall (aside from fabricated invented memories, e.g. so-called "memories" of past lives) are from the time approximately between the third and fifth year of age. No one has any memory of the first year of their life. In other words, we do not remember our lives as babies, and memories as a toddler are hazy or nonexistent. Current theorists believe that this is caused by some kind of natural amnesia, but this is based on their ignorance of how reasoning works. One has no memories of that period of early childhood because one was not conscious of anything during that period, and a memory that one consciously recalls is something one was aware of before that one recalls from memory to awareness again. The very young child is not conscious because a certain amount of unconscious reasoning must first take place in the brain in order to make awareness possible.

When a baby is born, at the precise point of birth, the potentiality of the mind to exist, which is all that exists before that point, begins to actualize itself. At that point, all that exists of the mind is the empty brain, a tool composed of matter in the womb during the process of fetal growth. The brain is empty in the sense that it contains no concepts, because it has not yet obtained perceptions from the external world out of which to process and derive concepts. The growth of the brain as a child is directed by genetic commands, the result of human genes which evolved an improved ability to respond to existence over many years of evolution. At the moment of birth, the brain is completely the product of genetic direction, and all that the brain has are whatever capacities are inherent in it as an organ, and whatever responses to stimuli which were hard-wired into it when the genes built it in the womb.

No sensory perceptions exist in the brain at the moment of birth. They could not, because until the moment of birth no external objects could act upon the sensory organs. After birth various stimuli act upon the sensory organs, such as light hitting the eyes and sound waves affecting the ears, which, as a result of the genetically created structure, send neural signals in response to having been acted upon to the brain. When the brain receives these sensory signals, the response which is a result of its human DNA genetically created structure begins. The response to sensory signals which the brain is born with is to exercise its ability to process the information it receives.

When does the mind begin to exist? To answer that question we must distinguish "the mind" from "the conscious mind". The mind is the brain. The conscious mind, which I also call the conscious part of the mind or the consciousness, is the brain doing certain actions. The conscious mind is the brain as the doer of the act of conceptual computation. So the mind is formed in the womb at birth, but when we speak of "our mind" we tend to mean the conscious mind, so in this sense the mind begins to form during the time immediately after the moment of birth when the brain receives perceptions, and the mind is not fully formed until conscious thought starts and we begin to think as human beings. We must also note that people can define the essence of "the mind" as either "the brain" or as the brain doing the act of thinking. In the latter sense, one might say of a person in a coma that they have a brain but do not demonstrate having a mind. In the sense that the mind is the brain which thinks, we might say that the mind begins to exist at the moment of birth, because the brain is empty at birth and something that can properly be called "thinking" can only begin once the brain receives perceptions. If one is careful to define essences then this is accurate, but it could be confusing, so it is best to simply define "the mind" as "the brain" and instead refer to the thinking brain as "the consciousness".

The activity of the brain during the birth of reasoning consists only of reasoning of the unconscious part of the mind, which I call "unconscious reasoning," since the computation of the sensory signals that the brain does is done without any awareness of what it is doing. I shall deal with this in detail in the next section, so for here suffice it to say that conscious reasoning is computation which the brain does that it is aware of doing while doing it, and unconscious reasoning is computation that the brain does without being aware that it is doing it. Later in this essay I will argue that the difference between unconscious and conscious reasoning is that unconscious reasoning processes perceptions and concepts, whereas conscious reasoning reasons from the objects of perceptions and concepts, and for the brain to be aware of something means that is has achieved knowledge of the thing in external reality by means of the perceptions and concepts in the brain, such that when a young child first becomes aware of reality this marks the point at which the brain's unconscious reasoning has figured out that the perceptions are coming from objects and graduates to conscious reasoning.

But the great length of detail necessary to explain that theory must wait for later. The beginning of reasoning, the brain processing the first sensory data that it receives, consists entirely of unconscious reasoning, since no concepts and therefore no consciousness has yet been created at this point. Because no consciousness exists at this point, consciousness cannot direct the mind during the birth of reasoning, so the response of the young child's brain to sensory perceptions is dictated solely by the directions that the genetic blueprint builds into the brain. No conscious mind, nor anything else, is present there to direct the reasoning.

In order to explain the birth of reasoning as it occurs in young children, first I must explain perception at is happens in children, then I must explain the nature of concepts, and then I must present my theory of unconscious reasoning, because first reasoning uses unconscious reasoning to reason concepts from perceptions. So here let me briefly return to the topic of perception, specifically as it relates to the brain and the birth of reasoning, and then complete the remaining tasks in the upcoming sections.

When, in the case of sight and hearing, an external means of perception, e.g. light or sound vibration, or in the case of the other three senses the perceived objects themselves, come into contact with an organ of sensory perception, that organ receives a message from the external object that consists of information about the object. This information is then sent from the organs by neurons to the brain. The brain then receives this information into itself, as the perception in the brain. The actual physical result in the brain produced by the message sent by the neurons from the organs is the perception in the brain, which is the brain's means of perception.

These perceptions are the means by which the brain knows the object of their origin, i.e. the knowledge of the perceived object is stored by the brain in the perception. The perception in the brain is the thing in the brain that puts the brain into direct contact with the object of the perception, in other words, the brain has access to the object because it can touch the perception of the object within the brain, the brain accesses the object by means of the perception of the object. The perceptions in the brain, and the signals sent to the brain by the nerves from the sensory organs, and the first neural signals the organs produce, and the light and vibrations and scents etc. emitted by the objects are the means of perception, and form the act of perception. The means of perception correspond to but are not identical with the thing that they bring knowledge of to the mind. The correspondence of a perception to its object consists of its having been caused by the being of the thing it shows in a precise way, so that the brain has knowledge of the being of the thing it shows from the being of the perception. Precisely how a perception corresponds to its object will be fleshed out later when I present two theories, the Math Hypothesis and the Brain Language Hypothesis.

An example of the correspondence of a perception to its object is a case of seeing something red. An object exists that has the property of being red. This property consists of doing a certain action. The act of being red consists of emitting only one frequency of light, the red frequency. The doer of that action, not the red light itself, is the red you see. If an object is red and it is reflective and is hit by light of more frequencies than red only, then it cannot emit and therefore must absorb all the other frequencies of light, consequentially of its essence. Light hits the object, is affected, and red light is reflected off the object. That light now corresponds to the object's color because the being of the light that shows color, its frequency, corresponds to the act of being red. Also, when a thing reflects light, while it is the thing's act of being a color that determines which frequencies of light it reflects, that does not actually cause the ray of light to be reflected from that location, rather it is the thing's act of being there, the thing's act of occupying space at that location, that causes the light to be reflected from that location.

Thus the ray of light's direction corresponds to the location of the last thing that acted upon it. The ray of light corresponds to the color and location of the thing that it shows, because that thing's color and location acted upon the light and caused those properties of the light that correspond to those properties of the thing. When the light hits the sensory organ, in this case the lens of the eye which brings the light to the rod and cone cells, it acts upon the rod and cone cells, and the rod and cone cells create neural signals whose properties are based upon the properties of the light that acted upon the organ's sensory neurons. The direction or angle of entrance, the intensity and the color of the light act upon the organ (the retina) and cause it to produce a certain neural impulse.

Those aspects of being of the nerve impulse sent to the brain from the retina that were caused by the aspects of the being of the light caused by certain actions of the original object are ultimately caused by that object and correspond to it. Nerve cells carry those neural impulses to the brain, where they act upon the brain. The brain stores the information it receives, those aspects of the being of the neural impulses, in the neurons of the brain's memory. There are specific neurons in the brain that are activated when the brain receives a nerve impulse from retina cone cells triggered by red light, and the neural activity of those neurons in the brain corresponds to red light. The brain alters in some way in response to the neural impulse it receives, creating something in the brain that corresponds to the neural impulse which the sense organ sent to it. It is that thing in the brain, whose being is ultimately caused by the action that the original red thing did to the light, whose being corresponds to the being of that thing, which is the brain's means of knowing the perceived thing, and it is also the mind's means of perceiving that thing. The thing in the brain that corresponds to the perceived thing is the brain's sensory perception of it.

A scientist might object that the human brain does not really work that way, in the sense that there is no one neuron that fires when red light is observed. I will briefly note that my description above is a vast oversimplification when I say that one cluster of neurons in the brain directly corresponds to red light. The brain has a complicated process of decoding sensory data from the eyes, which analyzes color, the movement and directions of objects in the field of vision, etc., and I am aware that the visual data is processed by hundreds of distinct brain regions and hundreds of thousands of neurons, each one attuned to a different detail of the perceived things. But the brain has sections which analyze visual data, e.g. the occipital lobe, and obviously the brain does learn to recognize colors, such as the color red. My point here is not to pretend that I am a neurologist or biologist (I am not) and say what the perception in the brain actually consists of at the biological level. My only purpose is to show that the perception must exist in the brain and it necessarily corresponds to its object. As a whole, within its entire analytic process decoding the sensory data, the brain contains a perception of redness which corresponds to the red thing.

Although one perception exists in the brain for each perceived thing, and the being of the perception corresponds to the being of the perceived thing as I described, that does not means that the perception in the brain is physically one thing, like one neuron or one section of brain. Rather the perception is information encoded and stored in the neurons of the brain in some way, and to know how would require knowing the language of the brain, which is variable in this line of reasoning, and must be discovered by biologists, not philosophers. For example, one perception of an apple is one thing in the sense that it is the thing in the brain that is the means of perceiving the apple, but the perception as it physically exists in the brain could involve thousands of neurons in the occipital and frontal lobes of the brain and the limbic system.

When people talk about information they often confuse the shown thing with the showing thing, as for example when they talk about the information that a computer sends and receives over the internet. Here let me specify that this information in the brain is the showing thing, not the shown thing. The reason why one can reason concepts of the things that one perceives from the perceptions is because the perceptions are the information encoded in the brain that corresponds to the perceived things, and the concepts are the information in the brain that corresponds to what thought-about actions the perceived things are doing. Since the information in the brain corresponds to the thing, the knowledge of the being of the thing is already in the brain in the form of the perception, so the only thing required for distinct knowledge of each of the thing's actions is for reasoning to decode the information and process the information in the perceptions. In other words concepts can be reasoned from perceptions because all the information that is in concepts is there in the perceptions already, it merely requires reason to bring the information that show each distinct action separately out. And this information must be there, because the information in the brain corresponds to the objects, and it is this information about the object which the concepts consist of.

The knowledge of the thing's total act of being is in the perception information, because it was the thing's total being action that affected the light/air/sense organ, as my windmill example demonstrates. For example, the concept of red is reasoned from perceptions of red things, and although these perceptions are of the total being that was perceived, a part of that being is being red, and because the perception corresponds to its object a part of the perception must be the perception of being red, i.e. the perception must contain the knowledge of redness. Similarly, as I showed in my last analysis of correspondence, the act of occupying space is perceived, and therefore it can be reasoned directly from perceptions, and, for the same reason, any act that a thing can be perceived to do can be reasoned from the perception of the thing, and therefore the concept of the act can be reasoned from the perception. A thing being solid consists of it preventing other things from occupying its space, and solidity can be perceived when one pushes on something, and the concept of solidity can be reasoned by the infant brain from the perceptions of the things it touches.

The motion of things can be perceived, and it can be reasoned from the perceptions of things that are perceived over time, and time itself can be reasoned from perception, from the rate of change of the perceived things, which the brain can measure and which shows the rate of change of the perceived things, which time is ultimately reasoned from. From this is should also be obvious that the mind has access to the perceptions and concepts because they are in the brain and the mind is in direct physical contact with them. The perceptions and concepts are the mind's means of knowing their objects, the perceived things and the thought-about things that one is aware of. How the means of knowledge enable the mind to be in direct contact with their objects should be clear from my explanation of correspondence and the fact that the brain's means of knowledge can give direct knowledge of the object by means of the means, since the means adds nothing of itself to the known thing.

Let me mention briefly the Brain Language Hypothesis and the Math Hypothesis. The Brain Language Hypothesis posits that, similar to the binary code of a computer forming the language of computer software, and also similar to the words in the English language corresponding to the objects shown by each word, the brain possesses some language of the neurons for the neurons to correspond to the objects in reality, and it is speaking in this language that enables the brain to know external objects by means of the perceptions and concepts in the brain. For example, to extremely oversimplify, if Neuron X's connection via synapse to Neuron Y shows to the brain that the paint is red or the apple is red, then XY is the brain's word for red in the Brain Language. Unconscious reason would process XY to reason about redness, and conscious reason would use XY as a tool to be aware of the color red and to think about redness itself. In other words, unconscious reason would analyze XY itself and conscious reason would analyze the object in reality which corresponds to XY by means of XY in the brain. It seems clear that something mathematically about the number and strength of neural synapses between the different neurons forms some sort of mathematical basis, similar to software's binary code, for the Brain Language, and the Math Hypothesis posits that the Brain Language is a mathematical language. The biologists must actually figure out what the Brain Language is, and as of right now they do not know what the Brain Language consists of, so they cannot figure out whether the Math Hypothesis is correct.

The biological level of the reasoning, i.e. the Brain Language, does not need to be known in order for scientific philosophy to proceed, since my reasoning in this essay depends only on the fact that the means of perception correspond to the perceived thing, which can be reasoned from the original thing affecting the light/air/sense organ and causing these to do something, and the organ sending a signal to the brain in response to being affected, i.e. being caused to send the signal by the thing that affects the organ, and the brain storing the signal it receives. In this case, the means of perception "corresponding" to the perceived thing means that the means of perception is caused by the perceived thing, and one can reason this with certainty from perceivable aspects of the act of perception. The details of how the correspondence works at the biological level of the neurons, and which neural signals and what in the brain shows which external things, i.e. what the language of the neurons and brain is, need not concern us further, and is variable in this reasoning. What is essential is that the perceptions in the brain are the brain's means of perception, that they correspond to the sensed things that they are the perceptions of, and that they contain the information collected by the sense organs.

From birth, the genetically created structure of the sensory organs causes them to receive sensory information and send it to the brain. Likewise, from birth the genetically created structure of the brain causes it to receive sensory information and store it as perceptions of the sensed things. After birth and the beginning of perception, the brain, again directed by its genetically created nature, begins reasoning by taking the sensory perceptions that it receives and, using its biological method for the computation of sensory perceptions, it processes and computes this information, and reasons concepts from the sensory perceptions. These concepts are reasoned from pure sensory information and nothing else. I call the reasoning of those first concepts in the brain from sensory perceptions without the use of any concepts the act of "first reasoning," and I call those concepts "first concepts."

Before I talk more about how first reasoning occurs and how reason develops, I must explain what concepts are.

Chapter Thirty Nine: Reason: Concepts

I call the brain's processing of concepts, whether rational or irrational, by the name "concept-computation" or just "computation" for short. The two kinds of computation are (1) reasoning, which I also call reason-computation, and (2) irrational computation. All reasoning consists of reason-computations. A reason-computation consists of the brain exercising its biological ability to reason by means of its neurons, to take information which consists of a certain set of perceptions and/or concepts and compute those perceptions and/or concepts, producing a concept which is the necessary product of that perceptual and/or conceptual information. A line of reasoning is a series of reason computations in which the concept that is reasoned in one reason-computation is computed with other information in the next reason-computation. Each reason-computation is one point in the line of reasoning, and a line of reasoning has a beginning, the initial premises, and reaches an ending, the conclusion. The reasoning draws a line which reasons from the premises to the conclusion.

A concept, which is the exact same thing as an idea, is the means by which you are aware of that thing which the concept is of, and it is also the means by which you think about the object of the concept. The object of a concept is the thing which that concept is of, which is the thing that one is aware of by means of the concept, which is that thing which the concept shows to the brain. All concepts are in the brain, and a concept is the brain's means of knowing that thing which is its object. Concepts are the means of awareness and thought. A concept shows its object to the mind's eye, which means that it brings a thing which is not itself to the mind by means of itself. The concept is in the mind, the thing that you are aware of is not. Whenever you are aware of a thing, you are aware of that thing by means of the concept of that thing in your mind.

Since concepts show things which are not concepts to the mind, it is a mistake originating from the confusion between the objects of knowledge and the means of knowledge to believe that the thing that one is aware is the concept in the mind. That would mean that one was aware of the means by means of itself, which is impossible given the essence of the act of showing. That error gives rise to saying that things like "beauty" and "truth" are concepts, when they are really essential things that one is aware of by means of the concepts of them. In other words, beauty is an essential thing, not an idea. The idea of beauty is the thing in the brain by means of which one is aware of beauty. The idea of beauty is not beauty itself. Beauty is the essential beautiful thing, and real beauty exists in objective physical reality as the things that are beautiful, while the idea of beauty is an actual physical thing in the brain's neurons.

People say "the concept of truth" or anything similar, or say "that's an interesting concept," etc. It sounds like they are talking about the concepts of these things, but they are using those words in order to talk about the things they are talking about, not the concepts in their minds of those things. When people talk about the concept of truth, the essence of what they are talking about is that it is truth, not that it is a concept, and if truth is not a concept but is an essential thing which is the object of a concept then they are, in fact, not talking about the concept of truth, but about truth itself. One should not equate essential things with concepts, nor equate the objects of concepts and concepts, nor equate words and the things to which they refer, nor equate perceptions and perceived things, unless one believes that the only things one knows are the contents of the mind rather than things known by means of the contents of the mind. And since one can reason that the brain is the means of knowledge rather than its own object from things that one perceives, is aware of, and talks about, since one can easily examine these and see whether or not they are the brain, e.g. this book looks like a book, not a mass of neurons, but your brain is the means by which you are reading this book, this mistake should be avoided.

To state this as a deduction, if (1) concepts are in the mind, and (2) the things that you think about exist in objective reality, then you can conclude that (3) the things that you think about are not concepts, after which an investigation of the matter would make clear that concepts are the means by which you think about things in reality. The belief that perceived things are perceptions and that thought-about things are concepts leads inevitably to the belief that the mind only knows the contents of the mind, not external objective reality, which can be called subjectivism (or in its most extreme form solipsism or panpsychism). This is a result of ignorance of the means of knowledge. If you believe in subjectivism then the "means" argument discussed earlier should be sufficient to persuade you to rethink your views.

The concepts in the brain are not "representations" and they bear no resemblance to their objects (the single exception being the concepts of concepts, but those only resemble their objects because they are the concepts of concepts, not because they are concepts). Nor do concepts need to resemble their objects for the brain to use them to know and be aware of their objects, because a concept corresponds to its object in much the same way as the perceptions in the brain, i.e. using the Brain Language. The concept shows its object because the being of that thing in the brain is ultimately caused by its object, so that in the language of its brain the concept refers to that thing which is its object, and the brain can deduce the object from its concept and be aware of it and think about it. A concept corresponds to its object, and the brain can use a concept to show to the mind's eye the thing itself which is the object of the concept.

The concept is in use as a means of awareness when one is aware of something, but the concept always remains in the brain's memory after it is created, and so long as the concept is in the brain it is the brain's means of knowledge of its object (as can be reasoned from the role of concepts in reasoning). The very same thing in the brain is the brain's means of being aware of something and the brain's means of knowing and thinking about that same thing. Once one knows of a thing one has a concept of it, for the computation of the brain by means of which one comes to know the thing creates the concept of that thing, and as one learns more of what the thing is the concept changes in response to this. The concept must change when one's knowledge of the thing changes, since the concept is the knowledge of the thing, and it also must change because the concept must correspond to its object. One can also reason this from the explanation of how conceptual reasoning works, which comes up later in this essay. The concept must remain in the brain after one ceases to be aware of its object, because one can recall things from memory and be aware of the same thing between intervals of non-awareness, and the concept would have to remain in the memory without one being conscious of it in order to be aware of the thing when one recalls it.

The objects of your concepts are what you see, touch, hear, think about, know, and are aware of. Since the essence of a concept is that it is the means of awareness of its object, it is impossible to be directly aware of a concept, that is, to be aware of a concept by means of itself. Everything can be the object of a concept. This must be so, since a thing(e) is the object of the concept of a thing, and everything is a thing. Everything that is perceived, known, imagined, or thought, everything that the mind is aware of in any way, is the object of a concept, since the mind is aware of it by means of the concept of it. These are what concepts are of, not what concepts themselves are. Conversely, only concepts are concepts themselves, i.e. only the actual physically existing biological means of awareness in the brain are concepts. The concept is in the mind consequentially of its essence, but whether or not one is aware of something outside of the mind or in the mind depends on what one is aware of. How it is possible to be aware of something outside of the mind is easy to understand if one grasps what the act of showing consists of. The concepts in the mind, if they are of external objectively existing things, show those very things to the mind's eye, and it is the external objectively existing things that one is aware of. The mind's concepts are the means of knowing reality, but the concepts do not constitute reality or create reality because the reality that they show exists objectively.

Above I have made references to essential things, real things, and perceived things. Essential things are different ontologically from real things and perceived things because the second two exist while the first is merely a tool of thought. Still they are all things, and one conceives of each thing as the doer of the being action that it does. A perceived thing does the action that it is perceived to do. A real thing does the act of being at a certain place and time in reality. And an essential thing does the essential act and no others. One can be aware of any of these, and one is aware of each by means of the concept of it. Let me introduce the term "thought-about thing" to mean a thing as the object of a concept, in other words a thing that one is aware of and that one thinks about.

One can ask what is the difference between perceiving a perceived thing and being aware of a perceived thing? And why would there be a concept of a perceived thing different from the perception of the perceived thing? The answer is that real things do a number of different actions, and some of these actions are perceivable, being red for example, and some of these actions are thought-about, being an apple for example. One can see a thing being an apple, so that being an apple is perceived rather than thought about, but I argue that there are actually two meanings of the word "perceives," one which refers to sensations entering the brain from the sense organs, and another from the mind's eye seeing an object. In the sense which distinguishes sensations from abstract essences, being red is perceived whereas being an apple is thought-about. In the sense that perception consists of the mind's eye perceiving objects, an apple is perceived. But in this second sense I think that "to perceive" actually reduces to "be aware of a thing or property that entered the brain via sensation."

Here it is useful to distinguish unconscious perception, which consists of the brain receiving signals from the sense organs, with conscious perception, which consists of the conscious mind actually perceiving and experiencing perceived things. What most people call perception, and what I call conscious perception, is actually the awareness of perceived things, which must not be confused with unconscious perception. When one is aware of the thought-about aspects of perceived things, one is aware of the perceived thing. A thing is not two different things, one perceived and one the object of awareness, rather it is the same thing both perceived and as the object of awareness, with the visible aspect (the thing as the doer of the visible actions) being perceived and the thought-about aspect (the thing as the doer of the thought-about actions) being the object of awareness. In reality, a human mind's experience of perceivable reality, i.e. the field of vision perceived by the mind's eye, consists of experiencing things with the perceivable and thought-about aspects both seen in the one object that one sees, in other words perception and awareness are blended together by the brain when one experiences something that the brain has both a concept and a perception of, similar to how the two different visions from each of the two eyes is blended into one perceived object seen by the mind's eye.

The brain's actual act of perception consists of a series of stages: (1) unconscious perception, where the brain receives a neural signal from the eye which received light reflected off a red object, followed by (2) unconscious conceptual identification, where the brain identifies the perception as being of a red object, which then leads to (3) conscious perception, where the mind's eye sees a red thing by means of the brain's perception of the red object, which might then conclude in (4) conscious essential identification, where the conscious mind thinks about the red perceived thing. Unconscious conceptual identification would identify that the red thing is an apple, so that conscious perception would see a red apple and not just a red blob of color, but it would require conscious essential identification to reason that this red apple was a specialty type of sweet apple that is for sale at a grocery store and costs five dollars, since that is an entirely abstract and intellectually complicated sort of property that could only be known by conscious reason and not by unconscious reason. Later I will explain the role of conceptual identification in conscious perception, which will make clearer and more precise the human act of perceiving and experiencing reality, e.g. what precisely it is that you see when you look around. Note that unconscious perception is done by the eyes, but conscious perception, i.e. your experience of seeing a red apple, is done by your mind, not by your eyes.

It is possible to be aware of things that objectively exist, and it is also possible to be aware of things that do not exist. Extended space and time exist independently of all minds, knowledge, and means of perception. The space and time which exists objectively is what most people call "reality" or "existence". To be in a certain location and time in objectively existing space is what most people mean when they say that a thing "exists" or "is real". When one is aware of something that exists at a certain place and time doing a certain total act of being, and objectively the thing is doing that total act of being at that place and time, then the thing that one is aware of exists objectively. When one is aware of something existing at a certain time and place, and outside of one's mind that thing does not exist, i.e. at that place and time in objective space the thing there is not doing the act of being that thing that one is aware of, then one is aware of a thing that does not objectively exist.

The act of being true is the act of showing to the mind's eye something that exists. The act of being false is the act of showing to the mind's eye a thing that does not exist. Anything that shows a thing to the mind, including words, sentences, concepts, and works of art, can be true or false. Perceptions can only show things that exist to the mind, so perceptions can only be true. If the showing thing is true, then both the showing thing and the thing that it shows to the mind exist. If the showing thing is false, then the showing thing exists, but the thing that it shows does not exist. Thus, if your brain has a false concept, then the concept in your brain exists, but its object, that thing that you are aware of and think about and possess knowledge of, does not exist. The thing whose being corresponds to the being of a false concept does not exist.

Chapter Forty: Reason: Unconscious Reasoning

Think back to your first memories, the very first things you can remember. Can you remember the first year of your life? No, because you were not aware between the moment of your birth and your first memories. That period of time was the period of your first reasoning, when your means of awareness, that is, your concepts, were first created. The accounts of people's womb memories or memories of birth or the first year of life can easily explained by the extensive work that has been done on the mind's ability to construct and believe false memories of the past based on what others have told them about what happened, or purely from their own imagination. The numerous psychological studies done on reconstructive memory is sufficient proof for this point. The reasoning that the brain does during the first two to four years of life, during the initial developmental stage of thinking, is composed entirely of unconscious reasoning.

The only kind of reasoning that most people are familiar with or know exists, the only kind of reasoning that most people are aware of, is conscious reasoning. That is the kind of reasoning that you, my reader, are doubtless using right now as you read this book, since you are aware of reading it and you are consciously thinking about what you read. Conscious reasoning is reasoning done by you in which you are aware of the things with which your reasoning is concerned. All conscious thought and all conscious figuring of things out, all reasoning which you are aware of doing as you do it, is of this kind. Indeed, what has classically been called "thinking" consists of conscious reasoning. Conscious reasoning is reasoning done in order to gain conscious knowledge of the things you are aware of, the objects of your concepts. If you have true concepts then by means of them you can be directly aware of reality, and the conscious reasoning you do has the goal of discovering the being of real things. Any reasoning that you are directly aware of doing is conscious reasoning, and throughout the history of philosophy various philosophers have equated reason with conscious reasoning. Typically, thought and reason are associated with conscious reasoning, whereas, to the extent that psychologists and philosophers recognize that the human brain performs unconscious actions, it is assumed either that all unconscious thought is irrational, e.g. Freud, or the unconscious is viewed as some sort of mystical intuition in contrast to reason, e.g. Jung. Because of this, conscious reasoning is probably the only kind of reasoning that you have ever been aware of doing, and the idea of unconscious reason is unorthodox and unfamiliar.

But there exists a second kind of reasoning, which I call unconscious reasoning, which functions in contrast to conscious reasoning. Two main differences exist between conscious reasoning and unconscious reasoning. The first difference is that you are aware of doing conscious reasoning when you do it, you are aware of the things that you are thinking about when you consciously reason, whereas you are not aware of doing unconscious reasoning when you do it, and you are not aware of the things that you are unconsciously reasoning about when you unconsciously reason. The second difference is that, in precise terms, the focus of conscious reasoning is the objects of perceptions and concepts, i.e. perceived things and thought-about things, while the focus of unconscious reasoning is the perceptions and concepts in the brain, without the brain being aware that they show anything.

Unconscious reasoning is the pure computation of symbols of Brain Language without any reference to what they show. I argue that conscious reasoning consists of unconscious reasoning that evolves in such a way that the computation of Brain Language symbols gains access to the knowledge that the symbols correspond to objects in the external world, and the essence of consciousness is that the brain accesses the objects in the external world by means of reason-computations. Also, conscious reason and unconscious reason are both done by the human brain, but conscious reason is done by the brain as consciousness, in other words by the conscious part of the brain or "upper brain", while unconscious reason is done by the brain as unconscious mind, i.e. by the unconscious part of the brain or "lower brain." These are not necessarily two different physical parts of the brain, but are the entire brain as the doer of two different actions, e.g. the occipital lobe is used for both unconscious and conscious perception. See my discussion of where consciousness and unconscious thought reside in the brain in my section on the mind as brain experiment.

Before I proceed to explain unconscious reasoning in detail, I must demonstrate the possibility of unconscious reasoning. First let me show you that you already know that unconscious reasoning exists without your even realizing it, and then let me address the classical belief that all reasoning is conscious by definition.

That every human being to ever live and think used unconscious reasoning, yet for me to be the first person to become aware of it, is easily explained, since, first, the essence of unconscious reasoning is that it is reasoning that one does without being aware of doing it, so that a person could easily live out their life and unconsciously reason constantly and yet never be aware of it. And, secondly, the many effects of unconscious reasoning have all been attributed to other causes as a product of people's assumptions about how the mind works and their ignorance of unconscious reasoning. What these effects are can be easily guessed once I list the causes which these effects are attributed to. These causes are feeling, intuition, instinct, and nothingness. Whenever someone knows something, and the knowledge of this thing enters their consciousness, and they do not know where it came from, they say that they "just felt it," or that their "instincts" told them, or that the knowledge came from their "intuition," or that the knowledge popped into their awareness from no source at all. But all of these are cases of the products of a person's unconscious reasoning entering their consciousness, and the person not consciously knowing where the knowledge came from and not being aware of having unconsciously reasoned it and attributing the effect to irrational causes.

When you can tell that a person is lying to you from their body language, or when you know the perfect words to say to someone without having thought about it beforehand, or when you are solving a mathematical problem and the answer pops into your head, or when you awake from sleep with some brilliant idea in mind, attributing the cause of this kind of knowledge to the above listed sources is like attributing one's bad luck to having broken a mirror. The belief in those things, e.g. intuition or instinct, arises from superstition and ignorance. A rational examination of how the mind works can reveal both the existence of unconscious reasoning and how it works. Once one is aware of unconscious reasoning then one can more easily tell when the knowledge that one consciously uses comes from unconscious reasoning. Given the kinds of beliefs that humanity has held for the majority of its time on this planet, unconscious reasoning would have seemed like magic to anyone who noticed it, and a belief in mystical sources of knowledge naturally developed from that.

Also, since one is only directly conscious of one's conscious thinking one can only reason the existence of unconscious reasoning from it effects, like reasoning the existence of a person only from seeing that person's shadow, so that if one had any assumptions about how the mind worked and these assumptions led one to attribute the effects of unconscious reasoning to other things before one consciously reasoned what had caused the effects from the effects themselves, then one would never consciously reason that unconscious reasoning exists, and the more one benefitted from it the more one would believe in intuition, instinct, or some other superstition. The so-called "gift" or "knack" that people talk about as a natural talent for doing something is purely the result of unconscious reasoning, the result of a large amount of unconscious reasoning at a young age in the case of gifted prodigies, and the result of a great deal of unconscious reasoning on how to do something from the experiences that a person has had in the case of a "knack" that one develops.

For a real life example, I, the author, am able to type at a speed of 100 words per minute, despite the fact that I do not consciously know the location of any of the letters on a keyboard, because my unconscious reasoning knows where the letters are and I type using my unconscious reasoning. And when you read these words, your unconscious reasoning assists you in reading, since you read normally and you do not stop and analyze each letter to consciously reason what the words are, e.g. when you read "the dog barked at the man" you read this sentence in a few seconds and you do not take five minutes to consciously reason "t and h and e spell the, d and o and g spell dog, this is a sentence about a dog," etc.

Unconscious reasoning can be used as an interpretation to understand how the lower brain works in relation to the upper brain in a human being. Unconscious reason directs one's skeletal muscle when one does things that one is not aware of doing, including habits. For example, if someone throws something at you and you catch it before you have time to consciously think about it or even be aware of what happened, unconscious reason caused and directed your action. Also, I would say that the neurological processes by which a human brain interprets sensory data in order to develop knowledge of the perceived things is rooted in unconscious reasoning. Before your conscious mind can see a baseball move through the air after a batter has hit it, the lower brain must process the perceptions, through a long series of neurons, some of which are excited when an object moves from left to right in that area of the retina's field of vision, some excited by a white object, some excited by a round object, etc., and it is unconscious reasoning which puts all the processing of sensory data together in order to reason that the perception is of a baseball, at which point in time conscious reasoning is then able to use the perception to consciously see the perceived thing.

Note that it is not true that the human brain is constructing the experience of perception, nor that the baseball is merely a representation, because pure empirical essential unconscious reasoning, particularly when it does conceptual identification, shows to the conscious mind the perceived thing that objectively exists in external reality. The unconscious reasoning is part of the means by which one consciously perceives, but it is not that which is perceived, whereas objective existence is perceived by reason.

Because one is never aware of doing it when one does it, it is easy for a person to not be aware of unconscious reasoning and to misidentify it. The only way to ever be conscious of it is to consciously reason that it exists, which still does not give direct awareness of unconscious reasoning when it is done. It is easy to understand why no thinker has ever realized the existence of unconscious reasoning, since when thought turns in upon itself and tries to reason what thinking consists of, this consists of conscious reasoning trying to reason what conscious reasoning is, and so unconscious reasoning would never be discovered or considered by the conscious examination of reasoning, if not from the observable effects of it mentioned above. Conscious thinking about thinking would never discover the existence of unconscious reasoning, although it could examine unconscious reasoning once it became aware of it. And the assumptions which religion gave to philosophy were sufficient to prevent any thinker from becoming aware of it.

The concept of intuition originated as a false concept of unconscious reasoning, and any philosopher who accepted the existence of intuition would attribute the effects of unconscious reasoning to that source, and so never reason the existence of unconscious reasoning and thereby become conscious of it. Until the understanding of how the brain operated and of the ability of computers to compute without awareness, the possibility of the brain unconsciously reasoning would not be likely to occur to anyone, and those are both recent scientific achievements.

The idea that reasoning is conscious by definition, while implicit in all the ancient philosophers, who took it for granted that all reasoning was conscious, was heavily strengthened by Descartes, who argued that the existence of conscious thinking was both the basis of self-knowledge and the basis of all certainty, which led subsequent thought into the equation of the reasoned with the conscious. Indeed, since ancient times the oppositional distinction between the mind and the body had led to an opposition between conscious thought, which was rational by nature, with the instinct of the body, which was an irrational by nature, since it was associated with the passions and emotions. This mind-body distinction between reason and instinct is the ultimate origin of the belief that reason must be conscious by nature, and that the body's unconscious means of knowledge is irrational by nature. This belief, which found its most vocal contemporary advocate in Freud and his psychology, is rooted in unscientific and unreasoned assumptions about how the brain operates. In order for the understanding of the human mind to advance, the assumption that all reasoning is conscious by nature would be best left behind.

Unconscious reasoning is reasoning done unconsciously. It is reasoning done by the brain, consisting of the reason-computation of the sensory perceptions and concepts in the brain, done without the mind's eye being aware of this reasoning or the objects of the sensory perceptions and concepts that are computed or the concepts which the unconscious reasoning produces. The difference between unconscious reason vs. conscious reason is the difference between the brain reasoning concepts from perceptions and concepts, on the one hand, and the brain reasoning from the objects of perceptions and concepts on the other hand. Unconscious reasoning is reasoning done from sensory perceptions and concepts to concepts, by means of reason computation. The concept reasoned is from the perceptions and concepts it was reasoned from. Although the perceptions and concepts are the knowledge of their objects no matter what, and unconscious reasoning is done by the brain in order for the body to better respond to perceived things, still at the level of the unconscious reasoning there is no awareness that the perceptions and concepts show anything, and the reasoning itself is only concerned with the perceptions and concepts, and with reasoning concepts that can be used to identify and respond to the perception.

Unconscious reasoning done by the human brain is somewhat similar to the reasoning done by a computer program, which analyses and responds to input without any awareness of what it is doing or the meaning of the signals it computes. Thus, although the unconscious reasoning is ultimately reasoning things about the objects of the perceptions, it is immediately only reasoning about the perceptions themselves, and its immediate goal is to reason concepts from perceptions solely in order to analyze the perceptions and respond to the perceptions in the brain, not the objects of perceptions, i.e. the perceived things. The unconscious reasoning has no awareness that the perceptions have objects, so it cannot be aware of the perceived things to want to respond to them. Unconscious reasoning responds indirectly to the objects of perceptions and concepts, and it responds directly to perceptions and concepts.

Chapter Forty One: Reason: First Reasoning

Here I discuss how unconscious reason works at the biological level, by reference to the Brain Language Hypothesis and the Math Hypothesis, after which I can complete my analysis of the birth of reasoning by showing how first reasoning works by unconsciously reasoning first concepts from perceptions. The examples in this chapter will tie together the previous explanations of unconscious reasoning, concepts, perceptions, and the Brain Language Hypothesis and the Math Hypothesis.

Recall that the Brain Language Hypothesis posits that concepts and perceptions are stored in the brain using a language which corresponds to objects in reality. For example, if Neuron X fires and sends neurotransmitters to Neuron Y when the eye sends a signal of redness to the brain then XY is the word for red in the Brain Language. My explanation of the Brain Language, which I call the Math Hypothesis, is the most plausible explanation, although it waits for confirmation by neurologists and biologists to understand exactly how it works at the level of neurons in the brain, i.e. what the math consists of. The Math Hypothesis says simply that the brain functions by using mathematical equations that describe the numeric data embodied in perceptions. Perceptions are data, stored as sets of numbers, and a concept is an algebraic equation. The perception is identified by the concept when the numbers fit the equation.

For example, if the rod and cone visual sensory cells in the eye's retina were mapped on a geometric Cartesian plane, and every stimulation of a cone cell by red light was mapped on the grid, then if an apple is seen by the eye the red apple-shaped thing will register as a certain set of numbers, and the unconscious mind identifies the apple by means of the mathematical equation which identifies an apple-shaped red object.

To make this simpler, if one's sight is a ten by ten grid, i.e. a Cartesian plane with 100 points, and the points $(1,1)$, $(1,2)$, $(2,1)$, and $(2,2)$ are filled by black points, then one could mathematically identify that one is seeing a square object with a width and height of two units defined by the points that are seen. Or, if one sees a horizontal line in one's vision that is five feet above the floor, the brain may analyze it as the equation Y=5. If each moment that the field of vision is perceived is stamped with a time signature of when it was seen, then the eye would continually generate a set of data in X, Y, Z, P, Q, with X being length, Y being height, and Z being the time when it was seen, as well as P being shade of white to dark and Q being color, and these sets of X,Y, Z, P, Q, data would then be graphed by the mind, and patterns would be identified by means of mathematical equations which match the data.

My theory of first reasoning asserts that perceptual data enters the mind of the young child and the child's brain sees the patterns which naturally emerge from the numbers and reasons the first concepts from these numbers without the use of any prior concepts. For example, suppose that the child plays with a square box lid. The box lid sends perceptions of 4:4 (i.e. a ratio of 4 to 4 in the length to the width of the sides of the shape, which would be seen from the difference in color between the object and its background and would be picked up by the neurons processing the data from the eyes) and also sends perceptions of 17:17 and 41:41 and 83:83 and 103:103 to the brain, these being perceptions of the same square object seen from different angles and distances. The child's brain could then reason the pattern in the numbers from the numbers themselves and reason that (X:Y where X=Y) is the concept of a square. This concept would match any perception of an object with a length equal to its width, i.e. a square with four equal sides. If the child then sees a new square object, e.g. the lid of a box of chocolates, and the perception of this new square is 25:25, then the numbers X=25, Y=25 will match X:Y, X=Y, and the child will conceptually identify a square using the concept of a square which was a first concept reasoned from perceptions.

For another example, the brain of a child receives perceptions of various shades of red, which show up as Z=10, Z=50, Z=60, Z=130, Z=210 (let us suppose in this example that being a multiple of ten is the pattern in the numeric perceptions of the color red; the actual property could be anything and would be known only once biologists discover it). First reasoning then reasons the concept of redness that Z=10Q. If the child later sees an apple whose color data is registered as Z=90 then it will be identified as a red thing by the child's conceptual identification. The brain could then take the first concepts and compute them, e.g. by reasoning that the equation

X:Y(X=Y)+(Z=10Q) is the concept of a red square by reason-computing the concept of a square plus the concept of a red thing. If the child then saw a red square the perceptual numbers data would match the concept of a red square. For example, if the child saw the side of a red toy car which was a red square, the perception might be X=16, Y=16, Z=20, and these numbers would match the concept of a red square and be identified as a red square.

As an example of computing using a concept combined with a perception, suppose that P is the sensation of solidity, and P registers in the brain's analysis of the sensation on a numeric scale of 1 to 100 with a higher P coming from sensations of harder surfaces, and the concept of hardness is P>50 (i.e. P is greater than 50). The child takes the toy car and pushes on it and it remains firm and the perception of this firmness which the child's nerve endings in his fingers send to his brain is P=80. Then the concept of this toy car would then be computed to be X:Y(X=Y)+(Z=10Q)+(P=80), from which, when combined with other sensations such as P=70, P=90, and also a contrast to distinguish it such as P=20, the child could abstract a concept of hardness and then reason a concept of a hard red square toy car as X:Y(X=Y)+(Z=10Q)+(P>50). This would all happen as a result of merely seeing the patterns which naturally emerge from the sensory data, without any prior concepts to assist the first reasoning. The patterns are there because the patterns are the aspects of the perceptions which correlate to the objects in reality that act upon the sensory organs.

Note that unconscious reasoning is the blind processing of perceptions and concepts and the brain is not aware of the objects of its concepts and perceptions when it unconsciously reasons. For example, the brain can unconsciously reason from the concept X:Y(X=Y)+(Z=10Q)+(P>50) using mathematical deduction to form new concepts which are useful for conceptually identifying new perceptions, but only conscious reasoning could use that concept to actually be aware of a red toy car and then think about the toy car as a thing in itself, e.g. the older child understanding that his mother gave him the toy car as a gift for the holidays. I will elaborate on this later, but my hypothesis of consciousness is that once unconscious reasoning develops concepts of the self and reality, which happens at around age 3 to 5 in a child, then unconscious reasoning reasons from the concepts of objective reality and the self and then it realizes that its concepts correspond to objects in external reality, at which point the unconscious computation reasons about the objects of concepts by means of the concepts and it becomes conscious reasoning.

The retina in the eye is a cluster of rod and cone cells which are excited by light of various colors or intensity. When a rod or cone cell is excited, it sends a nerve impulse to the brain, which processes the visual sensory data. Assume that the brain stores and codes this eye's data in this way. For each rod and cone cell, a number corresponds to the location in the field of vision of which rod or cone cell that the signal came from, which the brain knows because the location of the cell stimulated is the result of where the light came from in the field of vision, and the brain knows the location of each cell that sends a nerve impulse to it because one specific nerve in the nerve bundle is attached to each cell so a specific set of neurons will be hit by a specific rod or cone cell being activated by light.

Another number of visual perceptual data corresponds to the intensity or brightness of the light, known by the rapidity or frequency with which the cell fires. Another number corresponds to the color or shade, which the brain knows because rod cones pick up black and white, and each cone cell is keyed to a certain color, e.g. a cone cell will only fire when hit by red light. And assume that the brain time-stamps each impulse with a number showing when the stimuli was seen. So the eyes are continually sending the brain a mass of raw numeric data, which is the perceptions, and the concepts are the mathematical equations, originating in the ratios or patterns found in the raw numbers, which identify things by means of the equations being satisfied by the sensory data sets of numbers.

So, if one has a concept of a face, and one sees a face, the brain identifies the face by the numbers matching the equation. First reasoning, i.e. the reason of the brain of babies and young children, would come from the brain analyzing these sensory data numbers to see what patterns, ratios, and proportions can be seen in the numbers. A face will have the same proportions, e.g. a line of dark hair above the oval of the white face as a darker line above a lighter oval, although the specific numbers will all be lighter when seen in bright light than when seen in a darkly lit room. If the baby sees a face multiple times, from different angles, in different lights, then what it is about the face that is the same each time in these different circumstances will be picked out by the ratio or proportions in the different data sets, which the baby's brain learns to identify.

In this way, concepts such as square, red, face, apple, etc., could be reasoned from raw sensory data without any need for innate ideas or a priori intuition. This explains why it takes three to four years for a baby's brain to the develop to the point of childhood where speech and rudimentary thought are possible, namely, because the brain has no innate knowledge and must reason concepts of each thing from this unwieldy mass of raw numerical data, by looking at the sensory data and waiting for patterns to present themselves to the brain.

For example, if the concept of someone's face identifies a white oval (the face) below a black semicircle (the hair), then the concept of that face would be an equation which would be satisfied by the collection of points forming a white oval below a black line. For another example, from seeing a specific real square as graphed on a Cartesian plane by $S1=(X=1,Y=1)$, $(X=1,Y=2)$, $(X=2,Y=1)$, and $(X=2,Y=2)$, the brain could then reason to the equation $S(e)=(X,Y)$, $(X, Y+1)$, $(X+1, Y)$, $(X+1, Y+1)$, to reason the equation of a small one unit long by one unit wide square. From that the brain could then reason to the concept of a square as such, etc. Generally, this would also support the theory of variables in essential reasoning. Of course, the actual method by means of which the brain mathematically calculates visual data, or analyzes sound waves, or processes taste and smell, etc., is far more complicated, and science (and math) does not yet understand how it works in detail, but I think the general idea is probably accurate.

The Math Hypothesis is supported by the fact that, for example, computer graphics can model three dimensional visual objects using math, and define an object as a set of numbers in a matrix, which is how computer-generated images in the movies and TV work. If a computer can reduce a visual object to math then a brain can probably identify it using math. Generally, the fact that computers function by means of the mathematical processing of numbers coded as data in binary code suggests that unconscious reason could be reduced to an entirely mathematical process.

The Math Hypothesis is also supported by the fact that for most, perhaps all, mathematical equations, one can graph the equation on a Cartesian plane, so one would also find plausible the reverse, that any set of numbers embodied in the graph of an image can be analyzed by reference to which equations they satisfy. And the theory is supported by what is known thus far about how the brain actually processes sensory data, which is, that the sensory data is filtered through series after series of neurons which are excited by various properties in the sensations, at the end of which the brain somehow has awareness of objects, e.g. some neurons are excited by the color red, others by an object in the upper right corner of the field of vision, others by an object moving from left to right, etc. For there to be one concept of, for example, a square, the brain must have learned from processing various squares to reason some concept that could identify every square, and the mathematical equation of an essential square is the only thing that could identify a square in every set of sensory perceptions that contains a specific square.

During the period of first reasoning the brain reasons its first concepts from the perceptions that it receives. At the beginning of reasoning, the only thing that reason has to compute is the information sent to it by the sensory organs. All the signals that the sense organs send to the brain are received and form the collection of perceptions from which the brain can reason. Because of the nature of reason, the brain is able to deduce concepts from a very large number of sense perceptions without the use or need for any previous concepts. The brain has no concepts at birth, and all the concepts it ever has are ultimately the products of the computation of sensory perceptions. All concepts come from perceptions when concepts are the result of pure reasoning, because pure reasoning reasons only from perceptions.

But it is also true, in a sense, that corrupted reasoning originates from perceptions, because the assumptions that enter corrupted reasoning cannot have come out of nowhere, rather they are the result of the random or irrational combination of concepts that ultimately came from perceptions. For example, if one reasons a conclusion from the assumption "the sky is green," the separate concepts "the sky" and "green" came from perception, but were glued together irrationally to form an assumption and a false concept.

One can reason that it is impossible for the brain to have concepts before or at birth from the way that concepts are created.

As an adult one can discover the nature of reason, i.e. that reason computes concepts from perceptions and not from intuition or instinct, and if one can conclude that a baby or infant uses reason to form first concepts, then it follows that the mind is a blank slate at birth. After first reasoning produces those concepts that the brain naturally produces during its infancy from the sensory information it received, without the use of any prior concepts, the concepts can be used in the mind to begin to be aware of things, and consciousness and thought begins. Prior to consciousness, the child's brain can create concepts by reasoning them from perceptions, or by reasoning concepts from other concepts, or by reasoning concepts from a combination of perceptions and concepts. With enough concepts and perceptions to be aware of objective reality, consciousness naturally begins in the mind (as will become clear in the section on consciousness which follows). After this point both unconscious and conscious reasoning can create concepts, with conscious reasoning creating the concepts of the thought-about things which are reasoned as conclusions. For every line of conscious reasoning that analyzes perceived things and thought-about things, a parallel line of unconscious reasoning exists where there is a perception for each perceived thing and a concept for each thought-about thing, because this line of unconscious reasoning is the means by which the brain consciously reasons the conscious line of reasoning.

Above I have described how reason works. In contrast to reason, the brain has a way that it can produce concepts which is, to be precise, not reasoning at all, and this is the empirical associative reasoning I described earlier. In its most extreme form, associative reasoning consists of putting concepts together to form new concepts at random, as the minds of the insane do, or based on pure coincidence, such as belief in a superstition from good luck correlating to a chance event. The more common form of this is putting together concepts irrationally for a bad reason, such as putting concepts together to form beliefs in obedience to the teachings of a religious authority on blind faith.

My theory of first reasoning completes my analysis of the birth of reasoning, which I offer as an alternative to axioms as an explanation of the source of rational knowledge. The brain begins with perceptions, reasons first concepts from the perceptions using first reasoning, then reasons concepts of the various things in its experience, and once it reasons concepts of reality and the self it achieves conscious reasoning. The brain then consciously reasons about the perceived things, and the brain can reach the pinnacle of reason by using essential reasoning to reason from perceived things to the most essential truths such as "a thing is what it is" and "existence exists objectively," which are asserted by Aristotle and Rand to be axioms. Thus, pure empirical essential reasoning derives all knowledge from perceived things, which are self-evident. All subsequent knowledge follows from the self-evident properties of perceived things known by means of the act of perception as a consequence of the essences which are identified and reasoned from. Reason can be removed from the shaky, crumbling foundation of axioms and placed upon the stable stone pedestal of empirical essentialism.

To better understand reasoning, let me digress briefly on the subject of the evolution of the human brain as a tool for survival and the purpose of human reasoning.

Chapter Forty Two: Reason: The Evolution of Reason in the Human Brain

Let us for the sake of argument put aside the Aristotle vs. Darwin debate of purpose vs. randomness in evolution. Since most contemporary scientists believe in randomness, let us proceed from that theory, and put together an account of the evolution of reason. Evolution, on this theory, does not produce genes from a teleological purpose, as if the genes get together, make plans and decide how best to change. Genes do not have neurons, so they cannot think, and it is plausible to believe that there is no goal that evolution has in mind, although an individual animal or human may be conceived of as having the purpose of survival, either implicitly for an animal or explicitly for a thinking human. According to Darwin's theory, genes randomly change as a result of genetic mutation and the combination of genetic material through sex, and the genes which are created that just happen by chance to be good for survival remain in the species because the animals with those genes survive and do not die off and tend to reproduce, and the fittest animals survive because they have good genes. The genes that just happen to be bad cease to be in the species because the animals with them do not survive and they die off and do not reproduce, and they die because they have bad genes.

Animals seek mates with the best characteristics, which is one of the reasons why the good genes stay and the bad genes die. Natural selection explains how a certain genetic trait can increase in a way that seems like an intelligent response to some purpose-driven need, but it does not make evolution teleological. In general, the evolution and progress of animals and humans can be accounted for by a scientific explanation which is based on total randomness and does not contain any purpose or design, although this leaves aside consideration of the actions of contemporary humans, for example technological progress as a type of evolution, or for another example genetic engineering.

That being said, let us consider how animals survive. Each animal has a certain means of survival, which is the means by which the animal continues to exist, getting for itself the materials it needs to maintain its body while avoiding those things that will cause its death. For those living organisms which are capable of self-generated motion, responding to the things that they come into contact with, i.e. responding to their immediate environment, is their means of survival. For example, an eagle must respond to the sight of prey by attacking, a gazelle must respond to a lion by running away, animals in heat must respond to the mating signals of the opposite sex by mating, etc. Each stimulus has a certain response which the animal must meet it with in order to best survive. Thus, for animals survival consists of responding to stimuli.

What response an animal gives to a stimulus does not happen by magic. There are three kinds of responses to stimuli. All animals capable of motion have the first kind of response, which is the genetically determined response. Genes are capable of hard-wiring a certain response to a certain stimuli into the structure of an animal's body, or into a circuit leading from a sensory neuron to a motor neuron, or into the animal's brain. One example of this is the response to cold with shivering. When the brain identifies that stimulus, it automatically responds with the genetically determined response. A certain genetic response lasts because the animals with those genes respond in the right way and survive, and genetic responses are useful for animals in environments that do not change. But because genes cannot evolve as quickly as an environment can change, faster ways of developing responses to stimuli evolved.

The second kind of response, which humans and some other animals have, is the conditioned response. There has been too much research on how conditioned responses are created for me to explain it here, dating back to Pavlov's dogs salivating when the dinner bell is rung after the dogs are conditioned to associate the bell with food by hearing it repeatedly when they are fed. Suffice it to say that conditioning is a kind of empirical associative reasoning, and it has the same results as the empirical associative reasoning described earlier in this essay. An animal can be conditioned to respond to new stimuli as fast as the stimuli acts upon it, and because it associates older genetic responses with new stimuli it tends to get the right response for the new stimuli, but it is purely associative and so it is not necessary that it will produce the right response, i.e. the survival-maintaining response, for the stimuli. Skinner's pigeon, which was conditioned to jerk its head to try to get more food pellets because by random chance it had jerked its head when a food pellet was previously given, is one example of this. So conditioning as an evolutionary tool to react to stimuli had a flaw, and a new kind of response evolved in the early humans, the conceptual response, as the solution to the faults of genetic and conditioned reactions.

Unconscious reasoning evolved in humans as the next step in a method for developing responses to new stimuli. Through the use of empirical essential reasoning, the human could form concepts of the stimuli that entered its brain, it could then reason from experience the right response for that stimuli, and it could then identify the stimuli that entered its brain using concepts, by means of conceptual identification of perceptions, and automatically give that stimuli the response dictated by the concept through which it was identified. The advantage of this was that, because it used essential reasoning and conceptual identification, it could not be wrong like associative reasoning. Although the perceptions that entered the brain came from external sources, and the concepts referred to external essences, for all the brain knew it was just calculating the right response for the stimulus that entered it, in other words, the brain reacted to the perception in ignorance of the perceived thing, and the brain was then sending the message of how to respond to the stimuli that it had reason-calculated out to the muscles to act out the response.

Thus, although the perceptions and the concepts are of objects, and the unconscious reasoning does use the reason-computation of perceptions and concepts to respond to the things that ultimately cause the perceptions, still the unconscious reasoning consists of the processing of perceptions and concepts without the objects of the perceptions or concepts figuring into the reasoning. The unconscious reasoning itself is concerned only with the perceptions and concepts, not with their objects. Neither the act of awareness nor the objects of the perceptions and concepts are directly involved in unconscious reasoning, and the concepts are reasoned from concepts and perceptions without awareness of their objects. That is what makes the reasoning unconscious.

Because conceptual reasoning was so useful for survival, the genes which led the human brain to conceptually reason from perception after birth were strengthened, and the human brain developed as the primary means of survival for human beings. With the development of the brain it became possible to reason more concepts, and to reason concepts of more essential things. Note that first reasoning begins with knowledge of the completely specific, i.e. perceptions, reasons to the next more specific concepts, concepts of perceivable actions, and from there always reasons from the more specific to the more essential, since forming a concept of the essence consists of isolating an action and then ceasing to consider the rest, which requires the actions that one ceases to consider to have been in consideration prior to this for one to cease to consider them. For example, one reasons from one real ripe apple, to the essential ripe apple, to the essential apple, to the essential fruit, to the essential thing, always reasoning from specific to essential.

The more the brain developed, the more it could reason and create conceptual responses useful for survival. However, the human brain reached a point in its evolution where it came up against the limits of unconscious reason, because unconscious reason is mere automatic processing of signs and symbols of Brain Language, in ignorance of the external world, so it cannot use awareness of the external world to check and correct errors in its reasoning. So, naturally, consciousness and conscious reasoning evolved.

Chapter Forty Three: Reason: Consciousness

After birth the development of the human brain proceeds automatically during its first phase of intellectual development. The brain's physical growth comes from a genetic response, and first concepts are reasoned by what is actually a genetic response of the brain to perceptions entering the brain, and then the concepts which can naturally be reasoned from what was experienced are reasoned. Now, at some point it became part of this natural process of development for the child to reason the concept of the objective, and the concept of the self. The concept of the objective is what developmental psychologist Piaget called "object permanence," and he showed that completed development of this occurred at age 2. As one can see from Piaget's work, the reasoning of first concepts is a very lengthy process, and the concept develops from the reason-computation of a huge number of perceptions, with the concept being advanced as each new perception is added to the line of reasoning, until finally the concept of the objective is reached. It is surely similar with the concept of self also.

Once the concepts of the objective and the self are formed, it becomes possible for unconscious reasoning to reason about its perceptions and concepts as things that show objectively existing things outside the self. That reasoning is directly concerned with the objects of perceptions and the objects of concepts as opposed to perceptions and concepts. That reasoning is conscious of the objects of the perceptions and concepts, because it reasons about the objects by means of the perceptions and concepts, by reasoning about the perceptions and concepts as signs of objective things external to the self. By doing that, the reasoning becomes conscious of the objects of the perceptions and concepts, and the unconscious reasoning becomes conscious reasoning.

When unconscious reasoning reasons about the objects of the perceptions and concepts, it is conscious reasoning. How the brain operates is the same whether it does unconscious reasoning or conscious reasoning, and the difference between them is only in whether the concepts of the objective and the self are being used in order to reason about the objects of the concepts and perceptions, or whether the reasoning is purely concerned with the perceptions and concepts without reasoning about the external things to which they refer. It is accurate to say that conscious reasoning is unconscious reasoning that is conscious, i.e. unconscious reasoning about the objects of the concepts, and that unconscious reasoning is the substance of all conscious reasoning. From this it can be reasoned why every line of conscious reasoning about perceived things and thought-about things has a parallel line of unconscious reasoning about perceptions and concepts. The reason is that the conscious reasoning consists of that unconscious reasoning combined with reasoning about the objects of the concepts and perceptions as perceived and thought-about things, i.e. as external things, so that when one consciously reasons, one reason-computes all the same perceptions and concepts that the unconscious line of reasoning consists of, but the conscious line of reasoning consists of additional reasoning computations about the objects of the concepts and perceptions.

The act of awareness is the act of the brain responding to the objects of the perceptions and concepts, and being aware of a thing consists of unconsciously reasoning what the object of a concept is from that concept, so that the reasoning, and hence the brain doing the reasoning, becomes aware of the object. From this you can see how it is possible to be aware of things by means of reasoning using the concepts in the brain. The so-called "mind's eye" is the brain as the doer of the act of conscious reasoning. What the mind's eye sees are those things that conscious reasoning reasons about.

Note that the act of perception that most people talk about is not the perception I have been talking about previously. I have been talking about perception as such, but here we can distinguish between unconscious perception and conscious perception. Unconscious perception consists of the sense organs being affected by sensory stimuli and sending the signals to the brain that enter the brain and become perceptions. Conscious perception consists of reasoning what the objects of the perceptions are and thereby becoming aware of the perceived things. What you might call "perception" is actually the awareness of the perceived, and indeed you are seeing this page now only because your unconscious reasoning is taking the signals that are entering your eyes and going to your brain, and reasoning the object from the perceptions using unconscious reasoning from the concepts of the objective and the self and reality and your concepts of the essences of whatever the perceived thing is and those perceptions in the brain, in order to consciously perceive the words on the page.

When you consciously perceive something, you do so only because your unconscious reasoning revealed the perceived object to your brain such that your brain becomes aware of it. Your conscious reasoning takes the perception in your brain created by the act of unconscious perception and then reasons the perceived thing in objective reality from the perception of the thing in your brain so that you become aware of the perceived thing. For example, when you look at the word "apple" on this page, the word sends light to your eyes which sends nerve impulses to your brain (unconscious perception) which creates something in your brain (the perception) which is processed by the visual and verbal parts of your brain (unconscious reason) that identifies the perception using your concept of the word "apple," which then enables you to reason the perceived thing from the perception, at which point you see the thing itself in reality (conscious perception, i.e. awareness) which was reasoned by your unconscious reasoning, leading to the conclusion of your ability to think about the apple I am telling you about (conscious reason).

Conscious perception is actually a mixture of perception and awareness, specifically it is the awareness of perceived properties, but I prefer to distinguish between unconscious perception and conscious perception, rather than between sensation and perception, to be more precise about what does and does not involve awareness. When I have said that perception is infallible, but thought and judgment are fallible, I can now be more precise in restating what I mean. Unconscious perception in infallible in a human being with normal proper function of sense organs and brain (the Sanity Proviso), but conscious perception is fallible to the extent that it is influenced by false concepts.

For example, if your eyes see a red apple then your brain knows that it is red, but if your concept of this apple is that the apple is your dead uncle George reincarnated as an apple, then when you look at this apple you will consciously perceive the ghost of George as a red apple, you will not actually see the apple as a fruit. It is in this sense that the possibility of hallucinations, and the existence of dreams, in the context of my theory as a whole, does not disprove the infallibility of perception, because while perceptions are infallible, concepts are fallible, and concepts are used to unconsciously conceptually identify perceptions prior to conscious perception. One can consciously reason whether the red object is an apple or George, but the process of conscious reasoning would come after the unconscious reasoning had already led you to be aware of a red apple-shaped ghost of George.

You might ask: you can see the apple as George and be wrong, but can you be wrong that the apple is red? My answer is yes, because, along the lines of a scientific view based around an accurate understanding of how the brain works, the apple that is consciously perceived is the result of extensive unconscious reasoning to analyze that the thing is red, which includes the use concepts, e.g. the concept of redness, so conscious perception is fallible, although unconscious perception is not. This is supported by the facts, e.g. a person who hallucinates, such as when under the influence of psychedelic drugs, can consciously perceive unreal imaginary things and not be able to distinguish them from objective reality, despite their brain continuing to receive sensory signals from their eyes. Similarly, a person might think that they perceived a miracle or magic, for a similar reason. As I will explain in detail later, my ultimate conclusion that perfect infallible knowledge is possible from an empirical basis rests upon the understanding that one can only claim knowledge if one uses pure empirical essential reasoning. In the absence of reason, at both the conscious and unconscious levels, the brain can be completely wrong, including seeing a red apple as a pink elephant.

However, two reasons exist whereby we generally do not need to doubt that an apple is red when we consciously perceive a red apple. First, the Sanity Proviso states that a basic degree of sanity is necessary in order for a philosophy of reason to be useful for a person, and this enables me to proceed without worrying too much about such extreme misidentifications as would constitute insanity, such as seeing pink elephants or ghostly dead uncles. Second, the reasoning offered at the conclusion of this essay, which will bear the fruit of all the lines of reasoning offered herein, constitutes a proof that the physical scientific world that one perceives exists objectively, and proves that reason provides knowledge of the truth.

Once one has consciously reasoned that the apple is red, by means of pure empirical essential reasoning, then the conscious perception is proven to be infallible, although conscious reasoning can only be infallible because it is the result of perfect reasoning, and it can be fallible to the extent that it is the result of flawed reasoning, whereas unconscious perception is always infallible because the perception automatically corresponds to its object due to the nature of the act of perception.

The conscious reasoning of consciously identifying what are the things that one sees goes on constantly, and conscious reasoning forming beliefs about perceived things results in modifications and additions to the concepts of the perceived things, and those concepts changed by conscious reason are then also used in unconscious reason, so the mind as it really works is a continuous combination of unconscious and conscious reason. But for the purposes of this analysis we must distinguish the two in order to better understand each one as distinct.

Also note that conscious reason provides the brain with new information not contained in the unconscious reason because the objects of concepts and perceptions are different from the perceptions and concepts themselves. Indeed, conscious reason is more intelligent than unconscious reason precisely because it analyzes the things in the external world of objective reality which really exists, and it not mere automatic processing of Brain Language symbols with no awareness of what they mean. The brain can learn more about a red apple by thinking about the apple itself than by merely knowing which neurons react to the stimuli when light from the red apple hits the eye. Only unconscious reason can process the neurons in the brain, and only conscious reason can think about the apple itself and reason from the apple in objective reality.

The distinction between the perceived being and the thought-about being of things must be understood here. The perceived being is what the thing is that can be directly perceived, and the thought-about being is what the thing is that is reasoned from the being that is directly perceived. One is aware of both, but only the former is perceived. What one consciously perceives is the product of having conceptually identified the perceptions that entered the brain using one's concepts. This is not the "interpretation" of the perceptions or "imposing order" on the perceptions. Rather it is reasoning what the

313

perceived things are from the perceptions and then being aware of the perceived things. For example, when you see a face, all that you are directly perceiving are shape and color, but you don't consciously see shapes of color, you consciously see a face, because of the unconscious reasoning that identified the perception as being of a face using conceptual reasoning. You see the face of the person whom you are looking at, if you are aware of the person you are looking at, if you unconsciously identified the perception as being of the face of that person. Indeed, the act of consciously perceiving that person's face depends upon the prior act of unconsciously identifying what thing the perception is of using conceptual identification.

Awareness is not some mysterious magic that comes from some genie or soul living in your head. Rather it consists of the brain reason-computing perceptions and concepts to reason about things that are the objects of perception and awareness. When your brain reasons about the things that you perceive and the things that you are aware of, that is the very same act as your conscious perception and your awareness. You, the thing reading this book, the consciousness reading these words, the mind that is thinking about what I am saying to you, you are the brain in your skull doing the act of conscious reasoning.

By the way, as I describe it in this analysis, to think of something and to be aware of something is the same, and I say "a thought-about thing" and "a thing as the object of awareness" as the same thing. This is true, in the sense that to be aware of something is to think of the object of the concept that it is what it is, for example, when you are aware of this book you are thinking that it is a book, that it is this book doing the specific actions it does, and that it is the doer of those actions as the thing you are aware of. One might try to separate awareness and thought by saying that the objects of awareness do not necessarily change, but when one is thinking about something what you are thinking about changes as you think about it. But in fact each thing stays the same as you think about it, e.g. if you think about going to the grocery and then think about what you will buy, each food you think of stays the same while you are thinking about it, what changes is that your thought jumps from thing to thing. Another distinction between awareness and thought would claim that awareness is passive, and consists of an object acting upon the mind, whereas thought is active, and consists of the mind actively thinking about something. As should be clear from my analysis, I view awareness as an active act that the mind does to be aware of the object, so that distinction also rings hollow for me. In common conversation one might possibly distinguish thought and awareness, but I don't find a distinction between them useful here.

Once the reasoning of the brain makes it possible for the brain to reason about things rather than concepts and perceptions, then conscious reasoning can reason from perceived things to essences, and from essences to more essential essences, and then use essences to identify perceived things, i.e. conceptual identification, by means of the perceptions and concepts in the brain used to perceive and be aware of their objects. A temporally simultaneous continuous line of unconscious reasoning reasons from the perceptions and concepts to their objects which makes every line of conscious reasoning possible. With this reasoning it becomes possible to consciously perceive, to consciously reason essences from consciously perceived things, to consciously perform essential reasoning, to consciously identify consciously perceived things, and in general to reason from perceived and thought-about things to other things that can be reasoned as conclusions from the premises.

When one consciously reasons a new something one becomes aware of it, because the unconscious reasoning that parallels the conscious reasoning creates the concept by means of which one is aware of the thing that one has just consciously reasoned. The conscious line of reasoning's parallel line of unconscious reasoning is the means by which the conscious reasoning is done, i.e. the reasoning of perceptions and concepts done in your brain enables you to perceive the things you perceive and to be aware of the things you are aware of. In other words, when a conscious reason-computation computes two premises, either perceived things or thought-about things, to reason a conclusion, the unconscious reasoning must also process the perceptions or concepts in order to form the concept by means of which the conscious reasoning can know the object which is the conclusion of the conscious reasoning. In a sense one might say that conscious reason consists of unconscious reason that becomes conscious by reasoning objects from concepts.

The correspondence of perceptions and concepts to their objects, in a manner which can be read by the brain using the Brain Language, explains how it is possible for the brain, using the concepts and perceptions in the brain that the brain's reasoning is in direct contact with, to reason about and have access to the things that objectively exist outside of the brain, i.e. to perceive and be aware of the external world. The things that you consciously perceive are the things that exist outside your brain, the external objective real things, e.g. this book that you are looking at exists where you see it, not inside your mind. The things that you are aware of, if you are aware of them by means of true concepts, are real things that exist outside of your mind, objectively, and if the thought-about thing is the object of a false concept then it does not exist and you are aware of a non-existing thing. The unconscious reasoning, in which the mind uses the perceptions and concepts in the brain which it is in physical contact with, are the means by which the mind is able to perceive the external, objectively existing real world outside of the brain.

One is conscious of doing conscious reasoning when one does it, i.e. one is aware of thinking when one thinks, because one is conscious of the objects of thought and one can reason that one is thinking from the things that one is thinking that one is aware of. For example, from the apple that one is aware of, one can consciously reason that one is thinking about an apple, and from this one can reason that one is thinking. This line of reasoning is so easy that most people are conscious of thinking when they think. Indeed, you can become aware that you are thinking right now and verify how easy it is. However it is possible to consciously reason without being distinctly aware that one is reasoning.

In conscious reasoning one consciously reasons from the things that one perceives and is aware of, to a thing that one becomes aware of. My reader should easily be able to observe this himself or herself. I showed above how it possible to know things directly by means of the perceptions and concepts of those things. Given this, it is an easy step to say that the things that one perceives and is aware of are things in themselves, and that when one is conscious of a thing, it is the thing itself that one is in mental contact with, the thing itself that is directly known, by means of unconscious perception that gets the perception into the brain, and also by means of unconscious reasoning which processes the perception to form the concept that is

316

the means of awareness. From this one can reason that, since one can directly perceive and be aware of things in themselves, conscious reasoning can reason directly from the being of things, which it knows directly by means of the unconscious reasoning of perceptions and concepts. The things that conscious reasoning reasons from and reasons to, the things that it is aware of and reasons about, are the things in themselves, the things that are the objects of the concepts and perceptions. Awareness is a means of direct knowledge of the things in reality, and it is precisely the means of awareness, which consists of perceptions, concepts, and the unconscious reasoning I described, which makes this possible.

The things that one is aware of are not "mental representations" or some kind of artificial construct that the brain produces from the conceptual interpretation of the received perceptions, rather the concepts of what the perceived things are is reasoned from the perceptions, and the brain can then use the concepts of things to become aware of the things themselves in reality that the concepts are of. One can reason directly from things themselves, from their directly known being, because one is able to directly perceive and be aware of things themselves, and one is able to perceive and be aware of things by means of the sensory perceptions and concepts of those things which physically exist in the brain of the perceiver and are the means of perception and awareness. Thus, conscious reasoning reasons from things themselves to things themselves by means of the perceptions and concepts of the objects of the reasoning. The objects of your conscious reasoning are the things that you perceive and are aware of, and these must be external to the perceptions and concepts and the mind's operations, precisely because the perceptions and concepts are the means by which the perceived and thought-about things are known.

In other words, considering the planet Mars, while it is true that the planet Mars that exists in outer space and the planet Mars that exists in your mind are different, the planet Mars that you are aware of is the planet that exists in outer space, not the one that exists in your mind. "The Mars that exists in your mind" is merely a euphemism or imprecise name for your concept of the planet Mars, which is the thing in your brain that is the means by which you know and are aware of Mars. But you are not directly aware of "the Mars in your mind." A concept cannot be used to be aware of itself because a concept is a showing thing and the essence of a showing thing is that it shows something other than itself to the mind. The things that you are aware of are not in your mind at all, and the things in your mind are not the things that you are directly aware of. The "means" argument applied to perceptions and concepts, as described in the section on the Objectivity Proviso, is adequate proof of this, although it can also be reasoned from an examination of what the act of awareness and conscious perception consists of.

When a concept is used to be aware of something, the concept can add nothing to the thing that one is aware of, because the concept is that which shows the thing that one is aware of, to show is to show a thing to the mind which is not the showing thing, and therefore the concept cannot in any way be the thing that is shown, i.e. the thing the mind is aware of. A thing cannot both be the object of awareness and the means of awareness, because the act of showing consists of a showing thing being the means by which something besides itself is known, and the act of being a shown thing is the act of being the thing that the showing thing provides knowledge of. When one consciously reasons, one reasons from the thing itself that one is aware of, as you can see for yourself if you observe yourself thinking, and therefore conscious reasoning reasons from the thing itself, by means of the concept of the thing. One is aware of the thing by means of the concept, and one then reasons from the thing one is aware of. When one consciously reasons from a thing, one is reasoning from the actions done by that thing, not from the information in the concept of the thing, although it is done by means of the information in the concept of the thing.

It is only the ancient misunderstanding of the nature of the means of knowledge, and the collapsing of means and object into one in the works of many influential philosophers (most notably Kant) that leads people to incorrectly believe that it is not possible to reason from one thing by means of another and yet to still reason directly from that thing. This mistake leads naturally to subjectivism, to the belief that the means of knowledge necessarily influence, add to or create that which is known. But for the means of knowledge to be known by means of itself is metaphysically impossible, as I have demonstrated with essential reasoning. With how the mind operates more fully understood by us, it is easy to realize that, when one consciously reasons, one reasons directly from the things that one

318

perceives and is aware of, by means of the perceptions and concepts in the brain and the brain's use of them.

The essence of unconscious reasoning is that it is the reason-computation of perceptions and concepts done by the brain. The essence of conscious reasoning is that it is the reason-computation of the things that one is aware of and consciously perceives done by means of the reason-computation of their concepts and perceptions in the brain. When the brain computes concepts in order to learn about its concepts, it is unconsciously reasoning. When the brain computes concepts in order to learn about their objects, then that unconscious reasoning becomes conscious of the objects of its concepts, and the brain is consciously reasoning. That is the very difference in how unconscious reasoning and conscious reasoning work. From this it can be reasoned that unconscious reasoning is the means of conscious reasoning. The act of awareness is the act of the brain responding to the objects of concepts and perceptions. One can also reason that that thing which is aware, which has been called the consciousness or the "mind's eye" or the "conscious mind," is the brain doing the act of conscious reasoning by means of its act of unconscious reasoning.

The Freudian distinction between "conscious mind" and "unconscious mind" is highly erroneous, since not only are the two systems not in opposition to each other, nor does unconscious vs. conscious line up with irrational vs. rational, but they are actually not two different systems, but one brain doing two different actions, unconscious computation and conscious computation. Freud's oppositional distinction is based upon the assumption that conscious thought is rational and unconscious mental action is irrational, which has its origin in the same mind-body distinction I mentioned before, in which the non-physical soul is the seat of reason and sanity while the physical fleshy body is the seat of passion and irrationality. That theory is based on many assumptions and is not the product of empirical reasoning, which shows that the mind is a physically existing thing, the brain, and as such the mind is a part of the body.

It would be best to leave Freud's distinction behind, and instead consider the mind, as the brain, as having two parts, the unconscious part of the mind, which does unconscious reasoning, and the conscious part of the mind, which does conscious reasoning. Also note that it is quite possible in a real person for their unconscious mind to be rational while their conscious mind is irrational, as for example a person might be persuaded to consciously believe in religion by reading a book of irrational but persuasive arguments while at the same time having a "gut feeling" that religion is bogus nonsense. A conclusion is known true if it is reasoned, and it is probably false if it is the result of irrational computation, and this is true whether the reason which produces the conclusion is conscious or unconscious. Similarly, an emotion can be a rational or irrational reaction to a situation. For example, if your mother dies it is rational to be sad and cry, and if some minor unimportant irritation bothers you then it is irrational to get very angry. So emotions, like unconscious reasoning, are not inherently irrational in opposition to the conscious mind.

In passing, while discussing emotions, let me mention that an emotion is the brain's awareness of a thing that triggers in the brain a physical biological reaction to that thing. For example, sadness is the thing that makes you sad, as thought about by the brain, combined with the physiological biological reaction of crying. For another example, anger consists of the thing that makes you angry, which you are aware of, which then triggers the brain's aggressiveness rage physical reaction. Many emotional reactions in the brain trigger the human brain's biological stress reaction, i.e. the genetic "fight or flight" reaction, which gives energy to the muscles and promotes physical action at the expense of energy taken away from the cold logical reasoning part of the brain. This explains why emotions are historically associated with irrationality, although the belief that love and lust are irrational also contributes to the popular misconception. But the thing that you are aware of can either exist or not exist, and your brain's response to that thing can be pursuant to what it is or not pursuant, hence emotions can be rational or irrational.

Chapter Forty Four: Reason: Reason vs. Logic

The purpose of this essay is to prove that pure empirical essential reasoning is infallible and achieves knowledge of objective reality. I have already showed that perception is infallible, and, with the Brain Language Hypothesis and the Math Hypothesis, I have suggested how unconscious reasoning works, to the full extent that philosophy can explain it without the assistance of biologists and neurologists. Having established that perception gains knowledge of the external world, and that unconscious reasoning then processes perceptions to form concepts, it now remains to show how conscious reasoning works. Specifically, I need to prove that, where two premises are reason-computed to produce a conclusion, if the two premises are true then the conclusion must be true.

My argument here is completely contrary to the understanding of the Analytic philosophers. Generally, in the contemporary era philosophers are divided into the Analytics and the Continentals, and the Continentals do not believe in reasoning, so the Analytic account of reason has dominated philosophy, but here I offer a new model to compete with the Analytic account. The Analytics, in a mistake dating back to Frege, Russell and Wittgenstein, and perhaps originating with Aristotle, believed that logic is valid because a logical argument obeys a certain set of rules, and a logical inference which obeys the rules of logic are valid because it obeys the rules. These rules demand that a logical argument conform to a specific pattern or "form," hence the name "formal" logic. For example, if one premise is "All men are mortal" and the second premise is "Socrates is a man" then the conclusion "Socrates is mortal" is valid because of the structure of the argument as a logical syllogism, which conforms to the model "All X are Y, S is X, therefore S is Y". Another example would be "If X then Y, X, therefore Y". Analytic logic is called "symbolic formal logic" because it is concerned with the form of the argument, not the substance of the argument, and it deals in symbols which represent true or false statements, i.e. it analyzes symbols and not the things themselves which the symbols represent.

My argument, in contrast to the Analytic account, is that a conclusion of reasoning must be true if the premises are true, not because of the rules of logic, but because there is nothing in the conclusion other than the premises, because the conclusion is formed by reason combining the premises together, so that if the premises are true then the conclusion must be true because of the truth of the premises. I reduce reasoning to the analysis of essences, not to the rules of logic. In so doing, I restore meaning to reason and remove it from the realm of meaningless symbols computed by computer program-like unthinking automation.

For example, "All men are mortal" means that the consequence of the essence of being human is that the human is mortal, because to be human is to be a human type of animal, and the nature of the human animal is to die, either from accident or violence or illness or old age, after no more than about 100 years. This is essential reasoning. Then "Socrates is a man" is conceptual identification which identifies that the thing called Socrates possesses the attribute which is essential to being human, namely, having human DNA and being a human-type animal. Then "Socrates is mortal" is an extension of the essential reasoning, which reasons that because Socrates possesses the human essence he must also possess the human consequence, which includes being mortal. If you are aware of what it means to be human, and you are aware of Socrates, and you mentally combine the two things, then you can know that Socrates is mortal, because of the two thought-about things that you reasoned from, and the conclusion follows from the premises because the conclusion is the combination of the two premises combined by reason-computation.

But the Socrates reasoning is not a syllogism that is valid because of the rules of logic, rather it is valid because it is essential reasoning which analyzes the consequence of the essence and identifies which things do or do not have the essence, and then reasons from the consequence of the essence. Another example of the Socrates inference, phrased in a way to make clear that it is essential reasoning and conceptual identification, is: "all squares have four right angles, the lid of this box of chocolates is square, therefore it has four right angles in its four corners." This logical syllogism actually reduces to essential reasoning thusly: (1) the essence of a square is having four equal sides, (2) the consequence of the essence of being square is having four right angles, (3) the lid of this box of chocolates has four equal sides, therefore (5) the lid is square, from which we can reason the final conclusion that (5) the chocolate box lid has four right angles at its four corners.

The same essential reason analysis applies to the logical inference "If X then Y, X, therefore Y." "If X then Y" is something that can only be known by essential reasoning, where Y is the consequence of the essence of X, either from X causing Y, X requiring Y, or being X containing the act of being Y as a part of its total being action. "X" is conceptual identification, and "therefore Y" is the conclusion of the essential reasoning. But "If X therefore Y, X, therefore Y" has no meaning in the absence of the essences and things that X and Y consist of. This reasoning, when it really happens in real people's brains, is the analysis of what X and Y are. It is not the rote mindless computation of empty symbols by a computer.

To complete the analysis, the logical symbols "not", "and," "if then," and "or," and the various other symbols of symbolic formal logic, are merely ways that things can be analyzed. They are not symbols that enable learning about reality in the absence of the meaning of the symbols. Reasoning analyzes things, not symbols, and the Boolean operators are merely ways to frame words to assist thought in understanding things. Words and logic can be of assistance to thinking about things, but the things themselves are where our knowledge comes from. To say "not X" means that X is not true, or, to be more precise, it means that X does not exist. To say "X and Y" means that both X and Y are true. And "X or Y" means that either X or Y is true.

I am aware that logic as taught in contemporary philosophy departments at universities uses many complicated symbols, such as upside-down "A" and backwards "E." I studied the symbolic formal logic of Frege and Russell when I was a philosophy student in college, and I received a passing grade in a class on logic. But the details of which symbols are used does not change my fundamental criticism of logic as such. The needless complexity of logic merely shows how completely different logic is from the way that real people think and proves that logic has no practical usefulness for ordinary human beings. Logic's fancy symbols serve only to "muddy the water to make it appear deep," to quote Nietzsche (albeit in his discussion of Kant, not logic).

The symbols of logic are language techniques that can make it easier to reason by identifying with precision what we are thinking about. They can be very useful as a tool for mapping and diagramming the structure of an argument. And they play a special role in the logic of computer programs. But conscious reasoning does not look at meaningless empty symbols. It looks at meanings, i.e. the things to which words refer, and conscious reason looks at things that exist, not symbols in the mind or words on pages. Reasoning itself would look at the consequence of the essence of X, and what the thing Y consists of and what its total act of being is. If a premise is "X and Y" then conscious reason would look at X and Y and deduce what follows from both things existing and the two things being combined together. Reason can learn nothing merely from the processing of the logical symbol of "and".

A computer program can obey rules of logic, but conscious reasoning looks at reality, and existing things, not at symbols which are asserted to be true or false. As I said earlier, for a symbol to be true means that the thing which the symbol shows is an existing thing. Thus, logic as the analysis of symbols can show fancy trees and branches about the truth or falsity of symbols, but it cannot say that a symbol is true or false, because it looks only to words and not to things existing in objective reality. True to what we would expect, the Analytics have never had a good theory for how logic can know that a statement is true or false, and some Analytics, such as Wittgenstein, have even said that logic cannot achieve a principle of induction to show that X is true, and can only assume that X equals T or F and then process X using the rules of logic. For example, Wittgenstein, in a conversation with Bertrand Russell, was reported to have said that logic cannot prove the truth of the statement "a rhinoceros is not in the room," even if one is in the room and one looks around and doesn't see any rhinos there.

My theory, which states that reason looks at things and essences, not empty meaningless logic symbols, has a theory of induction, as well as a theory of deduction. If a rhino is in the room then my theory of perception and conceptual identification can prove that a rhino is in the room. My theory explains truth and falsehood, because words, signs, and showing things either show something to the mind's eye that exists, if the symbol is true, or show something that does not exist, if the symbol is false. The Analytic logicians talk of their symbols being true or false, but truth has no real meaning in Analytic logic, because Analytic logic's truth applies to symbols and has no reference to whether the symbol refers to something in reality that objectively exists. The statement "the rhino is not in the room" can be defined as equal to P, and logic can define that P equals T (or F), but logic is mere empty rhetoric unless perception can establish that P equals T, in other words, unless inductive reason can prove the truth of the statement "the rhino is not in the room."

Chapter Forty Five: Reason: Reason and Things in Themselves

Two Kantian transcendentalist objections to my theory of conscious perception can be asserted, and I will now refute them one by one. First is the objection that if the brain can transcend the perceptions that it receives to become aware of the perceived things that they show, then why can't the brain transcend the perceived things that it is contact with through conscious perception, to become aware of the noumena? The answer is that the brain does not transcend the perceptions to become aware of the perceived things, because to transcend experience is to go beyond it. Instead, the brain reasons from the perceptions it receives, from experience, to become aware of the perceived things that are consciously perceived. Perceptions show perceived things, so perceived things can be reasoned by the brain from perceptions, but perceived things do not show noumena. The perceived things are the things in themselves that exist, so reason cannot deduce any noumena from them, and, to phrase my philosophy in the language of Kant, the phenomena are the noumena. Kant confuses perceptions, which show things to the mind, and perceived things, which are that which is shown to the mind, to come up with his theory of phenomena and noumena.

For Kant, the things that one sees are not noumena, that is, things in themselves, instead they are phenomena, meaning mental constructs created by the mind's subjective experience of the things in themselves. I wholeheartedly reject that doctrine of the Kantian philosophy. The perceived things, and the thought-about things in reality, are the things in themselves. If you look at a computer, or a table, or an apple, or a book, each of those things is a thing in itself. The thing that one sees is not a representation created by the mind in response to the thing that acts upon the senses. Rather the thing one sees and the thing that reflected the rays of light to the eye is one and the same thing.

Second, one could raise the objection that because the brain reasons what the perceptions are of by means of conceptually identifying the perceptions, the concepts used by the brain in this process bias or influence the act of conscious perception, so that the consciously perceived thing is actually a mental creation in response to the perception, a subjective mental interpretation of the perception that is created by the unconscious reasoning, rather than the objectively existing thing itself. This argument basically says that because consciously perceived things are perceived by means of the conceptual identification of perceptions, they cannot be objective, because the means of perception, the conceptual identification of perceptions, must surely influence them or add something to them. This argument assumes that a thing in itself must be seen through an unmediated experience in order to be known, and a means of knowledge will inevitably distort the object of knowledge. If one does not assume the subjectivity of the means of perception, then we can examine the conceptual identification of perceptions and see whether it adds anything to that which is perceived or merely makes an objectively existing thing known to the mind.

Let us ask whether concepts are subjective or unbiased, and what that would mean. The concepts, recall, are the means by which you are aware of the things that you are aware of, and this includes when you are aware of perceived things, for you are only aware of the thing by means of the concept of it, formed from the concept which identified the perception, so that you are aware of the perceived thing being the doer of every act of which the concept identified that perception (e.g. if the concepts of red and square identified the perception, then you see a red square). If the concepts are the means by which you are aware of a thing that objectively exists then the concept is unbiased, which means that it shows something totally objective to the mind's eye without adding anything to that thing which you are aware of. If the concept is subjective, then what you are aware of by means of it is created by the mind or the mind adds something to it.

The answer to the question of whether concepts are unbiased or biased depends upon whether the being of the concepts is created by the brain or by the objectively existing things. I have explained how the being of perceptions is caused by the external objects which they are of, and how concepts are created from perceptions. From this one can reason that the being of the external thing, not the being of the brain, causes the being of the concept. The brain defines the concept in the sense that the concept is written in the Brain Language, e.g. the brain knows that the synaptic connection of neuron A to neuron

B means that the umbrella is leaning against the chair, but the being of the concept as information is defined by its object, not the subject, e.g. by the umbrella, not by the brain. Thus, if a concept is reasoned, then it is unbiased and will show an objectively existing thing to the mind's eye. These concepts are reasoned from perceptions, they do not come from within the mind, and so they add nothing of the mind to that which is reasoned. What is reasoned comes entirely from the perceived things.

The question then becomes whether the use of the unbiased concepts to be aware of the perceivable being of something, i.e. to consciously perceive something, is going to consist of simply being aware of the objectively existing thing or whether the brain will add something to the perceived thing or create the perceived thing. Since the thing that one consciously perceives is the thing which was reasoned from the perception, the thing reasoned from the perception is the thing which the perception shows, and (referring back to correspondence) the thing which the perception shows is the thing that originated it, the thing that it shows is an objectively existing thing, and since it exists outside the brain the brain cannot add anything to it nor create it.

For example, when the brain identifies a perception with the concepts of red and square, and reasons from it that it is the perception of a red square, and reasons the existence of a red square from it, and you consciously perceive a red square, the red square that you see is not something that your brain created, it is the very red square which produced the perceptions that entered your eyes, which exists outside your brain, which you become aware of by reasoning it from the perceptions that it produced. That the things that you see are representations created in response to mental contact with things in themselves has no basis in perception. When you look at a red square you see a red square, not a representation of a red square.

Kantian theory can only be believed if one reasons from an unreasoned assumption that perception is subjective prior to reasoning what you are seeing from the seen things themselves. Look at something and ask yourself honestly whether you are seeing the thing that you see, or something else? If you are seeing something, then the thing that you see is what it is, and you can see that this is so by looking at something and observing what it is that you see. A representation of a thing is not the thing itself. A mental representation of a cell phone is not a cell phone. So when you look at a cell phone, if you are seeing a cell phone, then you are seeing a thing itself and not a representation of something else. What enables the brain to consciously see is not its interpretation of the perceptions. The brain's conscious perception is enabled by reasoning what the perceived objects are from the perceptions. Interpretation is subjective and adds something. Reasoning is objective and adds nothing.

How the brain works is the very opposite of Kant's theory. The structure of the mind makes it possible for the mind to see things in themselves, to which the act of experiencing adds nothing to experience. The perceptions are the brain's means of direct knowledge of the perceived things. The brain does not take sensory perception information and from it create the things that one sees. Instead, it takes sensory perception information and reasons what is perceived from it, by reasoning the object of the perception from the perception, and the product of this is perceiving the perceived thing. Reasoning enables the brain to perceive the thing itself, which exists outside of the mind, which is the object of the perception. Once the brain reasons the object from its perception, the brain knows the object itself. This is how it is possible to reason directly from objectively existing things in themselves, i.e. things as such.

Chapter Forty Six: Reason as a Three Step Process: (1) Perception, (2) Essential Reasoning, and (3) Conceptual Identification, a.k.a. Induction, Deduction, and Practical Application

Reasoning, in its true sense, is pure empirical essential reasoning. Empirical reasoning is reasoning which reasons its conclusions either immediately or ultimately from perceived things, i.e. the conclusion either comes directly from perceived things, or from conclusions derived from other conclusions that were reasoned from a line of reasoning that is ultimately grounded in first reasoning from perceived things. Essential reasoning is reasoning which relies on the essences of things, by identifying the essence, reasoning the consequence of the essence, and then identifying things by essence. Reasoning which is not empirical essential reasoning, reasoning which is not ultimately based on perceived things and is based on assumptions, i.e. corrupted reasoning, and reasoning which does not reason from the essences of things and is based on coincidence or accident, i.e. empirical associative reasoning, are pale imitations of reasoning, they are not real reasoning.

Conscious reasoning consists of three chronological steps. The first step is reasoning essences from perceived things. The second step is reasoning the consequence from the essence. And the third step is identifying perceived things by essences. Reason as a three step process could generally be called step one perception, step two essential reasoning, and step three conceptual identification. Or, to use names more common in philosophy, the three steps could be called (1) induction, (2) deduction, and (3) practical application. I shall explain each of these three steps in order, one by one.

Chapter Forty Seven: Reason as a Three Step Process: Step One, Perception and Induction

Step one of reason as a three step process is reasoning essences from perceived things. How does one first gain knowledge of things and their actions in order to reason from them? This knowledge first comes from perception. I explained a plausible hypothesis of how first reasoning gives the mind concepts from perceptions. The unconscious reasoning of a baby or young child gets a mass of numerical data from the sense organs, and reason picks out the patterns and ratios that present themselves from within the mass of random numbers, in order to identify perceptions in the sensory data. This then forms equations capable of identifying trends in the data, which are first concepts. Secondary concepts are then reasoned from first concepts.

Concepts are the means of awareness. After the brain has concepts it can use its concepts to be aware of their objects. The means of the brain's awareness of things are perceptions and concepts. These are also the brain's means of conscious knowledge. The perception of a thing gives knowledge of the actions of the perceived things. The perception is the means of the brain's conscious knowledge of the perceived thing. It is the means by which the brain consciously perceives the perceived thing of which it is the perception.

One can consciously gain knowledge of an action from perceiving things doing that action. One can consciously gain knowledge of an action by observing a thing do the action and reasoning a concept of the action from perceptions, or, consciously, by being aware of the action that one is perceiving. For example, by watching a pen write, and seeing that action it is doing, and being aware of the pen that one sees, and then being aware only of that action and not the specifics that one perceived, one would be aware of the act of writing. In this way one can reason the act of writing from a perceived specific writing thing. When one does this one forms a concept by means of which one can be aware of the act of writing, and one can then consider the act of writing as essence and reason its consequence, and then identify things doing that act as writing things.

When one observes a thing one can become aware of all the actions that one can perceive it doing, and then reason about those actions. All conscious knowledge of thought-about actions comes (either immediately or ultimately) after one reasons them from consciously perceived things. All conscious knowledge begins with consciously perceived things, reasons from them, and then reasons from the thought-about actions that one reasoned from perceived things. In this way all reasoning is based upon empirical observation, which simply means that everything consciously reasoned is reasoned from consciously perceived things.

Let me briefly dismiss three objections to the claim that all reason is empirical. The first counterargument against me is that spiritual truths are not reasoned from empirical experience. Second, that math and logic, and perhaps philosophical revelation and insight and wisdom, is not reasoned from empirical experience. And third, that synthetic truths may be discovered empirically but analytic truths are derived from the mind's analysis of concepts.

In reply to the first objection, my philosophy is a philosophy of science, meaning that I only recognize physical objects as existing in reality, so spiritual truths do not concern me.

Second, math is reasoned ultimately from empirical experience, because the concepts of numbers are learned by children from observing things, e.g. a child learns the number "five" from repeatedly observing five apples or five playing blocks or five toy trucks, and all math that professors reason from as adults is grounded in the math education they had as children and teenagers, which is based on empirical experience. If math were a world of spirit and not based in the physical world then we would not expect science to be able to use math to understand the physical world. Reason and philosophical wisdom, to the extent that they are achieved by understanding how reason works, are also based on empirical experience and observation.

Third, all concepts are derived from experience, either by reason or irrational computation. I do not recognize a distinction between synthetic and analytic truths. The proposition "the clothing store is down the street" is learned by observing the local buildings, empirically. But the proposition "a bachelor is an unmarried man" is learned only if one has concepts of humans, marriage, being single, being a man, etc., all of which concepts are derived from empirical experience and the observations of human life. Truth can be discovered by the analysis of a concept, but only if that concept was itself reasoned from empirical experience. Generally, if a truth is deduced from an essence, it is an empirical truth, because essences are reasoned from perceived things, when reason functions properly.

By reasoning thought-about actions from perceived things, one becomes aware of the actions that one reasoned, and one can then reason about these. For example, the act of being red is perceived, and the act of being an apple is thought-about, and once one has reasoned the act of being an apple, one can reason about things that do that being action by reasoning about the act of being, by reasoning the consequence of the action. The act of being an apple is perceived, but not directly, i.e. it must be reasoned from the directly perceived actions, like being red, being a certain shape, having a brown stem and small seeds and white flesh, etc. Apples are directly perceived, but one is only aware of the thing being an apple because the concept identifies the perception that one received, and if one perceives an apple before one has a concept of an apple, then one can consciously reason the being of an apple from the perceived specific apple, by reasoning what action the perceived thing does.

In the case of an apple, to reason a concept of an apple from a perceived apple, one could reason a concept of the act of being an apple by only being aware of the act of being an apple which the specific apple you were aware of does and ceasing to consider the specific apple's other actions. However, to reason that an apple is a fruit, one would first need to be aware of trees and fruit, either from extensive observation of plants or from someone explaining to you how plants use fruit to disseminate seeds. One would then need to identify this specific apple as a fruit, and then reason that being a fruit is essential to this apple being an apple and that being an apple requires this apple to be a fruit, from which you could use this apple to reason that an essential apple is a fruit, and then conclude that all apples are fruit.

If one perceives a real apple, one can see it be red, and have seeds, and a stem, and flesh, and be apple-shaped, and taste like an apple. From all of this, one can reason that it is a certain kind of fruit, the type of fruit with these properties, namely, that it is an apple. Because the apple is this set of properties, which can be seen, once one perceives the thing, one can identify the act of being an apple. If you hold an apple in your hand, it is doing the act of being an apple. Although being an apple is a thought-about property (which I would define as being the fruit of an apple tree or as having an apple's genetics), being an apple causes a specific set of perceived properties in the thing which is an apple, which form the visual aspect of an apple, i.e. the visual properties of the apple are the consequence of the essence of being an apple. You can see this apple do this action's visual aspect, i.e. you can see it be the set of properties which form an apple, so when you see this your perception gives you knowledge of the act of being an apple. To mentally identify and distinguish this act of being an apple, as a fruit which is red and round and has white flesh and small seeds and tastes this way and smells like this, etc., from the specific real apple one is looking at, is to form a concept of its being as the act of being an apple.

This, precisely, is the moment when induction takes place. Induction is the process of perceiving a specific thing and then isolating one of its actions from the rest of its being in order to begin a line of essential reasoning from the thing as the doer of that action as essence, i.e. from the essential thing which does only that action and no others. One has knowledge of the essence because the perceived thing is the entire perceivable total being act of the specific real thing, and the essential act of being is something that the real specific thing does (or is), so the essence is in the perceived thing that one knows from perception, so that you can go into the perceived thing that you know and isolate the essence which is in it. In other words, the perceived apple is an apple so the essence of being an apple can be reasoned from the perceived apple. In this way you can reason essences from perception and inductive reasoning.

One has a concept of the thought-about actions because one unconsciously creates a concept of the action from perceptions when one consciously reasons the thought-about being action from the perceived thing. The reason-computations of concepts and perceptions that the conscious reasoning consists of produces a concept of the thing when one consciously reasons to the thing, and becoming aware of the thing is the act which produces the concept that is the means of awareness. After one has reasoned an action from a perceived thing and one becomes aware of and gains knowledge of the action from the perceived thing one can reason from the action that one is aware of.

So we have seen how one gains conscious knowledge of the actions that things do, and how it is possible for the mind to directly be aware of, to directly know, a being action, i.e. an act of being, e.g. to be red or square. This knowledge makes conscious essential reasoning possible. Every action is the essence relative to all the other actions done by a thing, and the other actions are specifics relative to the essential action, because if a thing does the action, then the thing is a doer of that action, as well as all the other actions that it does. To reason about an action as the essence is to essentially reason about the action, to reason what follows from a thing doing that action, without reasoning from the thing doing any other actions. The knowledge of actions, and reasoning about an action that you are aware of, makes essential reasoning possible. Once you know and can distinguish the action, considering that action, and reasoning the consequence of that action, is to reason about that action as an essential action. The essence of a thing is identified by identifying the distinct actions that a thing is doing and reasoning from one of the actions as the essence. Perception provides knowledge of the perceived being acts of the perceived things, and from there, once one knows the being acts, one can then proceed to step two, which is essential reasoning.

If one knows a thing by perceiving or being aware of it one must know its total being action, but how does one then distinguish the different actions of which the total being action is composed? Once one knows a thing, to be aware of the separate actions is to separate them. When one is aware of a total act of being, the total act of being consists of all those actions, so one is already aware of all the actions, and one can reason from one as the essence by considering that one and not reasoning from any of the others. Once one either perceives or is aware of a thing (by perceiving it in the first case, by reasoning it in the second case) one has direct knowledge of the thing's known actions, so one can reason from them, and one can reason from the different components of the total action by considering them separately.

For example, when one perceives a white wooden square box, one perceives something doing the act of being a white wooden square box, which is the total being action, which consists of the act of being white, and the act of being square, and the act of being made of wood, and the act of being a box. One can identify the act of being square from the perceived thing's shape, and the act of being white from the perceived thing's color. Unconsciously these can be distinguished because different aspects of the perceptions show the different aspects of the perceived thing, but these can be consciously distinguished because one consciously perceives the thing doing the two actions, so one can see the difference.

Because the thing is white and square, and one sees the thing, one must be seeing the total being act, and those two being acts are parts of the total being act, so one must be seeing the act of being white and the act of being square when one sees the box, and if one sees the actions then one knows them via perception in order to reason from them. A perceived thing is a real specific thing, and so the perceived thing must the entire thing that it is (although, note that while all of a perceived real specific thing's directly perceivable properties must be perceived, the thing's thought-about properties, such as being the jewelry box that belongs to your sister, must be inferred or conceptually identified). Because one is reasoning from the total being action of the things themselves that one perceives, one can mentally distinguish the different actions done by a known thing that one is aware of, and reason from each separate action as an essence, because the total being act is the sum of all its parts connected together.

Chapter Forty Eight: Reason as a Three Step Process: Step Two, Reasoning the Consequence of the Essence Using the Essential Thing

Step two of reason as a three step process is reasoning the consequence from the essence, after induction achieves a concept of the essential being action. Once one is aware of an action, one can reason the consequence of the action from the action. The consequence of an action is the actions that doing the action causes a thing doing the action to do. There are three ways in which doing an action can cause a thing to do a second action. The first way is for the first action to directly cause the doer to do the second action. The second way is for the first action to require the thing to do the second action in order to be done, i.e. the second action is the means of the first. And the third way is for the second action to be a part of the first action, i.e. to be one of the actions of which the first action consists. In each of these three cases, doing the first action will make it a matter of ontological necessity for the thing doing the first action to do the second action also. And if the second action is the consequence of the first action, and the first action is the essence of thing X, then all X's will be do the second action, because the act of being X will require them to do the second action, and so it will impossible for an X to not do the second action. If the second action is the essence of thing Y, then all X must be Y, and it will be a necessary universal truth that every X is a Y. Thus, the ability to reason what doing an action causes, requires, and consists of is necessary for the empirical essential reasoning of universals.

Reasoning from essential things is a tool for reasoning the consequence of the essence. One conceives of a thing doing only the essential action, and one reasons what actions doing that action requires the thing doing the action to also do, and the essential thing must do this because it does the essential action, so it must do the consequential actions, and one reasons the consequence of the essence by analyzing the essential thing. In other words, one reasons what doing the essential action causes the essential thing to be, and one is then aware of a thing being that thing and nothing else. When one identifies things by the actions that they do, one identifies them as that "kind of thing," i.e. a thing doing the essential action, and the essential thing is a thing doing that action. For example, one identifies a thing as "a horse," "a boy," or "blue," or "iron," by the acts of being which the thing does. If a thing does the act of being a horse, then it is a horse. When one says this of specific things one says that the specific thing does the essential action, but when one simply thinks of the thing doing the essential being act and no other specific acts, e.g. when one thinks about a horse, one is thinking of an essential thing, with all other properties are variables.

When one reasons something of the essential thing, and then identifies real things as that kind of thing, one can then apply the reasoning to the real thing, to know that it does the consequential action because it does the essential action. What is true of the essential thing must be true of all real things that do that essential action. The real things of kind X or of the type of X, the real X's, must do everything that X(e) does, including the consequence of the essence, because X=e(X) and e(X)=c(e(X)), therefore X=c(e(X)). To phrase this differently, an X is an X because it has the essence of X, the essence of X causes the consequence of the essence of X, and X must have the essence of X, therefore X must have the consequence of the essence of X. The essential thing only does the essential action and the consequential actions, so it is a very useful tool for essential reasoning, but essential things are not real things, they do not exist, because the place and time of the essential thing is variable, and real things always exist at a specific place and time.

If one gets confused and equates the essential action with the essential thing then many devastating mistakes follow. For example, Plato's theory of Forms rests upon the belief that the essential thing is being and causes the being of the perceivable things, and that leads one to believe that the essential things exist (which cannot be reasoned from perception, as Plato knew), and then the belief in ideal non-physical things vs. imperfect physical things follows, and the belief that the specific perceivable things do not exist, or that they lack the perfect complete existence of the Platonic Forms, follows. As can be seen, all of this begins by conceiving of essential things and then thinking that they are being and the cause of the being of things, whereas in reality the actions that things do are being and are the cause of the being of things, but the actions are not themselves things, they are actions, they are what things do.

338

Note that a thing is a set of properties, and to be precise, when we say that a thing does the act of being something, we are really saying that one property does the act of being another property to be a thing, e.g. when we say that the square is made of wood we are saying that the square is wood, that the act of being square does the act of being made of wood. Similarly, when we say that wood is a substance and a square shape is a form and the wood is square, we are saying that a wooden square exists, i.e. that a thing is wood and square. This is no different from saying that the property of being wood does the act of being square and vice versa, i.e. the wood does the act of filling up an area of shape which is square, and the square shape does the act of being full of particles of wood.

A thing is not a substratum that properties hang onto. In other words, we do not attribute predicates to subjects. Rather the thing itself is merely a set of properties, each of which is an act of being, which are united as the real specific physical thing that is all of its properties. If we discuss one property, e.g. being wood, then the property and the thing are identical. The thing is the property, i.e. wood is being wooden. Just as a thing and its properties are identical, which avoids the problem of knowing what a thing is as separate from its properties, I argue that a thing and its acts of being are identical, which also avoids the problem of knowing a thing as separate from the actions that it does. I speak of things doing actions and things having properties, but for things in reality no hard and fast difference exists between a thing vs. an act of being vs. doing an action vs. having a property. These terms should not lead to muddled philosophical confusion. A wooden square is a thing. It is the same thing as the set of the two properties of being made of wood and being square. That is the same thing as the total act of being composed of the act of being wooden and the act of being square, which is the act of being a wooden square. One frequently speaks of a thing doing an action, or a thing having a property, or a subject having a predicate attributed to it. But this is because all real things do multiple acts of being, since at a minimum they do an act of being at a place and at a time and having a physical substance and a shape. A thing doing an action is one property having another property or one action doing another action, e.g. a foot kicking a ball is the act of being a foot doing the act of kicking a ball.

In the "Euthyphro" Plato begins his conception of the Form of a thing as the basis of which a thing is or is not something, so let us examine this. If blueness is that thing which makes a thing blue, if things are blue on the basis of blueness, then blueness is the act of being blue, because on the basis of whether it does the act of being blue or not a thing will be blue or not. But if you tell a person to imagine blueness, then that person will probably imagine blue, the essential blue thing. For this reason, the suffix "-ness", as in blueness, can be confusing.

Let us consider one example of reasoning from an essential thing. This example will use reason to show that every X is Y because Y is a part of the act of being X. I will discuss essential reason using cause and effect, which is somewhat more complicated, in the next section. I can consider a red square(e). The thing that I am aware of is a red square, and I think nothing else of it, I think only that it is a red square. The thing that I am aware of is red. Because it is the essential red square, and it is red, a red square thing must be red, because a thing which does the act of being a red square and no other actions is red, so doing the essential action must have required it to be red, it did no other actions besides the essential action which could have required it to be anything. If the essential red square is red, then every red square must be red, because everything besides the essential action is variable in the red square. The red square is red because it is a red square. Therefore every red square is red. The act of being red was in the act of being a red square, and my reasoning analyzed the act of being the thing that I was aware of in order to identify its component parts; that is how the line of reasoning worked. This is an example of how one reasons from essential things. However, as this example shows, it is easy not to pay attention to a critical step, namely, how one reasons the consequential action from the essential action. In this case it was reasoned that the act of being a red square consists of the act of being red and the act of being square, which can be reasoned because from the essential red square thing being red and square, or from an analysis of the act of being a red square.

Reasoning from essential things is also useful so that one does not become confused as to whether one is reasoning from the specifics of a thing or from the essence of a thing. The only actions which all things of a kind certainly do are the actions that a thing does because it does the essential action, i.e. the essence and the consequence. If one reasons what a thing is doing, but one considers the specifics as well as the essence, then what one reasons is specific to things with those specifics. One must define the essence with care and precision, and maintain consistent definitions throughout the line of reasoning, in order to use essential reasoning correctly.

For example, many scientists reason things that are true of lab mice and try to apply this to humans, because human DNA is somewhat similar to mouse DNA on account of common evolutionary ancestors, but unless the scientists reason their conclusions from the mice doing the act of being mammals, or specifically from the DNA that mice and humans share, then the conclusion was reasoned from the specific act of being a mouse and it does not apply validly as a necessity to all things doing the act of being a human, although there may be some correlation. Only what is reasoned from a specific mouse (or any quantity of specific mice) from the mouse doing the act of being a mammal, or the act of being an animal, or the act of being a thing, or having whatever DNA was inherited from the ancestor animals that humans and mice share, can be applied to humans, because those are actions which both the act of being a mouse and the act of being human require, so that only the consequence of that essence will be done by all humans and mice. Only if the consequence is reasoned from an action that a human must do consequentially of being a human, can reasoning from mice be applied with universal certainty to humans. When a scientist learns that a substance is not harmful to mice, this provides no knowledge whatsoever as to whether or not it is harmful to humans, because the specifics of the act of being a mouse could have made it harmless, and a human does not do the specific actions of which the specific act of being a mouse consists. If the scientist considered the essential mammal in his/her reasoning, and did not allow the specifics of the mice to enter consideration, then the reasoning would apply to humans, and to all mammals.

When one reasons from specifics then the conclusion is not the consequence of the essence, but the reverse is also true, i.e. when something is reasoned from a specific thing but it is reasoned from the thing doing the essential action then it can be reasoned with certainty of all things of that kind, and of the essential thing. For example, if one considers a human doing something and reasons that

341

the person was able to do the action because the person was human (with a functioning mind), e.g. that person is able to think, then one can reason from this that a human(e) is able to do the action, and from this one can reason that every human is able to do this, and having identified oneself as a human one can reason that oneself is able to do this. But only if this is reasoned not from the specifics of the person, but from the person being human (and having a working mind) and only that action, is this line of reasoning valid as a proof that all humans can think. It is essential reasoning only if it is reasoned from the essence, and if it is reasoned from the specifics (or non-essential variables) then it has no universality or necessity.

One can distinguish between the specific actions and the essential action by distinguishing between the different actions that a thing does, and then reasoning something from one of the actions the thing does and not the others, e.g. if one reasons that having a brain and a spine and muscles makes it possible for a human to do something, like walk around, and having those are some of the actions of which the act of being human consists, then one can reason that all humans can do that, i.e. that all humans are able to walk on two legs, but if one forgets that one reasoned this from those organs functioning correctly, i.e. if one does not include all the specifics which were reasoned from in the action of which the consequential claim is made, then one might run into something which lacks those specifics, like a human who has a spine which does not function and is confined to a wheelchair, and then one's critics could claim correctly that what you reasoned was not true of all humans, because physically disabled humans do not walk around on two legs, so one cannot say "humans walk on two legs".

And the skeptic critics would be right, and your mistake would have been thinking that you were considering a thing doing the act of being human, that is, a human(e), when you were considering a thing doing the act of being human and having healthy organs, that is, (a human with healthy organs)(e). You reasoned from the thing being human, and from the thing having organs, and from the organs being healthy, but you then forgot about one of the specifics that you had reasoned from, and then thought that the specific was variable, as if it had not been considered to begin with. It is for this reason that non-essential reasoning can never be certain and always leads to mistakes, and a precise and thorough consideration of what actions the essential thing does can help to avoid this. One example is MPTP/MPP+, a substance which does nothing to rats but which kills humans, and which scientists tested on lab rats and then thought was safe for humans.

Chapter Forty Nine: Reason as a Three Step Process: Reason Using Cause and Effect

Scientific reasoning exists in two manifestations: description and explanation. Description says what a thing is and what exists. Explanation says why a thing is what it is and why things exist. All scientific reasoning begins with description and then proceeds to explanation. Description consists of perception, and also of reason which assists perception, while explanation consists of essential reasoning, especially by deploying cause and effect analysis. Description would consist of research and empirical scientific experiments aided by probability math to identify statistically significant findings. For example, description could consist of a study which showed that people who eat a lot of fatty and fried and sugary foods die of heart attacks more frequently than the people who eat mainly fruits and vegetables. Explanation would then explain the data. For example, explanation would proceed by analyzing the way the heart and the blood systems work to show that fatty foods cause increased bad cholesterol in the blood which produces clogs in the veins in the heart which trigger heart attacks. It is a cliché in science that "correlation is not causation," and this is true as far as it goes. But some scientists and philosophers get confused and think that science is only capable of identifying correlations and cannot prove causes. The truth is that science can achieve knowledge of cause and effect by means of empirical essential reasoning.

The "guess and check" theory of science, established by Karl Popper, asserts that a scientific theory is mere conjecture, which can always be disproven (i.e. falsified) by experimental refutation in the future, but cannot be proven true to the extent of perfect absolute certainty. For Popper, to say that an idea is scientific means that it is capable of being disproved by new evidence. On the contrary, the empirical essential reasoning theory of science asserts that science, in the form of reason and perception, can reason that a thing in physical reality is a cause which is causing the observed effect, as known from empirical induction and deduction, e.g. studying how heart attacks happen, such that cause and effect can be proven true.

In other words, explanations can explain descriptions, and scientific explanation is possible and can rise to the level of knowledge. If the explanation is based on essential reasoning, and reasons the consequence from the essence, then no set of facts can exist, and no observations from future experiments are possible, which can disprove the explanation, because essential reasoning achieves necessary truths which are universally true.

The theory of the consequence of the essence explains two elements of the scientific method as used in laboratory experiments. The two elements at issue are (1) statistical significance, and (2) controlling for variables. In both cases, scientists use the scientific method to identify the "because" relationship in physical objects. X because of Y means that X is the consequence of the essence of Y. Statistical significance uses probability math to establish that X is not caused by random chance, which justifies the belief that X was caused by Y. If only Y and chance are present, and chance did not cause X, then Y must have caused X. Controlling for variables enables scientists to reason the consequence of the essence directly from the essence as it exists in real perceived things, instead of using essential things. For example, suppose that the test group of lab mice is fed food plus water plus Chemical Z, and the control group of identical mice is fed the same food and water but not Chemical Z. Then the test group mice have effect R, e.g. their fur turns red, and the control group does not. The scientists can reason that the food and water did not cause R, because if it did then both the test group and the control group would have R, and the control group did not have R. But since the only factors affecting the mice were food, water, and Z, if food and water did not cause R then Z must have caused R. Thus the experiment turns food and water into variables for the mice which leaves behind Z affecting mice as the essence which has consequence R.

Now let us explain the role of cause and effect in essential reasoning. First we should examine how one reasons cause and effect from perceived things. The act of one thing causing another thing is the act of one thing acting upon another thing. When one thing acts upon another thing, it can change the action that the thing acted upon is doing. If it does change it then the first thing's acting upon the second thing is the first thing's act of causing the second thing to do something, causing the change of action done by the thing acted upon. To cause something's changing is to change something. One thing changing something else can be perceived, and the act can be reasoned from the things perceived doing the action of changing something else. For example, when observing the billiard balls hitting each other while playing pool, from seeing one ball move another ball by hitting it, which consists of one ball acting upon another and changing it, cause and effect can be directly perceived. The concepts of cause and effect can be reasoned from things perceived causing and being effected by other things, i.e. the existence of cause and effect can be reasoned from perceived causes and effects, when you perceive one thing acting upon another and causing it to do something. For example, if you open a door, then you are the cause of the door opening. When one domino knocks down another in a chain of dominoes, the one domino falling onto the next one causes the next one to fall. Thus, cause and effect can be reasoned via induction from empirical observation.

To change something is to act upon it when the action done to it changes the action that it does. This change of action consists of motion leaving the causing thing and entering the caused thing, when the two come into contact or through a means which carries the motion from the causing thing to the effected thing, when this motion combined with the actions the effected thing is doing produces a new action done by the effected thing or ends an action that the effected thing was doing. So all cause and effect consists of the transference of motion from one object to another. For example, when you push a motionless thing, it begins to move, the beginning of an action, or when you kick a soccer ball (a.k.a. a football) coming towards you, you do not necessarily change its motion, i.e. add to or subtract from its speed (although you might, it is variable), but you do change its direction, which consists of it ceasing to do the act of moving towards you and beginning to do the act of moving away from you after you kick it away from you. To create something is to change it (the thing as the doer of the new action is created), for example when one moves something from one place to another, one has created a new thing, since its location has changed, its act of being at a certain location has changed, and so the thing's total being action has changed. If you move a pot of flowers from the windowsill to the counter, then the flowers on the window have ceased to exist, and the flowers on the counter have been created.

346

However, obviously, as the doer of everything else it was doing it is still the same thing, unless the other actions were changed also, e.g. the pot of flowers still does the act of being flowers, so as the flowers it is one and the same thing on the counter which had previously been on the windowsill. Only when a thing no longer does any action that it was doing is the old thing totally gone, replaced by a totally new thing. If one changes every action, causing every old action to cease and causing entirely new actions, then one has created a totally new thing.

When the thing considered was the thing as the doer of a certain action, then when this action, which was the essence of the thing, ceases to be done, then the thing is destroyed, even if it continues to do some of the other actions in the same way that it did before. It is the same as the doer of those other actions but it is not the same as the doer of the essential action. For example, when a person no longer does the act of being alive, i.e. when a person dies, the body might still remain, in the coffin, with most of the organs and cells still intact, but the life no longer exists, because the substance is no longer doing that action which it did as a whole which was its form as a living being, i.e. the essential action of the living thing, the act of being alive. Also, the word which names creation is used to show two different acts. The first shown action (i.e. the first definition of the word) is to cause a thing to begin to exist. The second shown action is that to create something is to take a substance, or parts, which already exist, and cause it or them to be something as a whole. For example, when a group of people create a car they take different parts and cause them to do the act of being a car as a whole, by acting upon the parts to cause them to connect and do an action together, namely to drive, or when a sculptor takes marble and carves a statue, she created the statue from the marble by causing the marble to do the act of filling a certain area as a whole, causing it to be a certain shape which looks like something.

Thus, from the observation of one thing acting upon another, as for example one thing moving another, causing another to move, one can reason the concepts of cause and effect. The one thing causing and the other thing being effected are both perceived, and after perceiving causes and effects it is a simple matter of isolating those actions from the other being acts that the perceived things do. It is easy because these actions are directly perceived, as for example the act of being white and the act of being moved do not look like each other, they can be distinguished by perception. To be moved and to be caused are the same action. As for the question of how one knows that the one thing moved the other, the answer is that it is perceivable, for example one can see the motion transfer from one billiard ball to the next when one watches pool balls in a game of pool.

Having reasoned concepts of cause and effect from sensory experience, one can distinguish between the being of the cause, which is the act of being a cause, which is the act of causing, and the existence of a cause, which is the cause's act of existing. One can also distinguish between the being of the effect, which is the act of being an effect, which is the act of being effected, which is the act that the thing is caused to do, and the existence of the effect, which is the effect's act of existing. The act of existing is the act of objectively being at a specific place and time. The being of cause and effect are the actions that the things do, and the existence of the cause is the act of existing done by a thing that also does the act of causing, and the existence of the effect is the act of existing done by a thing that also does an act which it was caused to do. Because things can be caused to exist, and things can also be caused to do other actions, the existence of a thing can be caused, and each being action of a thing can also be caused, and the existence of a cause or an effect can be caused. And a thing can be caused to cause, as in the case of a domino in the middle of a chain of dominoes, in which case the thing which is caused to cause is the effect in relation to the thing that caused it to cause, and is the cause in relation to the thing that it causes.

One can also distinguish between instantaneous cause and continuous cause. The difference between them is that an instantaneous cause produces an effect which then exists independently of the cause, while a continuous cause continually causes the effect to exist so that the existence of the effect depends upon the existence of the cause. For example, a foot kicking a ball is an instantaneous cause of the ball's motion, because the ball continues to move after the foot is no longer kicking it, even though the kicking is the cause of the ball's motion. A person holding a cup is the continuous cause of the cup's not falling to the ground, because at each moment the hand's act of pushing up on the cup causes it to not be pulled to the ground by gravity, so that the act of holding continually causes the act of not falling, and the existence of the effect depends upon the existence of the cause and its existence would cease if the existence of the cause ceased, as can be seen if one drops the cup. Note that of instantaneously caused effects, the existence of the effect does depend upon the existence of the cause, but it depends upon the cause existing in the past, at whatever precise moment the effect was caused, rather than depending upon its existence at every moment in time when the effect exists.

Here I will refute one objection which may be raised against me, the exceptions argument. This is a common objection used by philosophy professors when they want to look clever and sophisticated. It has been presented by Bertrand Russell, as for example in Russell's "The Analysis of Mind," although many others have also used it. The argument is used against cause and effect and also against universality and necessity generally. It says that exceptions to every rule can be found, so we cannot know cause and effect with certainty, therefore causation analysis is mere guesses and approximations that can be wrong. For example, we can say that one domino causes another domino to fall, but if it is possible that one of the dominoes is glued to the floor, then our cause and effect reasoning will not be necessary and universal. More generally, this can also be phrased as a challenge to my theory of essences. For example, if I say that the consequence of the essence of being a mouse is to like to eat cheese, and someone shows me a mouse that hates cheese, then the necessary consequence of the ontological essence was not in fact necessary, and my philosophy teeters and crumbles. As applied to essential reasoning, the exceptions argument claims that for two things with the same essence one thing might be fundamentally different from the other thing despite them both being the same type of thing as the doer of the essential act of being.

I answer the exceptions argument with two replies, first, intervening causes, and second, precision and clarity in the identification of essences. When we consider cause and effect analysis, we only consider the essence of what we think about, and we do not consider as a specific that something else will enter the picture as an intervening cause. In essential reason, all specifics are held variable in order to reason from the essential thing. An intervening cause is something outside the chain of causation that acts upon it to alter the effect of the cause. In essential reasoning it is normally implied that the essence does not contain an intervening cause. If we redefine our essential things to include an intervening cause as a specific then our reason would deduce the consequence of this thing plus that intervening cause. So when we think about a chain of dominoes we have defined that we are not considering what would happen if one of them is glued to the floor. But if we did consider the essence of dominoes plus glue then our cause and effect analysis would be correct and we would reason that the motion will stop once it reaches the glued-down domino. The glue is not an exception to a general rule, rather it is an entirely different scenario than the one about which we reasoned our universal necessary general rule.

The identification of variables vs. specifically defined intervening causes relates to identification of the essence, which simply means that in empirical essential reasoning we must define the essence with precision and clarity. If we reason that all mice eat cheese, and we find a mouse that will not eat cheese, we must reconsider the essences that we reasoned from. If we reasoned that a mouse is a thing with a mouse-shaped body and organs and mousy DNA, and we reasoned that the mouse DNA will cause the thing to eat cheese (as a result of a biological genetic analysis showing that mouse DNA defines a trait of eating cheese), then the mouse who will not eat cheese must have abnormal or mutated DNA, so it is a mouse-like thing but it is not in fact a mouse because it lacks the precise set of properties which are the essence of being a mouse, which includes the property of having mouse DNA. In reply to the critics who say that two things with the same essence could be different, I say that essence is being so to have the same essence is to be the same thing as a thing with that essential property. Therefore the consequence of the essence identified by essential reasoning must be possessed by all things with that essence with no exceptions.

The exceptions argument points out a problem, but that problem is not the inability to reason the consequence of the essence in a way that is necessary and universal. Instead, the problem is that we must be very careful and intelligent in our choice of defining which essences to consider. For example, if we make a mistake in defining "mouse" then we could get confused about whether all mice eat cheese and be subject to exceptions. DNA in different animals tends to have slight variations, so we must proceed with caution in choosing what aspect of DNA to define as the essence of a species. Defining "mouse" is not a philosophy of language "move in the language game", i.e. I am not defining the word "mouse" to exclude exceptions which refute me. Rather I am examining the being that actually exists in objective reality, i.e. what it is for a real existing thing to be a mouse. If I know what a mouse is then I can say that all mice eat cheese and no exceptions are possible. We may also view the genetic mutation which causes the mouse-like thing to dislike cheese as an intervening cause which interfered with the chain of causation that being a mouse causes a thing to like cheese. Necessity and universality from essential reasoning are rescued if we distinguish the two essences, e.g. (a mouse)(e) vs. (a mouse-like thing with an intervening cause of a new gene)(e). This distinction enables us to say that the fact that "mice eat cheese" is necessary and universal with no exceptions.

Once one has reasoned to cause and effect from empirical observation, one can employ knowledge of cause and effect in one's reasoning. Three ways exist to employ cause and effect analysis in essential reasoning. First, by reasoning the being of an effect from the being of a cause. Second, by reasoning the being of a cause from the being of its effect. Following either the first or second method, one reasons the existence of the cause or effect from the existence of the effect or cause which it was reasoned to be the cause or effect of. The third method is to perceive the cause causing the effect, which means, to directly perception the causing action, from which one can reason that the cause causes this effect.

In order to reason the being of an effect from the being of its cause one must first know what the causing action is, and then reason the effect from it. Nothing more can be said of this universally, because in all cases one reasons the specific effect from the specifics of the cause. The effect is that action which a thing will do when a thing does the causing action (to itself or to another thing).

Here is an example of reasoning the being of the effect from the being of the cause. For example, you can reason that the glass will crack from a bat being swung and hitting a pane of glass. You would first need to know of everything of which the cause consists, by reasoning concepts of them from perception. These include glass, a

351

bat, swinging, hitting, breaking, the hardness and softness of substances, and the speed and force of motions. Each of these is directly perceivable and can be reasoned from perceived things. Once one has reasoned concepts of each of these things from perception one can consider them all together, and when combined all together they produce the effect that they cause as a whole. One can reason the combined effect of these things from the things together. Referring back to the windmill example (in the section on perception), if one knows what action the thing is doing one can reason what it will do in response to something, i.e. when it comes into contact with something. This is true because the act that a thing always does is the same act that it will do to the things that it comes into contact with. Thus if one has reasoned a concept of the action then one has sufficient knowledge to reason its effect.

When a bat is swung at a pane of glass and hits it, it breaks it because something harder than something softer will break it when hit with sufficient speed and force, and swinging a bat gives the bat sufficient speed and force. The latter can be reasoned from the act of swinging, which can be reasoned from perception of the act, and the former can be reasoned from the hardness and softness of substances and the act of breaking, which can be reasoned from perceptions of substances breaking. Observing a thing breaking, one can see that one thing breaking another consists of the one thing separating the other into parts, and the hardness and softness of things is the strength and weakness of the connection of their parts, so that when two things come into contact, if one thing has sufficient speed and force and its parts are hard enough to not be broken by the other thing, then it will break the other thing by separating the other thing's parts without itself being separated.

Thus, once all of the elements of the cause are understood, with this understanding coming from reasoning concepts of the things from perceptions of them, then one can reason from the essential actions of the cause, from what those actions consist of, and reason to what its effect will be. One can reason what its effect on itself will be from the cause acting upon itself, and one can reason what its effect on another thing will be can be from the essence of the cause acting upon another thing, from reason combining the cause and the other thing in the reason-computation. And if one considers the specifics of the thing that it acts upon, if one has already reasoned concepts of that specific thing, then one can reason what the effect of the cause acting upon that specific thing will be, from the being of the cause and the being of the specific thing.

In order for one to reason the being of a cause from the being of its effect, one must first understand cause and effect, and then know what the effect is, and from this one can reason what thing would cause that effect. If one learns of it in this way, then one's knowledge of the cause can only be as specific as one's knowledge of the effect. Also, one might reason that many things could cause the effect (that "could" being ignorant possibility), but if one knows all the specifics of the effect, or even enough of them, one will be able to reason what one thing would cause precisely that effect. For example, if one knows that this thing is a piece of clay, then one can reason that its substance was caused by any combination of chemicals which can form clay, but one can reason nothing about the cause of its form. If one gains more specific knowledge of it, for example if one knows that it is a piece of clay shaped like a human knight wielding a sword, one can reason that its form was caused by a human artist, and if one gains more specific knowledge of it, such as where it was found, one can reason in what general area the artist was, e.g. that the artist was a Renaissance era sculptor. The more specifics of the effect which are known, the more the specifics of the cause can be reasoned from them.

Here is an example showing that one can reason the cause from the effect without any perception of the cause whatsoever, if one has reasoned what the effect is from perception, and reasoned any other necessary knowledge which pertains to the effect from sensory experience. One can reason what the cause is from the effect, if one knows what the effect was. For example, one could reason what knocked over a tree from the specific effects of the cause observable on the tree, i.e. from the perceived tree one can reason what caused it to be knocked down. If one saw axe-marks one could reason that it was cut down, because only being cut by an axe would leave axe marks, and if there are beaver tooth-marks all over its base, one can reason that beavers knocked it over, because only beavers leave beaver tooth-marks, and if it snapped sharply but without marks of any causal agent at the base or along the tree, and an examination of the tree reveals that it was strong and did not snap from weakness, one can reason that wind or something else which leaves no trace

353

knocked it over, and an examination of the tree shows that the tree was weak and rotting, and the break shows marks of rot, one can reason that the tree collapsed by itself, and if one finds termites at the base of the tree eating it, and traces of termites along the break, one can reason that termites ate through the tree and made it collapse, etc. There will be traces of the being of the cause in the being of the effect (these traces consisting of the specifics of the effect which the specifics of the cause caused) and the being of the cause can be reasoned from these traces.

One can also learn what the cause of an effect is by directly perceiving the cause causing the effect. This is reasoned from perceiving the cause causing the effect, i.e. from the perceived cause and the perceived effect and the perceived act of causing, not by reasoning the cause from the being of the perceived effect. One can reason that a cause causes an effect from perceiving it. For example, by perceiving bees making honey one can reason that the bees cause the honey directly from the perceived things. Note that in this case one reasons that the bees cause the honey not from perceiving the bees and then perceiving the honey, not from seeing cause and then seeing effect, as Hume would have thought, but by perceiving the cause actually doing the action of causing the effect, e.g. bees making the honey, by actually seeing the bees perform the bodily act of creating the honey from flower nectar.

When one reasons what the effect of a cause is by perceiving the cause causing the effect, this only gives one knowledge of the cause or effect of the specific thing that one has perceived, unless one reasons that it caused or was caused by its effect/cause because it did an essential action, which would require an understanding of the essential action from the perception of it, via induction, and then reasoning what the essence would cause or be caused by from the essence. For example, if one ignites a rocket cap and watches it blow up, one can reason from this that that specific rocket cap is caused to blow up by being ignited, but if one can reason that it blew up when ignited because it was a rocket cap, in other words that its act of being a rocket cap caused it to blow up when ignited, then one can reason that all rocket caps will blow up when ignited, from one specific perceived exploding rocket cap.

In a line of reasoning similar to cause and effect analysis, one can reason what the creator is from the created thing, because to create is to cause a thing to exist. For example, one can reason that it was created by humans from the being of a computer, reasoned from two premises, (1) the computer being a machine on this planet, and (2) humans being the only things on this planet that create machines. If one can reason that only that thing could cause something to do the consequential action, then one can reason that the essential thing was caused by that thing consequentially of its essence, because the creator is required for the thing to do the essential action. For example, one can reason that only an intelligent thing can create a machine, because (1) reasoning is necessary to invent a machine, and (2) only intelligent things can reason, so one can reason from these two premises to the conclusion that a machine(e) is created by an intelligent thing(e), from what the essence of what a machine is, and so reason that all machines are created by intelligent beings.

One can reason the existence of a cause from an existing effect (in other words, from the existence of the effect), and one can reason the existence of an effect from an existing cause. Reason that an effect exists from perceived things, and reason what the being of the cause of an effect is from the being of the effect, and from this you can reason that the cause exists. Or reason that a cause exists from perceived things, and reason what the being of the effect of the cause is from the being of the cause, and from this you can reason that the effect exists. Both of these are essential lines of reasoning. The existence of the cause will cause the existence of the effect, because causing the effect to exist is what the essential act of causing consists of. And the existence of the effect requires the existence of the cause, because to be effected is to be created by the cause, so that the effect could not exist without the cause, and the cause must exist in order to create the effect in existence. In the case of reasoning the existence of the cause from the existence of the effect, the current existence of the cause can be reasoned if it is a continuous cause, and the existence of the cause at one time in the past can be reasoned if it is an instantaneous cause. Each of these lines of reasoning achieve knowledge, because the existence of the cause makes the effect exist consequentially of its essential act of being a cause, and the existence of the effect requires the cause to exist consequentially of its essential act of being an effect.

Chapter Fifty: Reason as a Three Step Process: Some Examples of Reasoning

Here are some examples of reasoning, intended to translate abstract theory into practical reality. The first, the glass example, is an example of reasoning the consequence of the essence from the essence requiring the consequence as the means of being what it is. The second, the sinking lead example, is an example of reasoning using cause and effect. The third, from chemistry, is an example of reasoning the effect from the cause. The fourth example, of wood and saw, discusses reasoning cause from effect. The fifth is an example of essential reasoning.

The fact that the thing that cuts glass must be harder than glass in order to cut it is reasoned from the act of cutting, a concept of which can be reasoned from the perception of things doing the act. One thing cutting another consists of the first thing separating the second thing's parts. The hardness of a thing is the strength of the connection of the parts, so the cutter must be harder than the cut thing, or else the cut thing would cut through the cutter, i.e. when two things come into contact with force in the direction in which one would go through the other's parts, the one with the stronger connection between parts has its parts cut through the connection between the other's parts, because the harder thing was stronger than the softer thing, i.e. the thing with a connection between parts stronger than the weaker had the ability to separate the parts with the weaker connection. Employing this template to a specific case, you can reason what will happen when a diamond scratches a pane of glass from one's knowledge of diamond, glass, and the act of cutting. Specifically, one can reason that the diamond will leave a scratch on the glass because diamond is harder than glass and is sharp.

Lead sinking in water can be reasoned from the act of falling due to gravity, a concept of which can be reasoned from the perception of things doing the act. Lead must sink in water, because lead is heavy and water is light, and weight is the mass times the pull of gravity. Lead is pulled down by gravity and the water does not hold it up so it sinks. Lead being heavy consists of lead being what it is in response to gravity (which consists of the mass of its atoms being what they are according to its identity as a chemical substance) and water being light comes from water being the chemical liquid that it is, so that if lead is lead and water is water, lead must sink in water.

We can also find other examples. For example, one can reason what molecular substance was created from the two elements that were mixed together and the proportion of the atoms and how they were mixed. For a specific example, if one combusts wood by fire then one can reason that it will produce the chemicals in smoke as a result of the chemical reaction. Or if one combines hydrogen and oxygen atoms at a ratio of two to one via a chemical reaction one can reason that water is created.

Reasoning cause from effect can be done, but it is more difficult that reasoning effect from cause because often many different causes could have produced the same effect. If one sees a block of wood that has a cut in it, and one sees a blade and wood shavings lying next to it, one can reason that the blade probably cut the wood. But, because it is unlikely but possible that a drill bit cut the wood and left the shavings and someone removed the drill and put the blade down, one would need to see something that would confirm that the blade, and not a drill bit or anything else, was the thing that cut the wood in order to know it with perfect certainty. If one examined the cut and saw markings on the wood that matched the pattern of the serrated edge of the blade, one could confirm that the blade cut the wood (although this too might be mere 99.9% certainty and not 100% certainty. A discussion of certainty and knowledge comes in a later section).

For the next example, suppose that you have a job interview at an office. The interviewer notifies you ahead of time that she has an allergy problem, and asks that you not wear perfume or cologne. You then wonder whether you should wear scented deodorant. It was not mentioned, so no explicit prohibition against it exists, but you suspect that it might be a mistake to wear it. You then need to reason whether deodorant is the same type of thing as perfume or cologne in the sense of the interviewer wanting to avoid it. Essential reasoning is what is used to answer such a question, because to be a member of a group defined by being a type of thing is to have an essence which makes a thing be that type of thing. You can reason that perfume and cologne are on the list because they are scents which can trigger allergies. Thus, the essence of what the interviewer wants to avoid is being a scent which can trigger allergy. You can then reason that your deodorant is a scent, which might trigger an allergy. Therefore you can conclude that you are better off not wearing it in order to avoid making trouble during the job interview.

Chapter Fifty One: Reason as a Three Step Process: Reasoning What Exists

Now let us examine how conscious reasoning reasons what does and does not exist from consciously perceived things, in light of what we discovered about perception and essential reasoning earlier. There are three aspects of this: reasoning what can exist metaphysically, reasoning what can exist mechanically, and reasoning what does exist. You can reason what things are metaphysically possible, what things can exist metaphysically, by reasoning what essences are mutually contradictory and what essences are not mutually contradictory. For two actions to be mutually contradictory is for a thing doing one of the actions to not be mechanically able to do the second action as a consequence of doing the first action. Mutually contradictory actions are the actions that it is impossible for the same thing to do at the same time, consequentially of their essence, and so it is impossible for any real thing to be that (because real things are things, and nothing can be that) so the doers of mutually contradictory actions are metaphysically impossible. The doers of actions which are not mutually contradictory are metaphysically possible.

The essential actions must be reasoned from perceived things prior to reasoning metaphysical possibility, so that one has a means of knowing the actions to determine which are and are not mutually contradictory. To be precise, metaphysically impossible things are not even things, they are not-things, because they cannot do the actions that they do and so they cannot be things. The best example of mutually contradictory actions is the act of being a thing and the act of not being that thing. To be a thing will cause the thing to not be not that thing, so the consequence of one negates the other. Another example which I discussed earlier was for an apple to be ripe, and for an apple to be rotten, because a ripe apple is edible, and a rotten apple is inedible, so an apple cannot be both at the same time.

What can exist mechanically can be reasoned from cause and effect and what does exist. What can exist mechanically is whatever things can be caused by what really does exist, so that once you have reasoned what does exist from perceived things, and once you have reasoned cause and effect and what are the things that can be created by existing things, then you can reason that all the things that the things that do exist can create are mechanically possible, and all the things that the things that exist cannot create are mechanically impossible. Because what exists changes in certain ways each moment, what is mechanically possible changes with each moment also, but so long as a thing continues to do the same action it will maintain the same things as mechanically possible or impossible because of the thing as the doer of that action which remained the same.

There are two ways to reason what is mechanically impossible in a certain area. The first way is to reason what the existing things in that area are, and then reason what they can and cannot cause. What they cannot cause is mechanically impossible. For example, if a group of children are playing games, and they do not have a baseball bat, then we can reason that it is mechanically impossible for them to play baseball. The second way is to reason what is metaphysically impossible, because the metaphysically impossible cannot be caused to be and so it is mechanically impossible. For example, if biology proves that water is necessary for life to evolve, then we can reason that it is mechanically impossible for life to evolve in planets which lack water. However, unlike reasoning mechanical impossibility from metaphysical impossibility, we can see that mechanical possibility cannot be reasoned from metaphysical possibility. The only way to reason what is mechanically possible is to reason what exists from perceived things, and then reason what the existing things can cause from the existing things. For example, the presence of water on a planet does not prove that life could have evolved on it, because the specific details of what life would have needed to evolve, such as for example the correct temperature or gravitational pull, might have been absent, and only an empirical analysis of that specific planet could prove that the evolution of life was possible.

One reasons what does exist from perceived things, and this is the only way to reason what exists. You can reason that the perceived things exist from the perceived things, and you can then reason the existence of other things from perceived things with essential reasoning. I have already explained how the existence of perceived things can be reasoned from perceived things, but I will restate this for the sake of clarity. Of course, the perceived thing being at a specific time and place is directly perceived, because the things are perceived to be when and where they are. For example, looking at something, you can see where it is, and you see it at the moment when it is. All other directly perceivable properties are perceived, as for example if a thing is blue, you see the blue thing being blue, and you know it directly by perceiving it. The thing as the doer of all directly perceivable actions is known by perceiving the perceived thing.

Whenever you perceive something, and you identify it, you can reason that the thing exists. Then, from the existence of the perceived things, you can reason other things. From those things that you perceived, you must reason the rest of what exists. The only way to do this is to reason those things whose existence can be reasoned from the things that you have perceived, i.e. those things which the perceived things' existence makes necessary, which can be discovered by essential reasoning. The continued existence of things that were perceived when they are unperceived, the causes of the perceived things, and the effects of the perceived things, are the three unperceived things whose existence can be reasoned with total certainty from perceived things. If these other things are reasoned essentially from the perceived things, and you reason that the perceived things exist from the perceived things, then you can reason the existence of these other things from perceived things, so that they are known through empirical essential reasoning. Essential reasoning is necessary to reason from the things that you perceive to those other unperceived things with absolute certainty. For example, if you use scientific reasoning to know that the Earth revolves around the Sun then you can know that the Sun exists on the other side of the Earth during the night even though at nighttime you are in darkness and you cannot perceive the Sun at that time.

The line of reasoning from the existence of perceived things to the existence of their causes and effects should be easy to understand, given what I explained in the section on cause and effect. Reason that the perceived thing exists, and then reason what its cause or effect is from it, and then from these you can reason the existence of the cause from that of the effect, or the existence of the effect from the cause. The existence of the cause makes the existence of the effect necessary, because the existing cause's act of causing (which is its essential action as a cause) causes the existence of the effect. And the existing effect's existence requires the existence of its cause to exist (and the act it was caused to do is its essential act as an effect). Thus the existence of the cause causes the effect consequentially of it doing its essential action, and the existence of the effect requires the existence of the cause consequentially of its essence also. This line of empirical essential reasoning gives total certainty in the existence of cause/effect from the perceived effect/cause, for that reason.

For example, you can reason the existence of New York without perceiving New York from the existence of perceived things that only the existence of New York could cause or make possible, such as people and things from New York. This consists, not of trusting it when other people tell you that New York exists, but, rather, of reasoning the cause from the words as an effect, e.g. Frank Sinatra would not have sang "New York, New York" if New York City had not existed. Another example of this is reasoning the existence of the parents from the existence of the child. Those are two examples of reasoning the cause from the effect, and in both of these one can see that the knowledge of the cause is only as specific as the knowledge of the effect. An example of reasoning the existence of the effect from the existence of the cause follows. Reason what will happen when you drop a rock and there is nothing underneath it to catch it. You can reason from the act of being a rock and the act of gravity that it will fall. Then, look at yourself holding a rock, and let go of the rock. Without looking down, you can reason from the cause that the effect exists, namely, that the rock fell to the ground, because you perceived the cause, although you did not perceive the effect.

One can only reason from things, so one cannot reason beyond the experience you have, e.g. if you do not know anything about Russia, then you cannot reason where a building in Moscow is. You cannot reason the location of a building in Russia from an apple. You cannot reason everything from anything. You can only reason what it shows from a thing. You can reason the existence of the tree from which it came from the apple, or what would happen if you threw the apple through a window from the apple (and the window), and you can reason the apple's existence and the parts of the apple, like what its insides are, from the apple. But you cannot reason where the Kremlin is from the apple, because that is not, is not a part of, is not caused by, is not a means of, and is not an effect of the apple. And you can only tell what can be reasoned from something, by reasoning what can be reasoned from it. Given that a person's experience is all the things that the person has perceived, reasoning is limited by the experience that it has to reason from. The more perceived things you can reason from, i.e. the more things you have perceived, the more you can reason. But this limit does not make reason fallible, it merely means that it is impossible to reason from experience you don't have, since all empirical essential reasoning is reasoning from your experience.

Reasoning what exists is the same as reasoning what real things are and it is also the same as reasoning what does exist. If a thing does the act of existing and does other acts also, then the doer of the total being action exists, that thing exists. When thinking about real things, one can consider the real things themselves and every action that they do. This is different from reasoning about essential things, because in the latter case one does not consider any action except the essential action, but in the former case the essential action of the thing is the total being action that the real thing does, so every action must be considered, although if one can reason something essentially from one action that a thing does (e.g. the consequence, if the thing does the essential action) then the real thing doing the consequential action can be known without knowing any other actions that the thing is doing except for the essential action, because anything which does that act will be required to do the consequential act by doing that essential act.

Real things are completely specific, as opposed to essences or essential things. Each real thing exists in the same existence as all other existing things. They must all exist in the same reality, since the essence of being in existence is to exist, and existence is the place where existing things are, therefore existence is the place where all existing things are, and therefore all existing things are in the place called "existence" or "reality". Because every real thing exists in one reality, you can consider the effects of all other things known to exist on the real things that you consider. Also, when thinking about a real thing, one can reason from everything that one knows about it, because it is the same thing. All reasoning about real things must be either directly from or confirmed by perception, i.e.

when reasoning about essential things all one needs to have done is to reason a concept of the essential action from perception, but when reasoning about real things one needs to reason from perception that the thing is doing the actions that one thinks it is doing, either by seeing it, or from a consequence of essence or cause and effect analysis. One must also have previously reasoned a concept of its action that one is focusing on from perceiving something doing that action, or from other perceived things, in order to have a concept to use as a means of thinking.

Chapter Fifty Two: Reason as a Three Step Process: Reasoning What Does Not Exist

What does not exist is each thing that does not do the act of existing. It is an absence of action done by the things that exist, not an active action done by nothingness. Only reality exists, and nothingness does not exist. To say that a certain thing does not exist, is to say that a thing does that act of being, and does not do the act of existing, e.g. to say that a purple elephant does not exist is to say that this thing does the act of being an elephant, does the act of being purple, and does not do the act of existing. A thing which is not objectively at a certain place and time does not exist. When you reason what exists at a certain place and time then you can reason that everything which is not that thing does not exist there. That is the only way to reason to the nonexistence of specific things from the existence of specific things.

One can reason that something does not exist everywhere in two ways, and both of these involve reasoning that it cannot exist. The first is by reasoning that it is impossible for the thing to do two or more of its being actions at the same time, i.e. reasoning that it is metaphysically impossible. Doing the one act will prevent the thing from doing the other act, so the thing will not do both acts and will not be that thing, whether it exists or not. The second is by reasoning that a thing cannot both do one or more of its acts of being and the act of existing, i.e. reasoning that the act of objectively being at a place and time consequentially prevents being that thing. Existing will prevent the thing from being that thing, so that thing cannot exist. Note that reasoning that something does not exist anywhere is the same as reasoning that it does not exist everywhere. Anywhere is the same thing as (a place)(e), and if you reason that a thing cannot exist at (a place)(e), then from this you can reason that it cannot exist everywhere, because every place is a place. In other words, a place is each place, and each place together is every place. This is why all reasoning from essential things with variable locations and times is universal.

For example, if I reason that only a computer and a printer are on my desk, from seeing what is on my desk, I can reason from this that an apple is not on my desk. For an example of reasoning what does not exist everywhere, I can reason that a thing cannot both be and not be itself and exist, from the inability of a thing to both be and not be itself. Or I can reason that one thing cannot exist in two places at the same time, because it would not be the same thing in both places, because being at a location is one of the being actions done by an existing thing, and the two different acts of being at a location would make the one thing be two different things.

One can only reason from nonexistence of cause to nonexistence of effect if the cause was the continuous cause of the effect, not the instantaneous cause of the effect. This is so because the existence of the effect depends upon the current existence of the continuous cause to exist, but the effect of the instantaneous cause does not. However, the existence of the effect of the instantaneous cause does depend upon the existence of the instantaneous cause at the time when it caused the effect, so if the instantaneous cause of the effect can be reasoned from the effect to have been at a specific time in the past, and one reasons that the cause did not exist at that time, only then one can reason the nonexistence of the effect from the nonexistence of the instantaneous cause.

Chapter Fifty Three: Reason as a Three Step Process: Imagination

Now let us consider the role of imagination in reasoning. In order to do this we must first reason what the act of imagining consists of. Drawing upon the distinction between perception and awareness made previously, it is clear from the examination of imagined things that one is aware of them, but that they are not perceived. One can reason from an examination of the act of imagining, which one can observe oneself performing, that it consists of the awareness of the directly perceivable aspects of things. That this is what imagination consists of is clear from examples of imagination, as for example to "imagine the situation of being in an elevator," or "picturing it." Thinking of being in an elevator can be distinguished from imagining being in an elevator, because thinking of it is just being aware of an essential thing, (yourself in an elevator)(e), but when you think up specifics for the elevator's interior, especially when you think of the visual details of the inside of the elevator, you imagine the interior of the elevator. To be aware of the directly perceivable specifics is to imagine a thing. Because of this, the more details you imagine, the more vivid the imagined thing becomes.

When one consciously thinks of a thing with specific perceivable qualities, that is the same act as imagining that thing. When one thinks of a human(e), humans do specifically perceivable acts, but the act of being human in thought-about, so it is not imagined. But being a certain color, shade, shape and size, a noise with a certain tone and loudness, etc., are specific perceivable actions, and thinking of a thing doing them is to imagine that thing. For example, to think of a person with a white, oval face with freckles and a long pointed nose and red hair, and to think of the physical qualities in detail, is to imagine that face. The act of imagining results in the mind's eye "seeing" that imagined face, but the act of imagination is actually awareness, not perception. If you misidentify the imagined thing as a result of a mistake in judgment and the evaluation of introspection then it might seem like you perceive it. The mistake of thinking that imagination is perception is an easy mistake to make because the very thing you are aware of is a thing doing directly perceived actions. But imagining a thing is not the same act as perceiving it. One can identify a mentally distinguishable difference (assuming the Sanity Proviso) between things that are perceived and things that are imagined.

The reason for the distinction between imagination and perception can be understood with reference to what was explained before. In terms of unconscious reasoning, the brain takes an equation for a thing and plugs as many specific numbers as it can into the variables where the sensory perception numbers would fit into the equation. But this is not the same thing as perception numbers entering the brain from the sense organs. The brain lacks the ability to create enough numbers to actually mimic the vividness and realism of perception, which is why the imagined thing is always identifiably vague and fuzzy compared to the perceived things, and the perceived thing is always identifiably clear compared to the imagined thing. The clarity comes because perception sends a complete set of numbers to the brain, whereas the brain when it imagines can only fill the perception variables in the equation with numbers, but the equation always has some variables left, because it is an equation.

Now that we know what imagination is, we can see how it is used by conscious reasoning. Imagination simply consists of thinking about the directly perceivable aspects of things. Wherever this is useful in a line of reasoning, imagination is useful in that line of reasoning. Imagination is a kind of essential reasoning, reasoning about things as the doers of specific directly perceivable actions. Any reasoning that involves reasoning about the directly perceivable aspects of things is imagination, thus imagination has no use as perception but as essential reasoning it is necessary for any essential reasoning about things as the doers of directly perceivable actions, and imagination is then the same act as reasoning about those things.

Two examples of how imagination is used in conscious reasoning may make this easier to understand. First, consider the famous example of Einstein's thought experiment, in which he imagined what it would be like to be traveling at the speed of light. This consisted, not of the use of intuition, but of his reasoning from his knowledge of light and motion (ultimately reasoned from perceived things) by thinking of the things that someone moving at the speed of light would perceive. Thus his thought experiment using imagination really consisted of empirical essential reasoning, reasoning his conclusions from the things that he was aware of.

Second, consider the example of someone who wants to tell something to someone else, and finds it useful to imagine the conversation taking place to try to figure out what to say. What does this imagination actually consist of? When he imagines the other person, he is thinking of the other person and so reasons from his knowledge of that person. He imagines the conversation taking place because specific words have perceivable aspects, and by thinking about the specific words he is able to judge which specific words to use or not to use on the other person. When he imagines saying something and the person's response, this actually consists of his reasoning how the person will respond to what he says from his knowledge of the person and the specifics of the imagined spoken words. Once he reasons the person's response he identifies it as good or not good, i.e. he either likes how the conversation in his imagination goes or dislikes it, and so he chooses which words are good. His act of imagining actually consists of his act of reasoning what he should say to the other person, and this reasoning relies upon concepts reasoned from perception (the concepts of the other person and the words), not on intuition.

A popular philosophical mistake asserts that the test to determine whether something is metaphysically possible is whether it can be imagined. This theory often uses the example of a square circle, which can't be imagined and is therefore impossible, compared with the Sun not rising tomorrow, which can be imagined and is therefore possible. However, contrary to this doctrine, the nature of imagination offers us no reason why the ability to imagine something is proof of its possibility. Imagination relies on the concepts used, and imagining the object of a false concept might enable us to imagine a thing that is metaphysically impossible. If one has a concept of a thing that cannot exist then one can imagine something that is impossible, either mechanically or metaphysically. Because mistakes in reasoning (e.g. associative or corrupted reasoning) can lead one to believe that both metaphysically and mechanically impossible things both can and do exist, it is possible to imagine things of both kinds by means of false concepts. Indeed, one can debate this, but it is plausible to assert that I actually can imagine a square circle or a circular square. It is a shape that looks like a square but with rounded corners and curved sides. And I can imagine the Sun not rising tomorrow morning, but this is not a possibility according to science and astronomy backed by empirical essential reasoning.

Imagined things come from concept-computation, they are not perceived, and it is possible to imagine things that are mechanically impossible and things that are metaphysically impossible. For an example of imagining something that is mechanically impossible, I can easily imagining my desk walking away on its legs, and this is mechanically impossible because it does not have the physical ability to do that. For an example of imagining something that is metaphysically impossible, I can easily imagine a solid bar of lead float up through the air, and this is metaphysically impossible because consequentially of being heavy a thing will do the act of falling due to gravity and being suspended in a substance which lacks resistance to hold it up, and lead is heavy and air is light, so the act of being lead and the act of floating in air are mutually exclusive. A bar of lead could be carried up through air, but for it to propel itself upward through air by means of the act of floating is metaphysically impossible. You and I can both imagine my desk walking away on its legs, and we can both imagine a bar of lead floating through the air, and this is sufficient to refute the theory that imagination is proof of logical possibility.

The theory that the imagined must be possible stems from Hume's Empiricist mistake of thinking that to imagine is the same as to perceive. Hume concluded that, since to be is to be perceived, and to imagine is to perceive, everything that was imagined could exist, in contrast to contradictions like round squares that cannot be imagined. However, as we have seen, we can imagine things that are impossible, and a person also might be unable to imagine something that is possible (as for example the ancient Romans probably could not imagine space travel), so imagination has no inherent connection to possibility.

With this theory refuted, let us ask why the theory was believed in the first place. We can see the origin of this theory in the philosophy of Hume. The theory that imagination is proof of "logical possibility" relies upon Hume's equation of essential things with one specific perceived thing that represents other things to the mind, from which he derives his theory that imagined things come directly from perception, and therefore must be possible. Specifically, from his claim that to think of a thing and to think of a thing existing are identical, he deduces the theory that anything of which one has a clear and distinct idea must exist, since if the thing thought of and the thing thought of existing are the same then to think of a thing clearly and to believe that the thing exists are the same, and he then claims that the clarity of ideas is the very source of knowledge of what exists, and perceived things exist because they are seen with great clarity and vividness. Thus, he asserts that things which are clearly imagined must exist.

Hume's theory implies that existence is subjective, since thinking of a thing clearly can require it to exist, and it is based upon two false assumptions. First, the assumption that imagination is a kind of perception, and second, the assumption that an essential thing is a specific thing associated with other things of its kind through mental conditioning. The origins of the first assumption can be seen in Hobbes' belief that imagined things were the impressions that perceived things left in the mind, like the ripples in water left by a stone that hits it, which relies upon ignorance of the existence of concepts and reason-computation. This is not even a valid metaphor for the creation of concepts, since the action of the brain is not in the picture. Hume uses his theory of imagination to argue against his opponents by claiming that he can imagine himself being right and cannot imagine his opponents being right, which is the most preposterously biased argumentation I have ever encountered (see Hume's Treatise Book I, Part II, Section IV).

The theory that imagination proves possibility was later capitalized upon by the symbolic logicians. The logicians claim that the logical inconsistency of sentences is their means of knowing what is impossible (i.e. the contradictory is impossible), but they have no good way to use their system of symbols to prove what is possible, so they took up Hume's theory and claimed that the things being imagined was proof of their logical possibility. The square circle example was useful for them because with it they could claim that nothing contradictory is imaginable. Thus, Hume originated this theory of imagination, and symbolic logic maintains its popularity. However, the inability of a person to conceive of or imagine something does not prove that it is a contradiction or that it is impossible. For example, a deeply religious person might be unable to imagine a world without God, but the person's imagination has no inherent or reasoned connection to what is actually physically possible.

Chapter Fifty Four: Reason as a Three Step Process: Step Three, Conceptual Identification

Step three of reason as a three step process is conceptual identification, which consists of identifying consciously perceived things. Specifically, conceptual identification perceives a thing, identifies it by its essential property, and then knows that the thing with this essence also has the consequence which had been previously reasoned. In other words, conceptual identification is the practical application of essential reasoning.

After one has reasoned a concept of an essential action from perception and reasoned the consequence of the action that one is aware of, the next step in reasoning is to identify the things that are perceived by the actions that they do, thereby identifying them as a certain kind of thing, and reasoning from the thing doing the essential action that it does the consequential action. This can be called the conscious identification of perceived things, or identifying perceived things by essence, or conceptual identification. This consists of two parts, first unconscious reasoning, i.e. the means by which one consciously perceived the things that one perceives, and second conscious reasoning, i.e. the act of identifying the perceived thing. We can distinguish the unconscious reasoning that makes conscious perception possible, which consists of identifying perceptions by concepts, and then consider how one consciously identifies the things that one consciously perceives using conscious reasoning. The difference between unconscious conceptual identification and conscious essential identification of consciously perceived things is that in the unconscious reason, the brain identifies the perceptions, and in the conscious reason, the brain identifies the perceived things.

By conscious perception I mean anytime when someone consciously sees, hears, tastes, touches, or smells anything that the person is conscious of. When unconscious perception happens a ray of light goes from the object to the eye, but when conscious perception happens a line of thought goes from the concept in the brain out to the object in the external world. Prior to all conscious perception, the brain must successfully undertake unconscious conceptual identification, which takes the perceptions that the sense organs send to the brain and identifies the perceptions unconsciously by means of concepts, which results in the conscious part of the brain becoming aware of the perceived things shown by the perceptions. It seems hard to believe that unconscious reasoning could happen prior to conscious perception and make it happen without ever being noticed, but it is unconscious, so there is no reason why one would be aware of it without having consciously reasoned its existence, and it happens so fast, taking place between the entrance of the nerve signal into the brain and the conscious perception of the thing, that no one previously noticed it.

Unconscious conceptual identification can be established, because if one looks at something, one is aware of the thing as the object of the concept that one has of it. For example, if one looks at a cup of brown liquid, and one's concept of edible fluid this shade of brown is that it is coffee, and one's unconscious mind identifies it as coffee, then one will be aware that it is a cup of coffee when one looks at it. But if one's unconscious mind has a concept of that brown as chocolate milk, and it identified it as chocolate milk, then one would be aware of chocolate milk. We look at things and immediately see things, without taking minutes or hours to sit and think and carefully figure out what it is we are seeing, therefore this identification must be done unconsciously. However, conscious thought can change our concepts, because when the conscious mind thinks that X is Y it creates a concept of X as Y, which is then used in later unconscious conceptual identification. So if one consciously thinks that a shade of brown fluid is not coffee and is actually chocolate milk, then one will alter one's concept, and moving forward one's unconscious reason will use the new concept and identify that color of fluid as chocolate milk.

One would be aware of the thing as a beverage by means of the concept of coffee or chocolate milk, but one would be aware of that color of brown by means of the concept of that color and also by means of the perception in the brain, which is the nerve signal from the eye's retina as stored in the brain, so one could use the perception and the concept of that color to know that color itself as separate from being either coffee or chocolate milk, in order to change the concept of that color from a concept of that color as coffee to a concept of that color as chocolate milk. The brain does have a direct infallible access to reality by means of its perceptions. The role of unconscious reason in conscious perception is not that difficult to believe, since no convincing alternative theory explains how the brain is capable of perceiving something.

The interpretation of sensory perceptions by the brain is unconscious conceptual identification, which is a precondition of awareness of perceived things. My theory is supported by science, in the sense that neurobiology has established that the nerve signals from the sensory organs are processed by the structures in the middle of the brain, e.g. the basil ganglia, and then are processed by the neocortical lobes in the sides, top and rear of the brain, e.g. the occipital lobe for vision, and so the perceptions are processed by layer after layer of neurons before they reach the conscious reasoning part of the brain, which is probably located in the frontal lobe.

Here I must mention in passing that, although "the conscious mind" is probably the frontal lobe, conscious reasoning is probably done by the entire brain, because there are many different functions in all different parts of the brain, such as vision, hearing, consciousness, thought, emotion, and memory, all of which are combined in conscious thinking. In particular, both the amygdala in the middle of the brain and the frontal lobe in the front of the brain play a role in making decisions, but the consciousness makes decisions, so the conscious reason must be built from different brain structures acting as a whole. Consciously thinking about what move to make in a game of chess might employ the amygdala, the frontal lobe, the occipital lobe, the hippocampus, and other parts of the brain, all in something that a conscious mind would understand as one decision made by one mind as seen by internal introspection.

Note that the reason why not every perception is identified as a showing thing and has its perceived thing reasoned from it is that when the brain becomes aware of a perceived thing and consciously reasons from it, this obviously employs a great deal of the brain's resources, so that unconscious reasoning identifies which perceptions are important and unimportant, and then reasons the perceived things from the important perceptions. This is why one becomes aware of things that one is concentrating on and tends to not become aware of unimportant things like background noise, because one's concept of the important will identify the perceptions of the things you consider important, and your concept of the unimportant will identify the things you consider unimportant, when unconscious reasoning tries to reason which perceptions to reason the perceived things from.

Let us examine the conscious identification of consciously perceived things. One consciously identifies what one is looking at by perceiving things doing certain actions. Identify a thing doing an action, and you have then identified that thing as the doer of that action. Essential reasoning can then reason from the thing doing that action to also knowing that the thing does the action which is the consequence of that action. When one perceives a thing and identifies it doing the X action, where the X act is the essence of X, one has identified that thing as an X, and one can then be certain that it does the actions consequential of the essential X act. In other words, if you identify a thing doing the X action, and you previously reasoned that X(e) is Y, then you can know that the X you perceive is Y. I call identifying things by their actions essential identification or conceptual identification.

For example, if you look at a thing and it is made of plastic and has keys with letters, then you can identify that it is a computer keyboard, because it does all the directly perceivable actions that that type of thing does, and only being that thing could cause it to do those visible acts, and so you can reason to it doing the thought-about act of being a keyboard, whose essence is the act of sending signals to the computer when a key is pressed. Once you identify a thing as a keyboard, you can know that it runs on electricity, because the essential keyboard runs on electricity, and therefore all keyboards run on electricity, so you can know that it needs a power supply in order to work without taking this keyboard apart and learning that about this specific keyboard.

Similarly, if you look at a thing, and you see it writing in ink, then you can identify that it is a pen, because it is a thing that writes with ink. If you see a thing and it is small and long and cylindrical and made of plastic and it has a cap on it and the name of a pen-manufacturer on the side, in the same way, from these perceivable acts that a pen does, you can reason that it is a pen because only being a pen would cause the thing to do those visible acts. Having identified this thing as a pen, you then know that you can write using it, without actually learning that from the experience of this particular pen. So conceptual identification identifies things by essence and knows instantly that the thing does the consequence, but it can also employ cause and effect reasoning, and identify things by essence by identifying the effect in a perceived thing and reasoning to the cause, where the essence causes the consequence. Cause and effect analysis enables one to see a consequence and reason that only that essence could cause that consequence, such that seeing the consequence of the essence of being a keyboard or a pen would enable one to identify the thing as a keyboard or pen.

If you perceive something and it has a bill and feathers and webbed feet and it quacks and flies and swims, the compound act consisting of all those actions is the act that a duck does, so from perceiving a thing do those actions you can reason that the perceived thing is a duck. Because of the prior reasoning of the essence, one has already reasoned that being a duck will cause a thing to do those actions, so one is actually reasoning to the thought-about cause from the directly perceived effect, only in this case it is not the cause and effect of one thing acting on another, rather it is the action that the thing does (being a duck, i.e. having duck DNA) causing it to do other actions (have a bill and feathers, swim and fly, etc.)

If you can reason that the thought-about action requires the directly perceived action by being an effect or a means or a cause of it, or that the directly perceivable action is a part of the thought-about act of being, then you can reason from perceiving a thing do something to what the thing is. This identification is not a guess. It is the product of empirical essential reasoning, and because it reasons from the consequence of the essence it is necessarily correct. The directly perceived thing is perceived to do the essential action or the consequential actions, and one can then reason that the consequence of the essence, which was previously identified, is done by the perceived thing doing the essential action, so that the perceived thing is then known to do both the perceived action (e.g. having feathers and a bill) and the thought-about action (e.g. being a duck), and the consequence of the essence (e.g. being able to swim).

Chapter Fifty Five: Reason as a Three Step Process: Reason as a Tool for Interacting with Reality

Now that the three steps of conscious reasoning have been explained, their combined use can be described. From what I have said above it is clear how one gains knowledge of perceived things, essences, the consequences of the essences, and how one identifies the things that one consciously perceives. What is left to be considered, then, is how this reasoning is combined.

When one reasons, one can reason from a thing that one has already reasoned, and then reason from that, and so on, so that each thing that one reasons becomes one point in a line of reasoning to some eventual conclusion. Through the use of concepts, and consciously reasoning the essences of things from perceived things, reasoning the consequences of essential actions, and essentially identifying perceived things, you can identify the things that are around you and what situation you are in, identify the world from one moment to the next, and engage in long-term reasoning.

This enables consciously responding to the things around you, for which conscious identification of the perceived is necessary. It enables planning, for which both identifying the same existence over a period of time, i.e. being aware of a continually existing existence, which can be identified because the place where existing things are can be identified whenever one perceives anything that exists, and knowledge of the essences which the plan involves, are necessary. And it enables long lines of conscious reasoning, i.e. thinking about things. When you think about something, you can go through everything you have perceived which you remember which pertains to it, including all existing things and all essential things, reason from that, and then reason from anything else you have reasoned from, to reason more about the thing.

Reasoning enables one to respond to the things that one comes into contact with. For an example of a rational response, if some food is on your plate, and you look at it, and from its directly perceived being you consciously identify it as a kind of food that you are allergic to, you can decide not to eat it. What this consists of is reasoning that it is that kind of food consequentially of being what it

378

is perceived to be, and reasoning that the food will make you sick consequentially of being that kind of food, and deciding not to eat it because eating that perceived thing will make you sick.

For example, you see a pile of little pea-green pea-shaped balls in the food. You identify that they are green peas because they have a visible being which is what peas look like. You had already reasoned that the consequence of the essence of being peas is that they will make you sick because you are allergic to them. This was reasoned from experience using cause and effect analysis, e.g. from having eaten peas before and having had an allergic reaction, where you knew you were not allergic to whatever else you ate at that meal so that only the green peas could have caused you to get sick, and/or medical tests by an allergist doctor. You identify that the food contains peas and you know the consequence of the essence of peas is making you sick, so you then know that you should not eat the peas. You decide not to eat them, and you don't eat them, and you don't get sick with a violent allergic reaction thanks to your empirical essential reasoning.

As this example shows, the practical application of reason is conceptual identification, which takes the first two steps, perception/induction and essential reasoning, and then applies them to reality. For an example of a long line of reasoning, if you are in a grocery and you want to buy the cheapest per pound food, you might have to go through several different brands and calculate the price per pound from the serving size, servings per container, and price of the package, each of which is one reason-computation, and then from all the prices per pound you could reason the ultimate conclusion of which brand was cheapest per pound, with the final reason-computation which reasons from all previous reason-computations.

The Aristotelian syllogism "All men are mortal, Socrates is a man, therefore Socrates is mortal" is not justified as symbolic formal logic, but it is justified as essential reasoning applied by conceptual identification. Perception can learn about humankind. Essential reasoning can then establish that all humans are mortal. Then, conceptual identification can identify that Socrates is a man. The conclusion of the line of reasoning identifies that Socrates is mortal because he is a man.

As one final note on the topic of the usefulness of reason, let me distinguish the use of conscious reason as compared to unconscious reason. Conscious reason is smarter than unconscious reason, because unconscious reason is the blind processing of perceptions and concepts whereas conscious reason actually knows reality and the things themselves which exist. In general, at the beginning of a task conscious reasoning figures out the truth or how to do something, and then, once you consciously figure out what to do, you do some behavior consciously over and over again in response to a situation. The repetition of the behavior is picked up by the unconscious mind and it becomes an automatic behavior or a habit. For example, if you consciously pick the peas out of your mashed potatoes repeatedly, then eventually this will become a habit and you will unconsciously pick the peas out of your potatoes automatically without being aware of it. This makes sense, because conscious reasoning applies its intelligence to figure out the correct behavior for a situation where you have no prior knowledge of what to do, and then once the hard work of reasoning what to do has been completed it becomes easier for unconscious reasoning to step in and use memorized habits to automatically do the right behavior.

Chapter Fifty Six: Reason as a Three Step Process: Why Unconscious and Conscious Reasoning's Inferences are Valid

Unconscious reason is the brain's analysis and processing of perceptions and concepts. Unconscious reason reason-computes a group of concepts and/or perceptions to produce a concept. One can reason that the concept is created entirely from the concepts from which it is reasoned, because its existence is caused by reasoning and the concepts that it is reasoned from, and reasoning cannot add anything to the created concept or it would be the object of awareness rather than the means of knowing the object of that concept, so the being must come entirely from the reason-computed concepts. The being of the concepts causes the being of the concept produced by their reason-computation. The act of reasoning does not cause the being of the concepts it produces, it only causes the concepts it produces to exist, i.e. reason-computation causes the existence of the product concept, and the concepts and/or perceptions that are reason-computed cause the being of the product concept.

Reasoning is a means, the means by which one gains knowledge of one thing from another. Because it is a means, a form of the means argument applies to reasoning. The means argument applied to perception says that a perception shows a perceived thing and not itself because the perceived thing is that which is perceived, and the perception is "of" the object of the perception, with "of" denoting the act of showing something else other than itself. The means argument applied to concepts states that the concept is the means of awareness of the object of the concept. The means argument applied to reason is similar. Like concepts and perceptions, the means argument applied to reason states that the means of knowledge cannot add to or constitute that which is the known thing, as a consequence of the essence of being a means of knowing.

This is not a circular argument, because the means argument is reasoned without any assumptions, and a loop argument always begins with its conclusion as an original assumption. I am not merely asserting that it is not assumed, because I have, in fact, shown how reason can deduce this from the observation of things in our empirical experience of reality, and the perceived things are self-evident and prove the arguments that derive from them. For example, if I know an apple from perceiving it, forming a concept of it, and reasoning about it, then the perception, concept, and reason are the means of knowing it, and the source of knowledge is the apple, not reason (in the sense of being directed at itself as the object of knowledge), not the concept of the apple, and not the perception of the apple, rather the perception, the concept, and reason were my means of knowing the apple.

Because unconscious reasoning cannot add anything, a group of reason-computed concepts will produce only one specific concept, and that concept has to be the exact product of the reason-computed concepts, its being must be determined by the concepts from which it is reasoned, because it is created entirely from the reason-computed concepts. True concepts produce true concepts, and false concepts produce false concepts, when they are reason-computed, because reason does not add anything to the concepts it computes, nor does it really create anything, reason merely produces from the concepts that it reasons from without adding anything or creating anything not created from the concepts. Perceptions always produce true concepts, because perceptions are always true. You can only reason what is potentially in the things you reason from, i.e. what they have the ability to produce when reasoned from. One can see that reasoning works in this way just by observing any line of reasoning that one has done, and examining what the act of reasoning from something consists of. Of course reasoning creates the concepts that it reasons to, but reasoning only causes the existence of the reasoned concept, whereas the being of the reasoned concept is entirely caused by the perceptions and/or concepts that the product concept is reasoned from.

Reason-computed true concepts can only produce true concepts, because the act of being true of the reason-computed concepts will cause the product concept to do the act of being true. And the act of being false done by a reason-computed concept will cause the product concept to do the act of being false. For every true concept, its truth is caused by the truth of the concepts it was reasoned from, back to the first concepts reasoned from perceptions. The truth of first concepts was caused by the truth of the perceptions from which they were reasoned, and the truth of the perceptions was caused by their objects, because the object caused the being of the perception, and the perceptions caused the being of the first concepts, the first concepts caused the being of the concepts reasoned from them, those concepts caused the being of the concepts reasoned from them, etc., so that the truth of all reasoned concepts is ultimately caused by the perceived things from which they are all ultimately reasoned.

Reasoning cannot create the act of being true or false in its products, that act of the product is caused by the act of the things which are reason-computed. And if the line of reasoning goes right back up to concepts that were reasoned only from perceptions (i.e. if the reason is pure), if all the concepts in the reasoning can ultimately be traced back to essential reasoning from perceptions, and the act of being true of the perception is caused by its object (because the being of the perception corresponds to the being of the perceived thing, because the perceived thing causes the aspects of the being of the perceptions which show the being of the perceived things), and the truth of the perceptions from which they are reasoned cause the truth of first concepts, and subsequent reasoning reasons only from three things, perceptions, first concepts reasoned from perceptions, and concepts reasoned from those concepts, then the concept which is the product of pure empirical essential reasoning can only be true. It must be true because the object causes the truth of its perception, the perceptions cause the truth of the first concepts reasoned from them, and then the truth of those concepts cause the truth of all later concepts reasoned from them. So perceptions, first concepts, and reasoned secondary concepts are all true, and reason reasons only from those sources, so reason must always be correct.

If true concepts and one or more false concepts are computed in the same reason-computation, then the product concept must be at least partially false, since it was caused in part by a false concept. And if a concept is partially false then what it shows does not fully exist, and since a thing either exists or does not exist (this can be reasoned from what the act of existing consists of), the product concept does not show a thing that exists, and therefore it does the essential act of a false concept and it is a false concept. Therefore if even one single false concept is reasoned from in any line of reasoning, all of the later products of that line of reasoning are false.

But the above description of truth-causation is not precisely true of conscious reasoning, because conscious reasoning reasons, not from perceptions and concepts, but from what they are of, the things that one perceives and is aware of. Showing things, i.e. perceptions and concepts, are true or false, while the shown things exist or do not exist. Conscious reasoning reasons from things themselves to things themselves, by means of concepts. Conscious reasoning reasons directly from the things that one is aware of, from the essences of those things. In all the steps of conscious reasoning, i.e. reasoning the essence from perceived things via induction, reasoning the consequence of the essence, reasoning the cause from the essence of the effect, reasoning the effect from the essence of the cause, reasoning what does and does not exist from perceived things and essence, in all of this, you can only reason from the things you are aware of, and you can only reason what the things that you are reasoning from show you, because you are reasoning from them. If all the things that you reasoned from exist, then the thing that you reasoned from them will exist. This must be so because the existence of the thing is reasoned from the essences of the things from which it was reasoned, and so the existence of the things which it was reasoned from makes the existence of the reasoned thing necessary.

For example, if you consider a whole pie and someone's having removed a slice (which is that person doing the act of removing a piece, to the pie, in the past), you will reason that a partial pie exists. And if the pie existed, and the person did remove a piece, then the partial pie necessarily exists, because the existence of the two things from which it was reasoned causes its existence. So if you baked an apple pie and you leave it in your kitchen, and then later you see your brother eating a slice of it, you can reason that the pie which remains is missing a slice, and the existence of the partial pie can be reasoned as necessary from the two premises that a pie existed and a slice was removed.

Reasoning is also necessarily correct if that which is reasoned from requires the existence of the reasoned thing by being an existing effect of it or by being something of which the conclusion is a part. If the two things you reasoned from exist then what you reasoned from them exists also. It must exist because the existence of the things that were reasoned from will require it to exist. If you reason from something that does not exist then what you reason from

384

it will not necessarily exist. And if the nonexistence of things reasoned from would cause the nonexistence of what was reasoned from it then the conclusion will necessarily not exist and your conclusion will be wrong and incorrect. For example, if you reason from the existence of ten dollars in your wallet to your ability to buy a book and no money is in your wallet then you will have reasoned from something that does not exist to something that does not exist.

In unconscious reasoning the truth of the perceptions or concepts causes the truth of the concept which is reasoned from it, while in conscious reasoning, the existence of the things reasoned from consequentially requires the existence of that thing which is reasoned from them. If the things reasoned from are reasoned from perception, then their existence is perceived directly, and if the existence of something is reasoned to exist consequentially of the existence of some perceived thing, then one has reasoned, first, from perception that the thing reasoned from exists, and second, that if the thing reasoned from exists then the thing reasoned from it must exist (because its existence is an essential consequence of the existence of the first thing). From this a person can consciously reason that the second thing exists from the perception of the first thing. And if a third thing is reasoned from the second thing, then the existence of the third thing is consequentially required by the existence of the second thing, the existence of the second thing is consequentially required by the existence of the first thing, the existence of the first thing was directly perceived, and so on for a fourth thing, etc.

The truth-causation of unconscious reasoning and the essential reasoning of necessary existence in conscious reasoning are similar. The only difference is that in the former the truth of what is reasoned from directly causes the truth of what is reasoned, while in the latter the existence of what is reasoned from requires the existence of what is reasoned, but does not necessarily directly cause it (as for example when one reasons the existence of a cause from the existence of an effect). Of course the other difference is that conscious reasoning reasons from the things themselves, and unconscious reasoning reasons from perceptions and concepts in the brain, and unconscious reasoning creates concepts, while conscious reasoning discovers things.

Chapter Fifty Seven: Assumptions: Does Reason Use Assumptions?

Now that how reasoning works has been explained in detail, we can return to the question of whether or not reasoning begins with assumptions. It was shown that all reasoning is based upon the first reasoning of first concepts, which all other concepts are reasoned from. The question, then, is whether or not first reasoning makes use of assumptions.

Let us first ask what an assumption is, and then see whether reasoning uses them or not. We should begin with definitions to make this inquiry easier. A premise is a thing that one reasons from. An assumption is an unreasoned premise. Reason is pure if it reasons only from perceptions as premises, and reason is corrupted if it reasons from one or more assumptions. With our definitions taken care of, we can examine premises and assumptions. Unconscious reasoning's premises are perceptions and concepts, and conscious reasoning's premises are things, i.e. the objects of perceptions and concepts, which I call perceived things and thought-about things. Every reason-computation has premises. Its premises are all the things that are computed to reach the conclusion. The things that are computed, unless they are perceptions in unconscious reasoning or perceived things in conscious reasoning or assumptions, also have premises, those things from which the premises were reasoned. I have already shown how it is possible to reason from perceptions and perceived things, and I will not repeat myself here. Because pure empirical essential reasoning reasons only from perceptions and perceived things, it does not reason from assumptions and so has no assumptions. First reasoning is reasoning of this kind. Two obvious objections to this theory can be raised, the first having to do with the status of perceptions, and the second with reason's reliance upon concepts, which must be dealt with to advance the assumption-free theory of pure reasoning.

The first objection against me is the "perception = assumption" objection. I reply that first reasoning has no premises, it reasons only from sensory perceptions. It does not make sense to ask where the "rules of logic" that first reasoning makes use of come from, because the ability of first reasoning is inherent in the biological nature of the brain, which is genetically determined. Also, since this ability to engage in first reason adds nothing to the conclusions that are reasoned from the perceptions in the mind of the baby or young child, it cannot be said to be a premise of the reasoning. The only things from which the first concepts are produced are the perceptions from which they are reasoned. The objection which would naturally be raised at this point is that the sensory perceptions from which first concepts are reasoned are things that are reasoned from which are not reasoned, and therefore perceptions are the assumptions of first reasoning.

To answer this objection, we must compare the source of the knowledge provided by perception with the source of the knowledge provided by assumption. If they are the same then it is true that pure empirical reasoning reasons from assumptions, whereas if an ontological difference exists between perception and assumption as sources of knowledge, then it would be accurate to say that pure empirical reasoning begins with perceptions but does not begin with assumptions.

The knowledge from perception consists of the perceptions in the brain, which are the means of knowing those things which are known by means of them, the perceived things. Thus, given what has been said earlier, the source of the knowledge from perception is the perceived things when they act upon the sensory organs. The perceived things cause the knowledge of them when they cause the sensory organ to send signals to the brain. Thus, the source of knowledge from perceptions is objectively existing things.

When the brain reasons from an assumption, it is reasoning from a thing that was not empirically reasoned. Physically no path exists through which external things could cause anything which can be reasoned from in the brain except through the sensory organs. Because the assumption was not reasoned using information from the sensory organs it is impossible for external things to be the cause of the assumption. Thus, the brain must create the assumption independently of sensory information. Since the assumption is not reasoned from perception, and there is no other means by which the brain could gain knowledge of external things, when an assumption about external things is made the brain's belief has no legitimate source, and the unreasoned concept must have come from irrationality or random computation. For one example, you could reason from an assumption whose source was a false belief that someone else told you which you believed in blind faith. For another example, empirical associative computation could cause an assumption, such as if you assume that all dogs are small from having only ever seen small dogs before, and you then reason from the premise that all dogs are small to the conclusion that no dogs are dangerous.

To see how it is both possible and easy for the brain to believe something for which it has no evidence, we must examine how assumptions are made, and, more generally, how irrationality and false concepts are possible. As has been proven, when one reasons from a false concept the conclusion is a second false concept, so that the origin of some false concepts must be reasoning from other false concepts. No false concept can have been purely essentially empirically reasoned or else it would be true, so the false concepts that are reasoned from to create more false concepts are assumptions, and this is corrupted reasoning. All corrupted reasoning reasons from assumptions, and this can lead to new mistakes that are reasoned from, which in turn makes more mistakes, if the assumption is false. But the first assumptions cannot be created by being reasoned from assumptions, or there would be an infinite regress and the corrupted reason would have no point in time when it began. The brain can create an assumption in the first instance, i.e. believe a thing which it has not reasoned and has no reason to believe, in three ways. These three ways are also the three ways that false concepts are originated.

1. Random computation: One can reason the first source of false concepts from the nature of concepts. The concept is the knowledge of a thing being what the concept is the concept of, and the brain has access to every concept which it possesses, so the brain can put two or more concepts together to create the concept of the thing being all things that the combined concepts are of. For example, the concept of red and the concept of square can be combined to form the concept of a red square. Reason combines concepts pursuantly to the reason-computed empirical perceptions, but the brain can put concepts together to form new concepts randomly without regard to reason, i.e. irrationally. If you have a concept of an apple and a concept of your kitchen table, your brain can combine the two

concepts at random to form the belief that an apple is on your table, even if this is not based on perception and reason, and even if in fact the table does not have an apple on it.

Indeed, since all concepts must ultimately be created from perceptions, the first false concepts from which other false concepts are reasoned must consist of concepts which were reasoned from perceptions put together randomly to form a concept of a nonexistent thing. I call this the act of random computation. The brain has the ability to put concepts together in a totally random way to form a concept of something, and can then reason from that concept as an assumption. This is what some of the symptoms of schizophrenia consist of, although the biological cause of the brain doing this is unknown. Sane people can also do this at will (this is obviously what happens in most dreams), and I believe that this is where many of the concepts of severely irrational people come from.

Two kinds of cognition consist of the brain putting concepts together for a specific non-random reason but in the absence of reason. The first of which is associative reasoning, and the second of which is irrational belief. Both are discussed below.

2. Associative reasoning: When one reasons to universal being from non-essential associative reasoning, the conclusion of the associative reasoning will be a believed thing which was not reasoned. For example, one can reason from a group of mice being unaffected by a certain substance that all mice are unaffected by this substance, and then reason from all mice being unaffected by this substance. But this merely associates, it does not reason from essence. For example, all of the mice in one's lab experiment were unaffected, but this was because of the specific conditions of your lab, such as the chemicals in the food you fed to the mice, and you did not reason from the essence of being a mouse as such (e.g. DNA), you did not control for the food in the experiment, so other mice out in the wild who eat different food would have been affected by the substance. One will have put concepts together to form a concept irrationally, forming a concept of a mouse(e) being unaffected by the substance from an experiment showing specific mice being unaffected. While you believe that this was reasoning, because the cause of the brain forming the concept will be the attempt to reason to essential being from specific non-essential being, I call this empirical associative reason, because, in a sense, it

draws conclusions from unnecessary associations between things, i.e. from accidents and coincidences. When that unreasoned conclusion is reasoned from it is then an assumption. When one has an accident and then develops a phobia from it, even though the thing is not really scary or dangerous, this actually consists of associatively reasoning the danger of the thing from the one case of the accident, and then unconsciously reasoning from that associatively reasoned concept of the thing. Associative reasoning consists of reasoning universals from coincidences, and anything reasoned inappropriately from a coincidence is associative reasoning and will produce a false concept.

3. Irrational belief: This consists of believing what you are told by someone else when it is not reasoned, e.g. from faith or irrational trust. When one is told something, and one believes it without having reasoned it, one puts together a concept of the thing as existing, and then reasons from this concept as an assumption. Faith can be classified as an error of this type. For example, if one is told that something is red, square, and floating in water, one then puts together the concepts of red, square, floating, and existing, and reasons from this unreasoned concept. If one is told that lead floats in water, then one combines lead and floating by means of the concepts of them, to form a false concept of existing lead that floats in water. One did not reason that the thing exists from empirical essential reasoning, rather it is believed, i.e. conceived of as being an existing thing, simply because one was told to believe it. One can believe what one is told for any irrational reason. When the thing that one was told is reasoned from it is an assumption.

As a practical matter it is necessary to trust what other people tell you constantly, but one must always be aware that one can be lied to, and a strictly rational person will always use reason to evaluate the likelihood that one can trust what one is told. Knowledge is not an either/or proposition, in the sense that one can hold a belief with a percentage of certainty. If the weatherman tells you it will rain tomorrow, you might believe it with a 25% certainty, or if your coworker tells you that she completed the report that you needed her to write for you, you might trust this with 99% certainty, if she is an honest and trustworthy person, but in the absence of pure empirical essential reason one cannot achieve 100% certainty, e.g. you need to see the report to confirm that she wrote it. Note however one more detail, namely, that it is possible to reason from words as the effect of a cause in order to know the cause. For example, if one can reason that only the assassination of Julius Caesar could have made Shakespeare and all the historians write about it as if it really happened, and a "big lie" is not a plausible possibility in the context of human psychology, then one can reason from the words as an effect to the cause, and one can know with 100% certainty that Julius Caesar was killed by the Roman senators. One can know that the Roman Empire existed in the ancient world, etc., using this type of reasoning. And one can infer that the laws of science and math are true without actually seeing their demonstrations, also using a similar type of reason.

It can clearly be seen that the source of the knowledge from perception is the perceived things themselves, and that the perceptions are caused by the perceived things, while the knowledge from assumptions actually has no source whatsoever, and the assumptions are caused by mistakes made by the brain rather than the act of empirical reasoning. In light of this we can achieve a refined understanding of what the act of reason consists of. Reason is the act by which the brain uses its natural abilities to gain knowledge of objectively existing things, this knowledge being stored as true concepts in the brain. Thus, the essential act of the brain reasoning consists first of the act of perceiving, and secondly of the act of the pure empirical essential reason-computation of perceptions, and third of the application of true concepts in identifying future perceptions. Reasoning consists of a three step action or the series of these three actions performed in succession over and over again. Once the precise functioning of reason is understood, it is correct to say that perception is one of the actions of which reasoning consists, although the act of perceiving without the reason-computation of the perceptions is not sufficient to constitute the act of reasoning as such.

I mentioned two objections, and here I address the second one. This is the objection that first reasoning must make use of prior concepts to identify the perceptions which it reasons from. It was explained in a previous section how first reasoning reasons concepts from perceptions with which to identify future perceptions without any need of prior concepts. First reasoning in the brain of a baby or young child merely looks at the sensory information that enters the brain from the sense organs, and sees the patterns or proportions and ratios that naturally emerge from the sensory data. When a baby learns to identify a toy car, the concept comes from the toy car, not from the baby's brain, in the sense that the toy car caused some sort of pattern to exist in the sensory data which the baby's eyes see, and the baby's brain learns to identify this pattern, by forming a mental equation that fits the numbers in the sensory data pattern.

This objection is based on thinking about first reasoning as if it were the first instance of consciously reasoning things. Given what was said earlier, we can see the importance of first reasoning being unconscious reasoning, since this clears up the most sensible idea as to why first reasoning would need assumptions, namely that the brain would need concepts by means of which to be aware of the first things from which all other things are reasoned, so that those concepts would have to be assumed. While it is possible to consciously reason things from perceived things without reasoning from anything else, one still must use concepts to be aware of the things that one consciously reasons from. Conscious reason from new experiences is not the same thing as unconsciously reasoning first concepts from perceptions during the period of first reasoning, which exists only the brain of a baby or young child. One must not think about first reasoning as if it were conscious reasoning or else mistakes will follow.

First reasoning does not require any previous concepts of the objects of the sensory perceptions to operate because it is in fact not concerned with the objects of the senses nor does it reason directly from or to them, it only reasons from the mathematical sensory data which the brain receives, to concepts which can identify the data, without any awareness of anything. The brain is born with the ability to unconsciously reason concepts from first perceptions without the use of prior concepts with which to identify the first perceptions, so the objection that concepts would be required for first reasoning to work is incorrect, and pictures first reasoning as conscious reasoning. The brain is not born with concepts already in it. Rather it is born with the ability to take perceptions which it receives from the sense organs and reason first concepts from those perceptions.

Let me conclude this section with a crucial distinction: the premises of a line of reasoning vs. the premises of an action. Confusion on this issue may lead people to think that reason is based on assumptions. The premise of a reason-computation is the things that it reasons from. The premise of an action is those things which must exist or be true in order for the action to be correct. For example, if I reason that I have five dollars in my wallet, the premise of the reasoning would be the perceived thing that I saw when I looked in my wallet, namely the five dollar bill. But if I am at a coffee shop and I choose to buy vanilla coffee instead of chocolate coffee, the premise of the action is that the vanilla is a better choice than the chocolate, even if I did not think or reason at all and made the decision on a whim. Or, if I eat a chicken sandwich, the premise of my thought may have been that I like the taste of chicken, i.e. this is what led me to conclude that I should eat it, whereas the premise of the action of eating the sandwich is that the sandwich is healthy and will give nutrition to my body. The premise of an action may also be called an implication, since it is implied in the action. If I drink a glass of milk, it is implicit as a premise of the action that the milk is fresh and has not gone bad, although this is not the premise of my reasoning if I did not actually look at the expiration date on the carton and reason about the milk's freshness when I decided to drink it.

This distinction can be employed in the analysis of first reason. The premises as reason of first reasoning are the perceptions which the baby or child's mind receives from the sense organs. However, one might argue that the premises of first reason as an action are that reason is correct, the objective external world exists, empirical essential reasoning is the best way to think, a thing is itself and contradictions are impossible, etc. If one confuses the premises of reason with the premises of actions, one might think that all reason assumes these propositions, perhaps as axioms. But, although reason has a wide array of things which are its premises as action, the only premises as reason of pure empirical essential reason, i.e. the things that are actually reasoned from to reach the conclusions, are perceptions and perceived things.

Chapter Fifty Eight: Assumptions: Corrupted Reason vs. Pure Reason

Now that is has been proven that reasoning does not reason from assumptions consequentially of its essence, let us consider the essence of corrupted reasoning, the product of reasoning from assumptions, that is, the reason-computation of unreasoned things/concepts. The implications of the essence of assumptions in reasoning becomes clear in light of what was said about the causation of knowledge of truth in the section which explained why reason's inferences are valid. In unconscious reasoning, the truth of that which is reasoned from causes the truth of what is reasoned. If a concept is false then reasoning from it will cause the conclusion which is reasoned to be false. In conscious reasoning, the existence of that which is reasoned from requires the existence of what is reasoned. If something that does not exist is reasoned from then the conclusion might not exist, and even if it does exist it will be believed for the wrong reason and so not be fully understood, so that the reasoned thing as the object of understanding as a whole will not exist. The truth of all reasoned truth is ultimately caused by the truth of all the sensory perceptions from which it was reasoned, and the act of being true done by a sensory perception is caused by the existence of the objects which produced the perception by acting upon the body's sensory organs.

Things reasoned from an assumption, if the assumption is false, will be false. If an assumption is false, every line of reasoning from that assumption can only produce more false concepts. If the assumption is true, things reasoned from it will be true, although they will be true accidentally and not as a consequence of the essence of the act of reasoning.

I call reasoning which does not contain any assumptions pure reasoning. I call reasoning which contains assumptions corrupted reasoning. Pure reasoning is ultimately from perceptions and empirical experience, while corrupted reasoning is ultimately from one or more assumptions. An assumption is an unreasoned premise that is reasoned from. Regarding its metaphysical possibility, an assumption can be true or false, but because it was not reasoned its truth or falsity cannot be rationally known. Thus, one cannot rationally know that an assumption is true, unless one takes the assumption and actually reasons whether it is true, at which point it becomes reasoned and therefore ceases to be an assumption.

Suppose that one has reasoned Y from assumption X. The truth of that which is reasoned from determines the truth of that which is reasoned, as described. Of unconscious reasoning, if X were true, then Y would be true, and if X were false, then Y would be false. Of conscious essential reasoning, if X exists, then Y must exist, and if X does not exist, then Y might not exist. Yet whether X exists or not is irrelevant, and what Y consists of is also irrelevant, because, consequentially of its essence as an assumed thing, X was not reasoned and therefore one has no way of knowing whether X exists or not, and therefore no way of knowing whether Y exists or not. Y can never be rationally certain and is therefore useless, since one can neither know it with any certainty, nor reason anything with any certainty from it, nor act as if one knows that it is true.

One could object that we have a way to know whether X exists or not, by reasoning whether or not X exists. But if one uses pure empirical essential reasoning to reason whether or not X exists, and one reasons that X exists, and then reasons certain knowledge of the existence of Y from X, one will then know X by means of reason, and if X is reasoned and Y is reasoned from X then Y will no longer be reasoned from an assumption but will be reasoned from a reasoned premise, and X will no longer be an assumption. If one reasons Y from assumption X, and one then reasons X from perceived things, then one can reason X from perceived things and reason Y from X, which means that one can ultimately reason Y from perceived things. If an assumption is reasoned then it is no longer an assumption, and the argument from the uncertainty of assumption no longer applies to it. Because the essence of an assumption includes not having been reasoned, one can never be rationally certain of conclusions reasoned from any assumed thing. However, in light of what I explained, it is permissible to use an assumption "for the sake of argument," i.e. to reason from an assumption only to see what conclusion it leads to, but this is only rational on the condition that one later goes back and reasons or verifies the assumption using empirical experience, to transform it from an assumption into something known by reason.

Both pure and corrupted lines of reasoning can produce knowledge of existing things, but corrupted reasoning can only achieve the truth by accident, while pure empirical essential reason achieves the truth necessarily. Reasoning from things that exist will produce knowledge of things that exist, and pure reasoning reasons all conclusions ultimately from perceived things, which necessarily exist. Corrupted reasoning reasons from things that either do or do not exist, and so can produce knowledge of existing things only if the things assumed to exist do actually exist. Only pure empirical essential reasoning can only produce knowledge of things that exist which is necessarily true, and thus only pure empirical essential reasoning produces true certain knowledge. The things that are reasoned to exist by pure empirical essential reasoning exist consequentially of their essence, because they were reasoned to exist consequentially of the existence of perceived things which are known directly through perception. Thus, because the things which are the conclusions of pure empirical essential reasoning necessarily exist, pure empirical essential reasoning can only be correct, must be correct, and is therefore always correct. Corrupted reasoning cannot be trusted, because it can be correct or incorrect and one can never know on the basis of assumption whether the premises of its conclusions exist or not. Of course, for the meaning of this to be understood we must examine what knowledge consists of.

Chapter Fifty Nine: Knowledge

In philosophy a widely held view of knowledge exists. This view, known as the "tripartite account," states that knowledge has three properties. First, it is a belief, second, the belief is true, and third, the belief is justified. This is also sometimes phrased differently, as follows. First, one is sure or certain that a belief is true, second, the belief is actually true, and third, one is right or correct in being sure and certain that the belief is true. If this is the model of knowledge which we accept, then the model can be refined via an essential analysis. Generally, it is best to define the essence of something by that property which causes or explains most of the other properties, because this makes essential reasoning as easy as possible. A property exists which a belief can have, and this property causes the three other properties in the tripartite account, such that those three properties are the consequence of this essence. The essence of knowledge is that the belief is necessarily true, in other words, that knowledge is a belief whose truth is necessary. If a belief is necessarily true, then it is believed to be true, it is true, and the necessity of its truth means that one was justified in believing it. This analysis may need one small further refinement, namely, that the belief must arise because of whatever it is that made the belief's truth be necessary, since only then would the belief be justified.

Thus, knowledge is belief that cannot be wrong. Based on this definition of knowledge, and the lines of reasoning offered earlier in this essay, we can reason to the conclusion that every belief which results from pure empirical essential reasoning is knowledge. Pure empirical essential reasoning produces true concepts consequentially of its essence. The three step analysis of reason is useful to understand this. As step one, reason gains knowledge of essences from perception. This knowledge from perception is certain consequentially of the essential act of perceiving, from the known objects causing the perceptions which correspond to them. As the second step, reason deduces the consequences of the essences from the being of the essences, and this cannot be wrong because the essence either causes, requires, or contains the consequence. The things reasoned from identifying perceived things by their essences exist consequentially of the existing perceived essences. As the third step, the mind perceives things and then identifies them using their essence and instantly knows for sure that the thing has the consequence of the essence, like seeing an unplugged computer and knowing that it needs to be plugged in to charge because of previously reasoned knowledge of the essential computer, without any need to dissect and study this particular computer's circuitry.

The truth of essential reasoning is necessary, i.e. it is true by necessity. It is necessarily true because the consequence follows necessarily from the essence and the existence of the essences is known and proved by perception. Pure reason is based only on perceived things and nothing else, and perception is necessarily accurate, so pure reason can only be right. Therefore beliefs which result from pure empirical essential reasoning are necessarily true. This necessarily true belief is what constitutes knowledge. Reason is right and correct consequentially of its essence. Pure empirical essential reasoning achieves beliefs which can properly be called knowledge.

The being of the reasoned thing cannot be other than it is known to be, and this is so consequentially of the act of being known by means of pure empirical essential reason, but it is not the act of knowing or reasoning that causes the being to be so, rather it is the being that causes the knowledge of it to be true. To define epistemological necessity further, for a thing to be unnecessary is for it to be able to be other than it is known to be. I have explained how this is the case when describing the actions which produce false concepts, and I have also noted previously which methods, e.g. associative reasoning and irrational belief, are capable of producing true or false concepts at random, so that the belief produced might be true but is not necessarily true, so that this is not belief whose truth is necessary and is not knowledge.

Since pure empirical essential reasoning reasons things which necessarily exist consequentially of being reasoned from perceived essences, and perceived essences necessarily exist in order to be perceived, reason necessarily reasons things which necessarily exist, and therefore all lines of pure empirical essential reasoning produce knowledge of that which necessarily exists, and it cannot do otherwise because this is a consequence of its essential act, therefore the knowledge reason produces is necessarily true knowledge, and therefore pure empirical essential reason achieves 100% certain knowledge, always, consequentially of its essence.

Chapter Sixty: Solving the Problems

Now, to come to the defense of reason, it remains only for me to refute the four arguments against empirical reason that were examined in the introduction to An Essay on Reason and Perception. These are (1) the deception argument, (2) the subjectivity argument, (3) the coincidence argument, and (4) the assumption argument.

The first argument is that the senses are capable of deception. The first version of this argument, that the sensory organs are flawed, was refuted point by point in the section on perception. The second version of this argument, of the kind of Descartes' claim that all perception could be a dream, begins with the assumption that the senses are capable of deception, and is therefore a corrupted argument, with an assumption that is disproved in the section on perception.

The second argument is the argument from subjectivity. This argument states that because your experience is your own, it is personally biased and subjective, and cannot be a means to knowledge of objective existence. This argument is refuted by the means argument and the Objectivity Proviso, which demonstrated that the means of perception do not bias perception, so that it is possible for perception to both be personal experience and have objective reality and the external world as its object.

The third argument is the coincidence argument. According to this argument, it is possible that all perceived things are arranged coincidentally, so cause cannot be reasoned from sequence or order, nor can any certain conclusions involving necessity be drawn from experience. This is refuted by the argument from empirical essential reasoning. Essential reasoning produces knowledge of necessity, as was explained, and empirical essential reasoning produces knowledge of necessary cause and effect, as was explained.

The fourth argument is the assumption argument. This argument states that because all reason begins with assumptions, sometimes called axioms, all reason is based upon propositions that cannot be rationally proven, all reason is based on the irrational, and rational certainty cannot be proven. It was shown that, while all lines of reasoning that begin with an assumption cannot prove or disprove that assumption, pure empirical essential reasoning begins with no assumptions and is therefore not subject to counter-arguments based on its assumptions.

It might be useful, as one final tangent, to consider the implications of this essay with respect to two influential philosophers, Hume and Kant. Let us consider their key arguments and see how they withstand the conclusions of this essay. Hume argued to be is to be perceived, and that therefore everything imagined is perceived and therefore is, and therefore everything imaginable is logically possible. We have seen that possibility does not work that way. He also believed that the knowledge of cause and effect was a conjecture of coincidence and accident, composed of mere association from perceiving a series of repeated events, and that, since in the future the event might be different, there could be no knowledge of cause and effect from observation. We have seen how cause and effect work, and we examined how it is possible to reason certain knowledge of cause and effect from empirical essential reasoning.

Kant's response to Hume was to argue for knowledge of universal laws governing experience, on the basis of these laws coming, not from the objects of experience, but from the subject of experience, that is, the mind laying down conditions that the objects of experience must obey in the act of constituting the elements of the perceivable world. The benefit of this is that if the mind causes all of the perceptions that it creates to obey the laws of science then no possible experience in the future can contradict the laws of science because the mind will cause the future experience to obey the laws with no exceptions. The disadvantage is that this makes all knowledge totally subjective, dividing the world into the objects of experience, i.e. phenomena, which are for the subject, and things-in-themselves, i.e. noumena, which are in objective reality, but which cannot be the object of direct experience.

With the reasoning presented in this essay, it can be seen that the subject cannot impose laws upon the objects of experience, nor is this belief useful in order to achieve knowledge which is necessary, because my theory combines the possibility of certain knowledge with the possibility of knowing and perceiving objective reality directly. While discussing Kant, we may make note of his use of the concepts of the "a priori", that which is known prior to experience, and the "a posteriori", that which is known from within or after experience. Kant claimed that certainty could only be had if it was a priori, because, from Hume's argument, certainty cannot be found from within a posteriori sense experience. We have seen that certainty can be founded upon sensory experience, so we can do away with the quest for a priori knowledge, and satisfy our desire for truth and certainty with a posteriori reasoning.

In conclusion, it has been proven, entirely on the basis of sensory evidence, that reason can analyze the sensory evidence that consists of perceived things, and, by means of essential reasoning, can deduce certain knowledge of universals and cause and effect. Pure empirical essential reasoning produces rational 100% certain knowledge, and it cannot be wrong about anything.

This conclusion leads naturally, not to dogmatism, which resists questioning and rejects criticism, but to the open-minded use of reason to develop knowledge, and to increased rationality, and to the increased desire to question things and use rational critical inquiry. Although we can assert that our beliefs are knowledge, and know that they are true, we must always be open to criticism and exploration in the interest of verifying our premises and checking our reasoning. Even if a line of reasoning is reasoned to be 100% certainly true knowledge, we can check and reconfirm the premises and reexamine the line of reasoning to look for flaws or mistakes over and over again, although such is not necessary in order for our belief to constitute knowledge. If our knowledge is justified and reasoned then we should have no fear of justifying it and proving its truth in the face of doubts or counterarguments.

Dogmatism is the assertion of knowledge which lacks a rational foundation and which leads the person who asserts their claim of knowledge to angrily attack anyone who questions, challenges or criticizes him. I believe that knowledge is possible, but I am not a dogmatist. If we know that our knowledge is the result of reasoned proof, then we need not be afraid of questions and doubts, since we have the arguments with which to persuade. We, the proud owners of reasoned knowledge, must not be dogmatists. Rather we must always be open-minded and intellectually honest about our limitations.

While on this topic I will explain the connection between pure empirical essential reasoning and the philosophical scientific method. The philosophical scientific method is a means of checking the premises of a line of reasoning for accuracy and achieving verification of reason's conclusions. The expectations entailed by a hypothesis tested by the philosophical scientific method, i.e. what we would expect to see in our experience if the hypothesis is true, are the consequence of the essence, especially the perceived things which are necessary effects of the essence as cause. So if you see precisely what you expected to see in the data and evidence then the hypothesis is necessarily true as a consequence of the essence. And if you don't see the expectations in your observations of experience then the hypothesis is necessarily false. The expectations of a theory are what an abstract theory looks like in flesh and blood practical reality.

Levels of abstraction are a useful idea for understanding the need to check the premises of a line of reasoning. Each reason computation which goes from specific to essential removes you from perceived things by one level of abstraction. Each reason computation is capable of error, so each additional level of abstraction makes it easier to stray from reality and get lost in abstract theory. The philosophical scientific method applied to pure empirical essential reasoning uses perceived things direct from reality at level one to correct all mistakes made at higher levels of abstraction. To clear up any confusion about whether I think that reason is fallible or infallible, pure empirical essential reason is infallible when done right, but a person can always make a mistake in using it and do it wrongly. Pure empirical essential reasoning as a method cannot be wrong, but people and ideas can always be wrong. So it is always necessary to check your premises and verify your conclusions. Knowledge obtained by pure empirical essential reasoning must be confirmed by means of the philosophical scientific method to remove the possibility of error. If a conclusion is the result of pure empirical essential reason and it is then confirmed by the philosophical scientific method then the conclusion is knowledge and cannot be wrong and should be trusted by rational people.

Chapter Sixty One: Other Scientific Theories: The Meaning of Math, Space Line Theory, and Reincarnation and the Creator

In this section, I will present several theories. Let me note at the outset and emphasize that unlike the rest of this book, I do not make any claims as to whether the theories in this section are true or false. I personally do not know whether they are true or false. Scientists, especially experimental physicists, need to test them to find out. I present them here in the hope that science will consider these hypotheses and experimentally test them and develop them to add to the body of scientific knowledge. I have nothing at stake in whether they turn out to be right or wrong. Also note that they are presented in descending order of certainty. I am fairly sure that my meaning of math theory is true, I think some reasonable likelihood exists that Space Line theory is true, and the creator and reincarnation theories are unlikely and probably incorrect but possible and worth mentioning.

A. The Meaning of Math

The history of philosophy has seen many philosophers who were also mathematicians, from Pythagoras to Descartes to Pascal, and Plato and Wittgenstein were also interested in math. All of these philosophers happened to promote a philosophy of mind vs. body, and they tended to assign math to the mind side of the mind vs. body conflict, which led them to the attitude that math and numbers are a spiritual, non-physical, mystical world that lives above the base physical world of matter. Many contemporary academic philosophers subscribe to this view, which they inherited from the famous philosophers of the past. In contrast to the philosophical establishment, I argue that math is an analysis of the physical world, and that math has no existence other than as embodied in the world of flesh and blood and guts and science. The point of connection between math and the physical world is something that I call "the meaning of math," which is an analysis of mathematical concepts in terms of their basis in physical reality.

I. The Number One and All Other Numbers. The number one is actually the essential thing, a thing as such. One is one thing. I will not repeat the entire discussion of essence and essential things from earlier in this book, but if you see that the number one is the essential thing with a focus on the number of the thing, then the entirety of math follows from essential reasoning. Numbers are simply the essential thing with the specified property of how many there are. One is one thing, five is five things, twenty is twenty things, etc. Because it is the essential thing and only the number has been specified, a number could be anything with that number. For example, if you reason something about the number five, such as that it is two plus three, then this is true of everything with that essence. Two apples plus three oranges is five fruits, two hammers plus three hammers is five hammers, three houses plus two houses is a group of five houses, etc. Thus, when math reasons conclusions about numbers, these conclusions describe all physical objects in terms of their property of the number of them.

II. Addition and Subtraction. Addition shows that a number is formed by adding its components together, and subtraction shows that a number is reduced by removing part of the group. A number is a group of parts, e.g. the number three is three ones grouped together. I will not repeat my analysis of how acting as a whole forms a group, but the most basic acts that form a number are either being in the same place and forming a group together, for example three people in the same room form a number three, or else doing some specific detailed act together as a whole, for example the nine baseball players on the field doing the act of playing defense or batting forms a baseball team, and this act combines the nine ones into one whole of nine, i.e. $1+1+1+1+1+1+1+1+1=9$.

To add is to introduce another object into the same group, either by adding it to the same location, or by adding it as a role into the same cohesive group action. When you have a bowl with three apples in it, and then you put two oranges into the same bowl, you have added two to three, which results in five, the five fruits in the bowl. Thus the mathematical functions are a reflection of what happens in physical earthly reality.

III. Multiplication and Division. I argue that to multiply X times Y actually means that the X thing is doing the act of being Y, in other words, that X times Y means that X has the property of being Y. Division consists of removing a property from the thing that one is considering. To understand the meaning of multiplication and division, let me illustrate with an example. The number six is three times two. Look at this: XXX + XXX. Thus, two groups of things, with each group containing three objects, is a total of six. This really consists of the number three doing the act of existing twice, i.e. it consists of three being two, of the number three having the property of being two. Now look at this: XX + XX + XX. Three groups of two are also six. Six is two threes and three twos because that is what six is. Six is both two threes and three twos as a consequence of the essence of being the number six because the total being act of being six includes the being of two and the being of three as component properties. I posit that many, perhaps most or all, numbers have the property of being other lesser included numbers. Six is three two, so in a sense six is three, and six is two threes, so in a sense six is two. To be two is to be two things, and a group of three things is itself a thing, so six does the act of being three and also of being two, in the sense of doing the act of being three twos and of being two threes.

When a greater number is a lesser included number in this sense, multiplication is the addition to a number of the property of being another number, and division is the removal of being another number. Six is three twos. When you remove the act of being three from the number six, this leaves behind only the number two. This explains why six divided by three equals two. Similarly, when you have the number two, and you add the act of being three to it, two being three is two things being groups of three, which is three twos, which is also six.

To demonstrate 2x3=6 visually, look at: X + X. This is two. When you add the property of being three to the number two, each of the two essential things then does the act of being three. So X + X as two becomes XXX + XXX when two is multiplied by three, which becomes XXXXXX, which is six. In other words, two threes are six. The principle remains the same for all multiplication and division. For example, 40 does the act of being half of 80, and when you remove the act of being half, you are left with 80, which explains why 40 divided by one half equals 80, which more generally explains why dividing by half has the same result as multiplying by two, since the essential one is one half of two.

IV. Zero, Positive Numbers and Negative Numbers, also Multiplying and Dividing by Them. The meaning of the number zero is that it is nothingness. It is the total absence of anything. Positive numbers are things that are there, and negative numbers are the absence of things, or the act of removing or destroying a thing that exists. A positive number times a negative number is a negative number because adding the act of destroying or negating something to a positive essential thing leaves the act of destruction or negation, because the act done by the positive thing becomes the act of negating. Multiplying a negative number times another negative number results in a positive number simply because it is a "double negative", i.e. destroying the act of destruction leaves behind a positive. Multiplying by zero always results in zero because it means that you have zero groups of the other number, e.g. 6x0=0 because six times zero is zero sixes, which is nothing. It is impossible to divide by zero because any other number is something and does not do the act of being nothing, so it is impossible to remove the act of being nothing from a number which does not do that act naturally as a lesser included act of its greater act of being the number which it is.

V. Pi and Circles. Under a meaning of math analysis, numbers like pi and other specific constants have a meaning which can be understood philosophically and conceptually as well as mathematically. The circumference of a circle is pi times the radius. Assuming that a circle is the set of points defined as equidistant to the center, the radius is the "width" of the circle, in a sense, and the circumference is the "length" of the circle. Since the radius is a straight line and the radius times pi forms the circle, it is reasonable to suspect that the number pi actually represents the act of revolving in a circle, since if you take the radius and spin it 360 degrees it defines the circle, and multiplying by pi achieves a result that looks exactly like this. If the number pi represents the act of spinning or revolving, pi's scientific and philosophical significance is that the number extends infinitely and never repeats. The meaning of the number falling into a pattern of repeating would be that, in a sense, the number has ended, whereas infinite continual change is an infinite moving forward. In nature, revolution is the basic action that forms the activity of most physical objects, such as the spinning of planets or the revolution of electrons around atomic nuclei. This suggests that pi is the act of revolution and that the significance of pi in the Universe is that action continuously moves forward and never repeats or ends.

VI. I and Fractals. Similar to pi, the imaginary number i has a meaning. I is the square root of negative one, meaning that i times i equals negative one. Thus it must be the same number twice which times itself equals negative one. Positive one times positive one and negative one times negative one are both positive one. Positive one times negative one equals negative one, but these are two different numbers. Thus, the meaning of i is probably that it is both one and negative one at the same time, such that it times itself equals negative one. Its philosophical significance would be either that it embodies contradiction, or else that it embodies something that is both itself and something else in a different sense. I tend to think that the most natural interpretation of its meaning is that it means contradiction. However, the fractals which it defines seem to be maps of things that have essences but different specifics, such that the essence is a continuous pattern that looks the same yet different over an infinite number of varying specifics. It remains the task of the scientists and mathematicians of the future to discover and prove the true meaning of fractals, and to understand the meaning of other significant math ideas, such as right angles or constants like phi and e.

VII. Exponents and Dimensions. Being squared is for one thing to have the same property twice in two difference senses, so that it is the same property but not in the same sense. For example, a 4 by 4 square is a length of 4 which has the property of being a width of 4. All other exponents reduce to multiplication. For example, five to the power of three is 5 times 5 times 5, which means that it is a thing which consists of five groups where each of the groups is itself a group containing five groups and each of those groups contains five things. That is five things being five things being five things. The four dimensions of objects in space-time, namely length, width, depth, and time, are all arrived at by multiplication, specifically by squaring. Width is length squared, in the sense that a width is many lengths side by side doing the act of being long in the direction of the width. Similarly, time is a series of many cubes side by side, with each cube being one moment in time.

VIII. Other Numbers. I argue that each and every number, equation, and mathematical construct, can be understood in terms of physical reality. This has implications for math and science as well as for philosophy. Math can be reduced to meanings, and meanings can also be stated as math. For example, based on my meaning of math theory, I can hypothesis the existence of a number "H", which means infinity. H times H equals H, H plus H equals H, H minus H equals zero, and H divided by H equals one. All of this makes sense and is justified only if one understands that H means infinity. Infinity is infinite, so it adds nothing to what it already consists of to add to it or multiply it, because it already contains everything that could be added by addition or multiplying. When everything is subtracted from everything, nothing remains. And when you remove the act of being everything from a thing, i.e. division, then a thing as such remains, and an essential thing is the number one.

B. Space Line Theory

Space line theory asserts that the concept of God is a false concept of space. First, I will note that all matter is merely protons, neutrons and electrons in different proportions with different speeds and directions, so it should be conceded that it is possible to imagine reality as consisting merely of mathematical relationships. Space Line theory holds that the universe consists entirely of space, comprised of a series of points in space defined entirely by their relative special relationship to one another, which looks like a three-dimensional Cartesian Plane. Space Line holds that energy consists of the motion of space, and that all particles of matter are waves in space, because space can move and a particle is a line of space that is vibrating. Just as sound is a wave in air, matter/energy is a wave in space. Specifically, a particle is a small space line that has two simultaneous motions, a revolving around an axis in a circular motion, and a forward motion in the direction of its axis. This can be visualized: first you imagine a segment of straight line on a Cartesian Plane, like a piece of string in a fabric, then imagine it contract and spin.

Space Line posits that space is not empty, instead, space is a physical substance, and space is full of points, which form lines. The points of space physically exist, and each line and shape and plane and three-dimensional segment of space physically exists. An object, which is a wave in space, gets its location in space from having the points of space which define its location forming its wave. Time is the rate of movement of one part of space relative to the rate of movement of another thing of space, e.g. if one particle, A, revolves five times for every ten revolutions of another particle B, then B moves twice as fast as A. Matter is space which is bent into itself to form the object, and energy is the movement of space.

The Einstein Relativity theory is explained by Space Line in two ways. First, when space vibrates, the space tugs at its edges, just like if you pull at a thread in a sweater it pulls a bit of string out, and gravity is explained by the spatial pull of matter pulling at the fabric of space around it. When space gets pulled, the objects in that space are pulled, which is how gravity pulls objects. Second, the relativity of time is explained because, while we only see forward motion, e.g. seeing an object move, the sense in which a thing in time slows down or speeds up when it moves closer or farther away from the speed of light is its internal revolving motion. The central postulate of Space Line theory is that for every particle, its two motions are shared, and when it moves forward faster its internal revolving motion slows, and when it revolves faster its forward motion slows. This explains the Relativity theory that as a thing's forward motion speeds up and approaches the speed of light, its internal time slows down. A thing's forward motion is what is commonly understood as motion, whereas the motion of its revolution is its internal time, which gets faster or slower as seen by Relativity.

Space Line goes further and says that the speed of light is the speed at which the particle becomes entirely forward motion and loses its internal revolution which decreases to zero, so light is a line of space consisting purely of forward motion moving down the line. And if a particle slows down so that it has no forward motion and all of its motion is in its revolving, then it becomes a Black Hole, which happens when a high concentration of mass causes forward motion to be stopped by gravity. The difference between matter and antimatter is posited that matter's space vibration rotates in one direction, e.g. clockwise, and antimatter rotates in the opposite direction, e.g. counterclockwise, so that when matter comes into contact with antimatter the two motions cancel each other out, and a motionless line of space is left behind. It is also posited that an electron's probability field, as known by quantum physics, is actually the location of the wave of the space line vibration, as can be seen if one looks at pictures of electron probabilities, which look like fields of vibrating lines. Just as light is both a particle and a wave, every particle can be understood as a small wave in space, and every big wave would look like a particle if it was seen from a far enough distance away. Space Line Theory is consistent with string theory, which inspired it.

From this it should be obvious the sense in which God is a false concept of space. Space is everywhere, it "knows" everything, it is everything, and it is all-powerful. Also, Space Line Theory asserts that reality consists of space, which physically exists, as a four-dimensional fabric-like substance that has length, width, height, and moves forward in time with each moment in time consisting of a cube of space. Every physically existing object is merely a collection of points and lines of space with specific movements. Energy is the movement of space, and matter is space that is moving, and time is the rate of movement of one part of space relative to another part of space. In the Math Hypothesis in an Essay on Reason and Perception, I posited that the brain stores data numerically, and that unconscious reasoning identifies things by fitting numbers into equations. Here I can extend this and say that a thing's being actually can be described by an equation because a thing is a set of points, lines, and units of movement, which can be mathematically described. Thus, the brain stores data as numbers and identifies things by matching perceived things to mathematical equations, but it is also true that a thing really is a set of numbers and a thing's being in objective reality can be entirely described by math equations.

For an oversimplified example, if you define the origin from which to measure space as a point on your kitchen table, and a glass marble is sitting on the kitchen table at a place which is 5 inches to the right of the origin and 3 inches up from the origin, then a mathematical equation of $(x=5; y=3)$ would describe it (in location only, but its molecular composition could also be translated into numerical values). If the marble is rolling towards the right side of the table at the rate of one inch per minute, its forward motion could be mathematically described as $f(x)=x+1$ (although alternative mathematical notation systems could be chosen to accomplish the same description of physical reality).

To make this clearer, I could say that reality is a Cartesian Plane, so every real thing is a geometric object and every real thing is made of math. A thing is merely a set of measurable quantities of

411

its basic fundamental substantive being in space. For example, redness is a quality that can be mathematically described quantitatively by the attributes of wavelengths of light waves. So, too, every quality can be reduced to a mathematical description. However, this is not "the word and the word made flesh", i.e. math is not a spiritual higher plane of existence above the physical world, because, to be precise, the mathematical equation is an essential thing abstracted from the raw numbers that are the things which exist. So what exists are things in physical reality that can be described by math equations, not math equations embodied in space.

Space Line Theory asserts that space is the fundamental physical substance which does the act of being things and having properties, and multiple properties unite in the same thing because the same area of space does the act of being all the properties, e.g. being red and being round are done by the space which forms an apple. However, the physical substance, the thing, and the set of properties, are all the same thing. Also note that in answer to the question of whether the location of the Universe is absolute or relative, space is absolute in the sense that it exists objectively, but one location can only be measured relative to another location, e.g. (x=10; y=5) is measured relative to the origin or frame of reference at (x=0; y=0). In order to think that the Universe could have been at a different location we would need a frame of reference outside of our Universe with which to measure the location of the Universe, and physics has not yet identified whether anything exists outside the border of the Universe. Positing that something outside the Universe might exist in a thought experiment in the absence of hard science showing that something really does exist there is mere Analytic Unreality and is not scientific.

C. Fractal Reincarnation

A fractal is a pattern, graphed using a mathematical equation based on imaginary numbers (i.e. the square root of negative one). The pattern has a design, and the pattern has details which consist of an infinite number of smaller similar yet different designs, and the design itself is one among an infinite number of bigger similar yet different designs. So when you take a fractal and magnify it, you see a smaller fractal which looks the same yet different. Also, each point has a corresponding graph that reflects its place in the design. Here I am thinking specifically of the Mandelbrot set and the related Julia sets. It is difficult to describe this but it can be seen easily by looking at fractals.

The scientific method must learn, by scientific experimentation, whether fractals are merely pretty designs, or whether they have a scientific significance. One theory, which might be true or false, is that the fractal patterns reflect the nature of essences, specifically, just as the square root of negative one both is one and is not one (i.e. it is both positive one and negative one at the same time, since one times negative one is negative one and the square root of negative one when multiplied by itself is negative one), the theory is that imaginary numbers represent essences, since if you take two bottles of orange juice, one both is like the other (they are both bottles of orange juice) and is not like the other (e.g. one has a red label, the other a green label).

Or the significance of fractals may be that the Universe is in a fractal cycle. The Universe would be born in the Big Bang, would expand, life would evolve within it, then it would reach its point of maximal expansion, then it would contract, and the Universe would eventually collapse into a single point, and all life would die. Then, the pressure of being in one point would cause the Universe to explode again, creating a new Big Bang, and the Universe would live again, with contents similar to the past but also different and new, and the fractal pattern describes the life cycle of the Universe. If this is true, it would provide a scientific basis for reincarnation, namely, in a previous Universe you might have lived before, in a life that was the same yet different, and after you die, the Universe will collapse and explode, and you might live again in a life that is the same yet different. But this is merely a strange idea until science develops a better understanding of what happened before the Big Bang and where the Universe will end up in the future. This is not a religious idea, it is a scientific hypothesis.

The third possible significance of fractals would be that, although science thinks that Planck's Constant is the smallest possible physical existent, reality is actually infinitely large and infinitely small, and humans really are fractals in the sense that we are made of smaller fractals consisting of our cells and atoms and subatomic particles, and we are living inside of a bigger fractal. Specifically, subatomic particles, or whatever is smaller than them that forms them, would resemble us or our Universe in some way, and the Universe would be merely an atom in a huge organism or thing composed of hundreds or perhaps billions of Universes that form it. But, just as the big and small would be different, they would also be somehow similar to life on Earth, along some sort of mathematically defined pattern. Note that these three interpretations of fractals are all mere hypotheses in need of testing.

If reality is infinitely big and infinitely small, and extends infinitely in all directions, and extends back and forward in time infinitely, then one can posit the "one above and one below" theory, that a limit exists to how big or small a thing can be for us to know it or interact with it, and a thing can be so big or small that it cannot know us or affect us. For example, sub-sub-subatomic particles might each be a tiny Universe, but it might be unable to know or affect us, and it might be so small that we cannot possibly see it or interact with it by any means available to us. Similarly, our Universe might be merely an atom in a much bigger world, filled with giant beings made of billions of universes acting as a whole, but we are so small that we could not see or interact with the bigger world, and we are so small that they could not notice us or direct some action to a small enough focus to affect us.

D. The Purpose of Evolution, the Plan of the Universe, and the Creator of the Universe

This theory posits that no logical basis in science exists for saying that it is religious, and not scientific, to say that the Universe has a purpose for humans. Evolution is asserted to consist of random mutations, but the evolutions that make survival easier and better are the ones that natural selection promotes. Thus, scientific biology can say that it looks like evolution has an objective purpose, to reach an organism that has some end goal, namely to survive and flourish and succeed in this environment. The end goal would be to evolve a species that could achieve the perfect ability to interact with the environment. For example, humans evolved the ability to think not due to random chance but due to the fact that the nature of the Universe is such that reasoning is better able to cope with the environment than the coping mechanism of stupid instinct. Thus, thought is intended by the Universe to the extent that the nature of the Universe caused thought to evolve.

Evolution's successes and failures are defined by whether an organism's DNA is such that it succeeds in its environment, on planet Earth and in this Universe. In a sense, the more an organism "knows," i.e. matches and conforms to its environment and what that environment requires, the better it survives, and the more it is encouraged by evolution. It seems clear that the mind and reasoning is an evolutionary step forward towards the end goal of evolution, and that thinking is a superior form of adaptation as compared with genetic reactions or Pavlov-type conditioning. If this is true then thought and reason are not mere accidents, although the genes that enabled them were the result of random mutations and natural selection. They are the tools which are defined by the environmental conditions of Earth as necessary for evolutionary success, and therefore thinking and reasoning beings like humans were called for by Earth and the Universe.

From this I propose an idea: mutation is random in the sense of the genes changing, but the scope of possibility of what DNA can evolve into is defined by the DNA that already exists, which comes from the nature of the chemical components of the cells of living organisms. Evolution actually has a purpose, to evolve an end goal species that will possess minds and reasoning, which will be able to leave planet Earth and spread around the Universe. What specifically the Universe wants us to do is at this point unknown, and science would need to make progress in order to understand that. One possible goal is for the Universe to know itself through our minds, or to achieve virtue and ethics and goodness through us, or to experience pleasure and be happy. The Universe has no way of doing these things other than through organisms made of the Universe's physical matter. In a sense, the Universe acts through us. It is also possible that the Universe has some specific plan for life that evolves intelligence, such as physically colonizing the planets and developing the physical condition of the Universe in some way. Just as the fertilized egg grows into the person guided by the plan in the DNA, it might be our purpose to transform the raw materials of the Universe's planets into some sort of designed evolution guided by science.

I will conclude by discussing the scientific possibility that the Universe was created by a creator. This idea has been so completely warped and twisted by religious theologians that both the faithful and the scientists find it impossible to unwind it from the religious baggage which surrounds it, but I shall try to unwind it. The religious believers will think that by my saying that the Universe could have had a creator that I accept and concede that God might exist. Such confusion comes not from me but from the irrationality of the faithful. If the Universe had a creator then it would not be God. Instead it would be some sort of physical event or physical being. For example, the Big Bang might have created the Universe. Or the Big Bang and the origin of the Universe could have been caused for some purpose by an alien being or a race of technologically advanced alien beings who existed prior to the Big Bang. This is not science fiction and I am dead serious when I say this. The origins of the Universe must be discovered by physics, not theology. If the Universe had a creator as opposed to having existed forever and extending infinitely back into the past, then the creator would not be omniscient, or omnipotent, or loving, and would not have any relation to the God of the Bible, and would not be the father of Christ.

Indeed, if the Universe had a creator to whom we owe our existence, the proper attitude towards the creator would be the same attitude that they would have towards us if we were to build a race of artificially intelligent androids. We the creators would be physical, imperfect, ignorant beings, despite having given life to the robots. It would be insane for the androids to worship us and revere us the way the religious worship God. It is similarly insane for us to worship the creator of the Universe. The same sort of argument applies to a lesser extent to the fact that it would be irrational for a child to worship his or her parents religiously (although there are some cultures where ancestor-worship is common).

Science can test and confirm or refute the idea that the Universe was created by some sort of intelligent powerful being or species of beings who existed prior to the Big Bang. But the scientific understanding of the creator would learn what the creator was from actual hard science and would look to physics and computer modeling of the Big Bang to try to see what existed before the Big Bang and what caused the Big Bang. In addition to what existed before the Big Bang, we also need to know what will happen in this Universe in the distant future, and we need to reach the borders of our Universe and figure out what is on the other side.

Let me repeat an important point for clarity. The creator as known by science would be a physical being, e.g. some sort of very large alien or species of technologically advanced aliens that existed prior to the Big Bang. The creator would not be omniscient, omnipotent, or loving, and it would be absolutely nothing like a white bearded man in the sky, and would bear no resemblance to the God of the Bible. The God of human religions has been imagined by the religious prophets who knew nothing of science and looked to their hallucinatory daydreams for knowledge.

Science can test the hypothesis that a physical being or group of beings intentionally created the Universe, and designed evolution for the purpose of evolving certain types of species and organisms, which have a purpose in the plan of the Universe's creators, just as science can test any other theory. This theory might be true or false, and only the scientific method can tell. But this is not a religious theory. We must not allow religion to steal theories away from science by painting them with the brush of the taint of irrationality. Only theoretical physics can answer the question of the origin of the Universe by developing methods to enable us to look back upon what actually happened during the early stages of our Universe on the basis of astrophysical empirical data.

The religious fanatics are more insulted, and frightened, by this idea that God could be known by reason and science, than they are by atheism itself. They try their best to confuse and undermine this idea, saying that if science studies God then it must do so with the methods and assumptions of religion, and if the world had a creator then it must be the God of the Bible. In the study of God, the scientific method would be prior to God, and all religious truths which were not confirmed by science would be thrown away, and science would inform our discovery of whatever caused the Big Bang, not the other way around, not with our religious beliefs informing science. Science only neutrally tells us what exists, it does not validate our wishful thinking and what we want to believe, so science, as applied to the study of God, runs contrary to religious methods. It is a matter that is decided by objective reality itself, and not by us, as to what actually, really existed before the Big Bang. If you have any particular belief about this that you want to be true, as separate from a neutral scientific discovery, then you have already abandoned the scientific study of these issues.

Chapter Sixty Two: Conclusion: The Apple Metaphor

The principle of induction is essence. To be a type of thing is to have an essential property, and a thing's essence is the being which makes a thing be what it is. Pure empirical essential reasoning uses perception and induction to form concepts of essences and then achieves knowledge of necessary universal truths by reasoning the consequence of the essence, which is the set of properties that are caused or required by a thing's essence.

The mind is the human brain and consciousness is awareness of the things in external reality as known by means of the perceptions and concepts in the human brain. So when we dissect a brain we find the perceptions and concepts in the brain, but we do not see the objects which you and I are aware of inside the physical brain because the things in our conscious experience exist out in external reality and not in our brains. We can believe that the mind consists entirely of the physical human brain and does not contain any non-physical spiritual soul. We can do so without collapsing into the foolish idea that consciousness does not exist.

Knowledge, which is belief whose truth is necessary, comes from pure empirical essential reasoning and the philosophical scientific method. Philosophy can apply the scientific method to answer philosophical questions by means of the philosophical scientific method. The philosophical scientific method is a principle which explains the proper method of thinking for both science and philosophy. A complete understanding of perception proves that perceived things exist in objective reality, which means that they are independent of and separate from perception and the mind. And the things in experience are things in themselves, because for a properly functioning mind the world of experience is the external world and is not a world of subjective belief and interpretation. This is true because the mind is the means of perception and awareness but is not itself the thing that is perceived and thought about.

I will not engage in pointless repetition and restatement by repeating here the details of the various ideas and arguments offered earlier. Instead, I would like to use the remainder of this conclusion to discuss the metaphor of the apple and what the apple symbolizes.

418

Throughout this book I used an apple as an example to illustrate my ideas. The choice of apple as symbol was intentional and deliberate. Everyone who has read Dan Brown's novel "The Da Vinci Code" should remember the apple as the symbol of science, as the fruit which hit Isaac Newton in the head to enable him to perceive gravity and thereby inspire Newtonian physics. Indeed, Newton's apple is the perfect symbol to represent my theory of empirical essential reasoning. Newton directly perceived a specific real thing which did the act of falling from gravity, he achieved a concept of gravity from the apple via induction, and then he deductively reasoned his physics equations which described gravity from the essence of gravity. His mathematical reasoning ultimately rested upon the concept of gravity reasoned from the apple which he had perceived in his empirical experience.

But the symbolism of apples is deeper and richer than Newton, as Dan Brown knew. Historically, medieval European Christian art used an apple to symbolize sin, especially the sin of sexuality. This artistic usage derives from the popular theological idea that the fruit of the tree of the knowledge of good and evil in the Garden of Eden, described in the Biblical Genesis, was as apple. Adam and Eve ate this fruit and thereby knew their sinful nakedness and covered their genitals with fig leaves. Eating the apple incurred God's wrath and caused Adam and Eve to be cast out forever from the Garden of Eden.

The original sin of Christian theology is understood as the sin which led Adam and Eve to be thrown out of Eden, so the original sin was eating the apple. However a deeper analysis of the Christian metaphor reveals that the Christian original sin is actually sex, i.e. knowledge of the flesh. Sex is the origin of all human beings, in the sense that a person's parents having sex and resulting in conception of the fetus is the origin of the human being's physical existence.

My desire in using an apple as an example throughout this book is to reclaim the apple as a symbol of the physical existence of human beings which should be celebrated, not hidden in guilt and shame. My point does not relate to sex as such, but rather to the fact that a person is a physical human body and the mind is the physically existing human brain and nothing more. The messy, bloody, fleshy moment when a baby is born is the point at which that human being's thinking mind begins to function and perceive reality and form concepts. Birth and early childhood are the birth of reasoning where children learn the basic concepts and principles that form a foundation for all higher learning. The birth of reasoning is a better explanation than axioms to explain where reason comes from. All knowledge comes from a posteriori experience and no knowledge comes from an a priori soul. For example, the most abstract ideas of theoretical mathematics have their origin in concepts which the mathematicians and math professors formed as young children. Forming a concept of the number "five" from observing five apples or five toy cars or a picture of five dogs is the precondition of using the number five in advanced calculus or linear algebra.

We live in physical reality and achieve knowledge through empirical experience, e.g. we learn what an apple is by looking at it and eating it and tasting it. I seek to popularize the apple as a symbol of science not only because of the Bible and artistic symbolism but also in tribute to Newton's moment of scientific revelation. The apple is the emblem and insignia of science in its quest for progress and its ability to move human civilization beyond the outdated traditions of religion and into a future where technology will make life longer and more enjoyable for everyone.

The metaphor of the apple of knowledge means that a person can gain knowledge by eating an apple, because sensory perceptions and empirical experience analyzed by reasoning can achieve knowledge of reality. To extend the metaphor, when one eats the apple of knowledge one ceases to hear the voice of God, which is the voice of abstract theory detached from any basis in practical reality or experience, and one instead hears the voice of science and leaves the Garden of Eden in favor of life on Earth. One experiences the apple directly by tasting the apple itself and knowing what it tastes like.

When a person bites into an apple to confirm or refute his or her belief about the apple, he or she unites abstract theory with practical reality by using the experience of the apple so that reality holds theory accountable. If you think the apple will be sour and you taste it and it is sweet, then your idea is refuted and reality held your theory accountable, by means of the philosophical scientific method. If you think the apple is ripe and then you bite into it and it tastes good then your hypothesis is proven true by experience, also using the philosophical scientific method. And if you reason that your apple is edible because it is ripe, then you can reason that all ripe apples are edible, by means of the consequence of the essence and pure empirical essential reason. As a final concluding remark let me note that, as the examples throughout the book make clear, these ideas are more than mere abstract theory. The theories of the philosophical scientific method and pure empirical essential reasoning are a useful practical tool for how to think and how to interact with reality.

I will end this book with a one sentence summary that captures the flavor of my ideas: "True knowledge can come only from experience."

About the Author

I hope you enjoyed reading this. If you liked this book, won't you please tell a friend about it, or write a review of it online? Thanks!

Also please be sure to take a look at my other books:

Fiction:

The Golden Wand Trilogy

The Prince, The Girl, and the Revolution

Rob Seablue and the Eye of Tantalus

Project Utopia

The Office of Heavenly Restitution

Nonfiction:

Golden Rule Libertarianism

XYAB Economics

What They Won't Tell You About Objectivism

At the age of fifteen, Russell Hasan read "Atlas Shrugged" and decided that he wanted to become a philosopher and novelist when he grew up. Since then, he has written several books, with a focus on fantasy and science fiction, libertarian politics, and philosophical epistemology.

He majored in philosophy at Vassar, holds a law degree from the University of Connecticut, and at different points in his life has worked as a lawyer and as a software engineer. His distinctions include having written for the famous libertarian magazine Liberty for several years, and having read over three hundred fantasy and science fiction novels during his lifetime.

He lives in the United States, and is an obsessive New York Yankees fan.

Copyright Details